One Health Concepts and the Aquatic Ecosystem

_____

One Health Concepts and the Aquatic Ecosystem

Edited by

Laura D. Urdes

The World Aquatic Veterinary Medical Association, Romania; Faculty of Veterinary Medicine, University Spiru Haret Bucharest, Romania; Faculty of Management and Rural Development, University of Agricultural Sciences and Veterinary Medicine, Bucharest, Romania

Chris Walster

The World Aquatic Veterinary Medical Association, UK

Julius Tepper

The World Aquatic Veterinary Medical Association, USA; Long Island Fish Hospital, Manorville, New York, USA

CABI

CABI is a trading name of CAB International

CABI
Nosworthy Way
Wallingford
Oxfordshire OX10 8DE
UK

CABI
200 Portland Street
Boston
MA 02114
USA

Tel: +44 (0)1491 832111
E-mail: info@cabi.org
Website: www.cabi.org

T: +1 (617)682-9015
E-mail: cabi-nao@cabi.org

A catalogue record for this book is available from the British Library, London, UK.

ISBN-13: 9781800623224 (hardback)
         9781800623231 (ePDF)
         9781800623248 (ePub)

DOI: 10.1079/9781800623248.0000

Commissioning Editor: Jamie Lee
Editorial Assistant: Theresa Regueira
Production Editor: Rosie Hayden

Typeset by Exeter Premedia Services Pvt Ltd, Chennai, India
Printed in the USA

Contents

Contributors

Patricia Aguirre Mejía, Facultad de Posgrado, Universidad Técnica del Norte, Ecuador

Donald M. Broom, St Catharine's College and Department of Veterinary Medicine, University of Cambridge, Madingley Road, Cambridge, CB3 0ES, UK

Devon R. Dublin, Ocean Policy Research Institute, Sasakawa Peace Foundation, 1-15-16 Toranomon, Minato-ku, Tokyo 105-8524, Japan

Gratiela Gradisteanu, The Faculty of Biology, University of Bucharest, Romania

Pablo Jarrín-V., Dirección de Innovación, Instituto Nacional de Biodiversidad, Ecuador

Stephen J. Jordan, 25 Wayne Memorial Drive, N. Dartmouth, MA 02747

Alfonso Molina, Instituto de Investigación en Zoonosis (CIZ), Universidad Central del Ecuador, Ecuador

Iasmina Maria Moza, The Faculty of Biology, University of Bucharest, Romania

Ricardo Muñoz Cisternas, Observatorio de Políticas Públicas del Territorio, FARAC, Universidad de Santiago de Chile, Chile

Julia Prado-Beltran, Faculty of Engineering in Agricultural and Environmental Sciences, Universidad Técnica del Norte, Ecuador

Kestrel O. Perez, Department of Biology, St Joseph's University, New York, 245 Clinton Ave., Brooklyn, NY 11205

Konstantine J. Rountos, Department of Biology, St Joseph's University, New York, 155 W. Roe Blvd., Patchogue, NY 11772

Gabriela N. Tenea, Biofood and Nutraceutics Research and Development Group; Faculty of Engineering in Agricultural and Environmental Sciences, Universidad Técnica del Norte, Ecuador

Julius Tepper, The World Aquatic Veterinary Medical Association, USA; Long Island Fish Hospital, Manorville, New York, USA

Laura D. Urdes, The World Aquatic Veterinary Medical Association, Romania; Faculty of Veterinary Medicine, University Spiru Haret Bucharest, Romania; Faculty of Management and Rural Development, University of Agricultural Sciences and Veterinary Medicine, Bucharest, Romania

Chris Walster, The World Aquatic Veterinary Medical Association, UK

Preface

In addressing current global challenges, this textbook on One Health and the aquatic ecosystem offers an integrated exploration of One Health—a conceptual framework linking human, animal, and ecosystem health.

Beginning with foundational perspectives in Chapter 1, the work introduces the intertwined concepts of One Biology, One Welfare, and One Health, establishing a critical understanding of how these interconnected disciplines influence welfare, environmental resilience, and public health. The subsequent chapters delve into specialized topics across aquatic ecosystems, biodiversity conservation, pollution mitigation, and epidemiology in aquaculture. In Chapter 2, pressing issues, such as marine pollution, sustainable resource management, the impacts of environmental degradation on both terrestrial and marine systems, and the challenges of disease transmission are presented with concrete real-life examples. Chapter 3 focuses on biodiversity conservation and the ecological significance of genetic, species, and ecosystem-level diversity. This is complemented by Chapter 4, which further explores environmental pollution and presents cross-sectoral strategies for mitigating its widespread effects. Together, these chapters emphasize the inextricable link between environmental health and the resilience of human and non-human life. Chapters 5 through 7 explore applied health concerns in aquatic ecosystems, where epidemiology and biosecurity in aquaculture are investigated, offering tools for disease management and economic sustainability. Public health challenges, such as antimicrobial resistance and microbiome interactions, are examined in Chapter 6, while Chapter 7 addresses the human–wildlife interface and the risks of zoonotic transmission in changing ecosystems. These detailed investigations highlight the significance of multidisciplinary approaches and advanced technologies in fostering sustainable practices and robust biosecurity measures. Chapter 8 offers a forward-looking perspective on food security, novel food systems (e.g. insect protein, seaweed farming) to ensure food security, and environmental health. Chapter 9 is a concluding chapter, which synthesizes research on human–wildlife interactions, innovative food production, policy, and practice, providing a comprehensive blueprint for safeguarding public health while promoting sustainable environmental stewardship.

Collectively, this textbook serves as both a foundational academic resource and a call to action, promoting collaborative, cross-disciplinary solutions essential for sustaining planetary health in the 21st century.

Acknowledgments

_____

This manual is a collaborative effort to which colleagues and friends contributed. The editors would like to acknowledge the following:

Dr. Katinka de Balogh, former One Health focal point at the Regional Office for Asia and the Pacific Food and Agriculture Organization of the United Nations, is acknowledged for her constructive criticism of the structure of the manual.

Teresa Collins Tepper is acknowledged for kindly providing the photo for the book cover.

1 One Biology, One Welfare, One Health Concepts and Implications

Donald M. Broom*

St Catharine's College and Department of Veterinary Medicine, University of Cambridge, Madingley Road, Cambridge, CB3 0ES, UK

Abstract

This chapter accommodates various definitions of One Health, explaining why epidemiology, biosecurity, and other concepts addressed in the other chapters are included in this book. The chapter includes definitions and various uses of One Health in practice as well as explanations of one welfare and one medicine. It sets the scene for the next chapters, where the One Health concept is taught upon the idea that wetlands are the point where human, animal, and environmental health meet.

1.1 Introduction

Understanding the living world—and aquatic ecosystems in particular—requires that many areas of human knowledge be brought together (Kasper *et al.*, 2022). In order that harms to life in the world can be minimized, the decisions that have to be taken require that a range of government departments utilize these various areas of knowledge. In drawing together the topics considered in this book, the way in which the One Health concept has encouraged cross-disciplinary and interdepartmental approaches is emphasized. Some of the problems discussed are a consequence of too human-centered an approach, which has already led to negative consequences for humans and for the living world in general. Science has been progressing rapidly, but many attitudes, policies, and actions have not kept up with these changes in understanding. Some of the concepts discussed in the different chapters of this book are introduced

here. Key points are that life is dependent on water, and much of the surface of the earth is water. Many people assume that aquatic animals are less complex than those that live on land, that cold-blooded animals do not have feelings, emotions, and sophisticated brain function, and that humans do not need to consider the welfare of aquatic animals, but all of these assumptions are false (Broom, 2007a; Sneddon, 2015; Kristiansen *et al.*, 2020).

1.2 One Biology

As the DNA of more and more species is sequenced and described, it has become clear that the genetic differences between humans and other species of animals are small and the similarities are large (Boffelli *et al.*, 2004). It is also now known, especially from epigenetics research, that, while all characteristics of organisms depend on genetic information,

*Corresponding author: dmb16@cam.ac.uk

DOI: 10.1079/9781800623248.0001

each step in the translation of genetic information into proteins, and hence characteristics of individuals, can be altered by environmental factors (Alexander, 2017; Broom and Johnson, 2019), so nothing is completely genetically determined, nothing is instinctive, and nothing is innate. Humans belong to the animal kingdom and have hardly any ability that is not also possessed by other animal species, at least to some degree (Lerner and Berg, 2015). Those qualities that some people have presented as exclusive to humans, such as language, emotions, the notion of culture or society, cooperation, altruism, tool use, empathy, and a concept of the future, have been described in various groups of animals (Broom, 2003, 2014, 2022; Premack, 2007; Clayton and Emery, 2015; de Waal, 2016; McBride and Morton, 2018). Despite some differences in brain gross anatomy, there are similarities in the functioning of high-level cognitive analytical systems, pain systems, and other emotion analysis mechanisms. The pathogen-combating mechanisms of the immune and other systems differ little across humans and many aquatic species (Urdes *et al.*, 2021). It is for these reasons that other species can be used to better understand humans.

Since the fundamental biological processes in the living cells and systems of all animals, including humans, are the same (Striedter, 2019), it is logical to say that there is only one biology. As the biological processes in humans and non-humans are the same, the concepts of health and welfare mean the same in humans and all other animals (Tarazona *et al.*, 2020). If that is the case, the concept of 'zoonosis' seems anomalous. This term is mainly used to refer to human disease that can have an origin in a different animal species. However, this is a very human-centered usage, and the One Health approach might lead to any disease that can infect more than one species being called zoonotic. Another term is 'anthropozoonotic,' used for pathogens passed from humans to other species. Terminology often overemphasizes humans and should be changed if a biologically balanced terminology is adopted. The remainder of this chapter explains further the concepts of one welfare and One Health and introduces some of the implications of these approaches.

1.3   One Welfare

Welfare assessment evaluates how well an individual is faring or going through life. The welfare of an individual is its state as regards its attempts to cope with its environment (Broom, 1986). Coping means having control of mental and bodily stability. In order to cope, a nervous system and other mechanisms are needed, and the state, and hence the individual's welfare, can be scientifically measured (Broom and Johnson, 1993, 2019). We can consider the welfare of humans and of any other animal because they have a nervous system, but we cannot refer to the welfare of plants, bacteria, viruses, or inanimate objects. It is the individual whose welfare can be assessed and considered, so, while we cannot speak of the welfare of a group or population, we can refer to the mean welfare in a population. Well-being means the same as welfare as does quality of life, except in relation to very short timescales, but welfare is more often used as the scientific term (Broom, 2007b). Welfare can be positive or negative, good or poor, and the position of the welfare of an individual on the continuum from very negative to very positive can be evaluated scientifically. Measures of positive welfare and evaluations of the balance between positive and negative are required for this (Boissy *et al.*, 2007; Lawrence *et al.*, 2018; Broom, 2023). All animals, aquatic or terrestrial, have needs, and it is important to understand that if these are not met and the individual is not able to control its interactions with its environment, welfare will be poor. More complex environments that fulfill more needs are just as important for aquatic animals as for terrestrial animals (Zhang *et al.*, 2022).

The many different coping mechanisms, and hence the wide range of measures of welfare, include most aspects of brain function and many bodily functions with cellular, physiological, and behavioral components. Feelings are a key part of welfare in sentient animals. The physiologically describable aspects of feelings are referred to as emotions (Broom, 1998, 2014; see also Rolls, 2005; Boissy *et al.*, 2007; Paul and Mendl, 2018). Feelings and emotions are adaptive mechanisms and, like other adaptive mechanisms, they promote survival (Broom,

1998; Fraser, 2008). Health is also a key part of welfare. Like welfare, health can be qualified as good or poor and varies over a range. It refers to body systems, including those in the brain, that combat pathogens, tissue damage, or physiological disorder, so health can be defined as the state of an individual as regards its attempts to cope with pathology (Broom, 2006).

It has always been known that the terms 'welfare' and 'health' have the same meanings when applied to humans or to other animal species. A forerunner of the one welfare and One Health discussions was a Dahlem Conference entitled 'Coping with Challenge: Welfare in Animals including Humans' (Broom, 2001), which considered in detail the similarities in physiological, immunological, and clinical research on stress and welfare in humans and a range of other species. It was emphasized that human psychiatry and medicine could learn from farm animal and other welfare research and that animal welfare scientists should use more information from human studies. Progress was greater if each area of study was also applied to the other. Recently, the one welfare approach has emphasized that the concept of welfare is identical when applied to humans or to non-human animals (Colonius and Earley, 2013; García Pinillos et al., 2015, 2016; Broom, 2017; García Pinillos, 2018). This approach can be incorporated into teaching of animal welfare (McGreevy et al., 2020). Some of the relevant evidence shows that, when the welfare of individual humans or non-human animals is poor, there is increased susceptibility to disease. Improving welfare generally reduces disease. Disease effects are similar in humans and other species, so those with a medical background and those with a veterinary or other biological background benefit from exchanging information. As an example, Daigle (2018) describes similarity between postpartum problems in pigs and humans. In order to utilize this approach effectively and provide good care for people and good care for animals used by people, all should be considered as individuals. While this statement is unsurprising to many practitioners and researchers, some still have the idea that herd treatment is always sufficient and individual animals used by humans do not require consideration. This view is more extreme if the animals are aquatic.

1.4 One Health Concepts

Although, as mentioned above, it has long been assumed by many scientists that basic biology, welfare, and health are concepts relevant to all animals, publications on One Health have drawn the attention of many to what this means for medical and veterinary scientists and practitioners. The publications about One Health explain, first that health means exactly the same for non-human animals as it does for humans (Lerner and Berg, 2015), and second, that One Health is a worldwide strategy encouraging interdisciplinary collaboration and communication in relation to all aspects of health care for humans and non-human animals. It also promotes viewing the individual human, or other animal, in relation to its interactions with the environment. A resolution promoting the similarity of human and non-human animal health and the great potential for collaboration between the human medical and veterinary researchers and practitioners was adopted in 2007 by the American Medical Association and the American Veterinary Medical Association. The concept is further explained by Monath et al. (2010) and Karesh (2014), while an example of application of the principles to aquatic animal production is discussed by Urdes et al. (2022, 2024).

1.5 One Health Implications

The One Health approach requires a wide range of areas of investigation in terrestrial and aquatic habitats, both in those occupied or obviously affected by humans and in those that are not. These areas of investigation, some mentioned in the following incomplete list of examples, are the subject of the various chapters of this book. The methods used by epidemiologists are of importance in trying to understand the distribution and movements of pathogens from individual to individual, species to species, and environment to environment. Disease surveillance techniques have to be used in single species and multi-species groups. They take account of diseases passed to others by infected animals and by those that are not themselves infected. Biosecurity procedures can have great effects on when and how pathogens spread. Social interactions in human and non-human groups require study in order to understand the circumstances in which pathogen transmission might occur. The importance of such studies is clear since almost one fifth of the world population is involved in livestock production or the processing and marketing of food of animal origin (WOAH, 2024). All of these methodologies are important for aquatic vertebrates and invertebrates used by humans. Some of the most numerous farmed animals are invertebrates, for example prawns and shrimps, so the study of their welfare, including especially their health, is of great importance to humans and to the whole world (Alcivar-Warren *et al.*, 2023). Food security for humans and other species is a necessary investigation area in relation to many pathologies.

Antimicrobial resistance is one of the most important problems facing humanity at present. Misuse of antibiotics has facilitated the development of strains of bacteria that are resistant to some or all antibiotics, and antimicrobial resistance is estimated to lead to the deaths of five million people per year (Murray *et al.*, 2022) as well as the deaths of many other animals that cannot be treated successfully for infections. Monitoring of pathogen responses to antibiotics and other disease control methods are of great value in attempts to deal with antimicrobial resistance (Brown, 2015; Foyle, 2022; Walster and Foyle, 2023). There are several other important actions that can combat antimicrobial resistance in both terrestrial and aquatic environments: (i) control the sale of antibiotics and other antimicrobials, making it illegal to sell antibiotics unless a medical doctor or veterinarian has prescribed them and illegal to sell them over the counter (OTC); (ii) educate the public not to use antibiotics for viral or other non-bacterial diseases; (iii) educate the public that they should always complete a course of antibiotic treatment; (iv) educate the public so that nobody ever disposes of unused antibiotics into sewage etc. and unused antibiotics are returned by clients to a medical or veterinary practice; (v) never allow antibiotic use for growth promotion; (vi) minimize prophylactic use of antimicrobials, reduce total use, and use alternatives when available and appropriate; (vii) avoid farm animal housing that leads to poor welfare as better welfare reduces disease; improve human housing of poor people for the same reason.

Another point discussed in this book is that disasters can have a major impact on disease spread, and different control methods may be needed for some diseases.

The sustainability of actions and systems is an important concern in relation to disease control measures. Sustainability has a much wider meaning now than it had in early writings on the subject. Some usage of the term 'sustainable' has been too narrow as sustainability is not solely concerned with whether or not sales of a product can continue or what the carbon footprint of an activity is. For example, the ethics of the production method are now included, and a system can be unsustainable because of negative impacts on the efficiency of use of world resources, on human welfare, on the welfare of other animal species, or on various aspects of the environment. A definition of sustainability is: a system or procedure is sustainable if it is acceptable now and if its expected future effects are acceptable, in particular in relation to resource availability, consequences of functioning, and morality of action (Broom, 2014). In order to comply with this definition, the system or procedure should be ecologically sound, economically viable, socially just, and humane (Appleby, 2004). It brings together aspects such as health, welfare, environmental protection, productivity, food safety, food quality, and efficiency (Pethick *et al.*, 2011).

Health, as defined in Section 1.3 above, is a term that refers to individuals with systems for combating pathology. The mean health of a group or population of individuals can also be considered and evaluated. When reference is made to environmental health, the meaning is normally the environment in relation to the health of humans in that environment. The environmental problems for humans may involve accumulation of discarded materials or effluent. The environment itself cannot have health because it has no system for controlling pathological harms that might affect it. Similarly, the world ecosystem, or 'the planet,' does not have health as it is not a self-regulating system. Negative consequences for environments or for the world resulting from human actions will not be put right by those environments or by the world. This important scientific principle is explained further in several chapters of this book. Negative consequences for the environment may involve reduced stability, reduced biodiversity, or increased risk of disease among individuals in that environment. The One Health approach, including an understanding of principles of epidemiology, can be beneficial in considering how to deal with some of these problems.

Specific topics need to be addressed in relation to this One Health approach to water-based environments. Life is based on water, so aquatic environments are of crucial importance to understanding disease transfer in the world. Water moves around the oceans and can carry living organisms, including pathogens. A disease-causing organism that enters a body of water in one place can be transported to another place, and if the body of water is an ocean, the distance traveled may be very great.

One key area when protecting and restoring water-related ecosystems is the physical and chemical monitoring of oceans and other water bodies. It is also important to be able to identify and assess the various kinds of pollutants and to understand the movements of materials in currents. Populations of living organisms, including pathogens and proportions of affected hosts, have to be monitored. The impacts of harmful human actions and of conservation efforts can be assessed by various methods. The tools of One Health that are used to detect environmental pollution indicators that could threaten aquatic species include molecular ecology, population genomics, proteomics, and epigenetic epidemiology (Urdes and Alcivar-Warren, 2022). Aspects of such work are the treatment of wastewater and the efficacy of the maintenance of services, including the provision of safe and affordable drinking water. These and other relevant topics are considered in this book.

References

Alcivar-Warren, A., Bateman, K., Clinton, M., Foyle, L., Lewbart, G. *et al*. (2023) Aquatic invertebrates. In: Urdes, L., Walster, C. and Tepper, J. (eds) *Pathology and Epidemiology of Aquatic Animal Diseases for Practitioners*. Wiley Online Science, Hoboken, New Jersey, pp. 1–80.

Alexander, D. (2017) *Genes, Determinism and God*. Cambridge University Press, Cambridge, UK.

Appleby, M.C. (2004) Alternatives to conventional livestock production methods. In: Benson, G.J. and Rollin, B.E. (eds) *Production Animal Pain and Well-Being: Theory and Practice*. Blackwell, Ames, Iowa, pp. 339–350.

Boffelli, D., Nobrega, M.A. and Rubin, E.M. (2004) Comparative genomics at the vertebrate extremes. *Nature Reviews Genetics* 5, 456–465. DOI: org/10.1038/nrg1350.

Boissy, A., Manteuffel, G., Jensen, M.B., Moe, R.O., Spruijt, B. *et al*. (2007) Assessment of positive emotions in animals to improve their welfare. *Physiology and Behavior* 92, 375–397. DOI: org/10.1016/j.physbeh.2007.02.003.

Broom, D.M. (1986) Indicators of poor welfare. *British Veterinary Journal* 142, 524–526.

Broom, D.M. (1998) Welfare, stress and the evolution of feelings. *Advances in the Study of Behavior* 27, 371–403. DOI: 10.1016/S0065-3454(08)60369-1.

Broom, D.M. (ed.) (2001) *Coping with Challenge: Welfare in Animals including Humans*. Dahlem University Press, Berlin, Germany.

Broom, D.M. (2003) *The Evolution of Morality and Religion*. Cambridge University Press, Cambridge, UK.

Broom, D.M. (2006) Behaviour and welfare in relation to pathology. *Applied Animal Behaviour Science* 97, 71–83. DOI: org/10.1016/j.applanim.2005.11.019.

Broom, D.M. (2007a) Cognitive ability and sentience: Which aquatic animals should be protected? *Diseases of Aquatic Organisms* 75, 99–108. DOI: 10.3354/dao075099.

Broom, D.M. (2007b) Quality of life means welfare: How is it related to other concepts and assessed? *Animal Welfare* 16, 45–53. DOI: 10.1017/S0962728600031729.

Broom, D.M. (2014) *Sentience and Animal Welfare*. CAB International, Wallingford, UK.

Broom, D.M. (2017) *Animal Welfare in the European Union*. European Parliament Policy Department, Citizen's Rights and Constitutional Affairs, Brussels. DOI: 10.2861/891355.

Broom, D.M. (2022) *Broom and Fraser's Domestic Animal Behaviour and Welfare*, 6th edn. CAB International, Wallingford, UK. DOI: 10.1079/9/9781789249835.0001.

Broom, D.M. (2023) Can positive welfare counterbalance negative and can net welfare be assessed? *Frontiers in Animal Science* 4, 1101957. DOI: 10.3389/fanim.2023.1101957.

Broom, D.M. and Johnson, K.G. (1993) *Stress and Animal Welfare*. Chapman & Hall, London.

Broom, D.M. and Johnson, K.G. (2019) *Stress and Animal Welfare: Key Issues in the Biology of Humans and Other Animals*, 2nd edn. Springer Nature, Cham, Switzerland.

Brown, L. (2015) Antibiotics in aquaculture. *The Aquatic Veterinarian* 9(3), 15.

Clayton, N.S. and Emery, N.J. (2015) Avian models for human cognitive neuroscience: A proposal. *Neuron* 86, 1330–1342. DOI: 10.1016/j.neuron.2015.04.024.

Colonius, T.J. and Earley, R.W. (2013) One welfare: A call to develop a broader framework of thought and action. *Journal of the American Veterinary Medical Association* 242, 309–310.

Daigle, C. (2018) Parallels between postpartum disorders in humans and preweaning piglet mortality in sows. *Animals* 8(2), 22. DOI: 10.3390/ani8020022.

de Waal, F.B.M. (2016) *Are We Smart Enough to Know How Smart Animals Are?* W.W. Norton & Company, New York.

Foyle, L. (2022) Epidemiology of aquatic animal diseases. In: Urdes, L., Walster, C. and Tepper, J. (eds) *Fundamentals of Aquatic Veterinary Medicine*. Wiley, Hoboken, New Jersey, pp. 135–150.

Fraser, D. (2008) *Understanding Animal Welfare: The Science in its Cultural Context*. Wiley Blackwell, Chichester, UK.

García Pinillos, R. (2018) *One Welfare: A Framework to Improve Animal Welfare and Human Well-being*. CAB International, Wallingford, UK.

García Pinillos, R., Appleby, M.C., Scott-Park, F. and Smith, C.W. (2015) One welfare. *Veterinary Record* 177, 629–630. DOI: 10.1136/vr.h6830.

García Pinillos, R., Appleby, M., Manteca, X., Scott-Park, F., Smith, C. *et al.* (2016) One welfare – a platform for improving human and animal welfare. *Veterinary Record* 179, 412–413. DOI: 10.1136/vr.i5470.

Karesh, W.B. (ed.) (2014) *One Health. Scientific and Technical Review*, Vol. 33 (2). World Organisation for Animal Health, Paris.

Kasper, S., Adeyemo, O.K., Becker, T., Scarfe, D. and Tepper, J. (2022) Aquatic environment and life support systems. In: Urdes, L., Walster, C. and Tepper, J. (eds) *Fundamentals of Aquatic Veterinary Medicine*. Wiley, Hoboken, New Jersey, pp. 1–29.

Kristiansen, T.S., Fernö, A., Pavlidis, M.A. and Van de Vis, H. (eds) (2020) *The Welfare of Fish*, Vol. 20. Springer International Publishing, Cham, Switzerland.

Lawrence, A.B., Newberry, R.C. and Špinka, M. (2018) Positive welfare: What does it add to the debate over pig welfare? In: Špinka, M. (ed.) *Advances in Pig Welfare*. Woodhead Publishing, Cambridge, UK, pp. 415–444. DOI: 10.1016/B978-0-08-101012-9.00014-9.

Lerner, H. and Berg, C. (2015) The concept of health in one health and some practical implications for research and education: What is one health? *Infection Ecology and Epidemiology* 5, 2530. DOI: 10.3402/iee.v5.25300.

McBride, S.D. and Morton, A.J. (2018) Indices of comparative cognition: Assessing animal models of human brain function. *Experimental Brain Research* 236, 3379–3390. DOI: 10.1007/s00221-018-5370-8.

McGreevy, P.D., Fawcett, A., Johnson, J., Freire, R., Collins, T. *et al.* (2020) Review of the online one welfare portal: Shared curriculum resources for veterinary undergraduate learning and teaching in animal welfare and ethics. *Animals* 10, 1341. DOI: 10.3390/ani10081341.

Monath, T.P., Kahn, L.H. and Kaplan, B. (2010) One health perspective. *ILAR Journal* 51, 193–198.

Murray, C.J., Ikuta, K.S., Sharara, F., Swetschinski, L., Aguilar, G.R. *et al.* (2022) Global burden of bacterial antimicrobial resistance in 2019: A systematic analysis. *The Lancet* 399, 629–655. DOI: 10.1016/S0140-6736(21)02724-0.

Paul, E.S. and Mendl, M.T. (2018) Animal emotion: Descriptive and prescriptive definitions and their implications for a comparative perspective. *Applied Animal Behaviour Science* 205, 202–209. DOI: 10.1016/j.applanim.2018.01.008.

Pethick, D.W., Ball, A.J., Banks, R.G. and Hocquette, J.F. (2011) Current and future issues facing red meat quality in a competitive market and how to manage continuous improvement. *Animal Production Science* 51, 13–18.

Premack, D. (2007) Human and animal cognition: Continuity and discontinuity. *Proceedings of the National Academy of Sciences* 104, 13861–13867. DOI: 10.1073/pnas.0706147104.

Rolls, E.T. (2005) *Emotion Explained*. Oxford University Press, Oxford, UK.

Sneddon, L.U. (2015) Pain in aquatic animals. *Journal of Experimental Biology* 218, 967–976. DOI: 10.1242/jeb.088823.

Striedter, G.F. (2019) Variation across species and levels: Implications for model species research. *Brain Behavior and Evolution* 93, 57–69. DOI: 10.1159/000499664.

Tarazona, A.M., Ceballos, M.C. and Broom, D.M. (2020) Human relationships with domestic and other animals: One health, one welfare, one biology. *Animals* 10, 43. DOI: 10.3390/ani10010043.

Urdes, L. and Alcivar-Warren, A. (2022) A comparative study on metals and parasites in shellfish of freshwater and marine ecosystems. *Journal of Shellfish Research* 40, 565–588. DOI: org/10.2983/035.040.0313.

Urdes, L., Simion, V.E., Talaghir, L.G. and Mindrescu, V. (2022) An integrative approach to healthy social-ecological system to support increased resilience of resource management in food-producing systems. *Sustainability* 14, 14830. DOI: 10.3390/su142214830.

Urdes, L., Walster, C., Tepper, J. and Foyle, L. (2024) How one health and one welfare can strengthen the evidence of a management procedure - A case study of eyestalk ablation in farmed shrimp. *Scientific Papers. Series D. Animal Science* LXVII(1), 818–829.

Urdes, L., Walster, C. and Tepper, J. (eds) (2021) *Fundamentals of Aquatic Veterinary Medicine*. Wiley, Hoboken, New Jersey.

Walster, C. and Foyle, L. (2023) Biosecurity: The use of risk assessment, surveillance, outbreak investigation, modelling disease outbreaks. In: Urdes, L., Walster, C. and Tepper, J. (eds) *Pathology and Epidemiology of Aquatic Animal Diseases for Practitioners*. Wiley, Hoboken, New Jersey, pp. 382–403.

WOAH (2024) *Animal Welfare: A Vital Asset for A More Sustainable World*. World Organisation for Animal Health, Paris. DOI: 10.20506/woah.3440. (accessed 19 June 2025).

Zhang, Z., Fu, Y., Zhao, H. and Zhang, X. (2022) Social enrichment affects fish growth and aggression depending on fish species: Applications for aquaculture. *Frontiers in Marine Science* 9, 1011780. DOI: 10.3389/fmars.2022.1011780.

2 Toward a Sustainable One Ocean: Evaluating Life Below Water Through the Lens of One Health (Ecology Introduction)

Kestrel O. Perez[1]*, Stephen J. Jordan[2] and Konstantine J. Rountos[1]
[1]Department of Biology, St Joseph's University, New York, 245 Clinton Ave., Brooklyn, NY 11205; [2]25 Wayne Memorial Drive, N. Dartmouth, MA 02747

Abstract

The chapter describes ecology from the perspective of One Health, emphasizing the importance of sustainable marine ecosystems in the implementation of One Health. United Nations Sustainable Development Goal 14: Life below water (SDG 14) and its objectives are detailed in this chapter as listed below, with real-life examples impacting One Health.

2.1 Introduction

As global human populations grow in the next century, more than two billion people are expected to inhabit the world's coasts (Reimann *et al.*, 2023). Human populations have caused a variety of direct (e.g. pollution, overharvesting) and indirect (e.g. ocean acidification, harmful algal blooms) impacts, which have impaired coastal marine habitats and organisms (He and Silliman, 2019). When these multiple stressors are coupled with the effects of climate change, additive, synergistic, or sometimes mitigating effects can manifest, depending on the ecological scale evaluated (i.e. organisms, functional group, food chain, or ecosystem) (Gissi *et al.*, 2021).

The relationships between human civilizations, animals, and the marine environment are multifaceted and deeply intertwined. While humans have benefited socially, economically, and ecologically from these aquatic ecosystems and their organisms for centuries, the stability of these relationships is reaching or exceeding tipping points (e.g. Heinze *et al.*, 2021; Rockström *et al.*, 2024). In the context of One Health, examining the intersection of humans, animals, and the integrity of marine ecosystems reveals established impacts, persistent challenges, and promising directions forward. This chapter aims to synthesize these relevant concepts and interconnections from a marine ecology perspective. Specific emphasis will be placed on (i) the effects of marine pollution (i.e. marine debris and eutrophication), overharvesting, habitat degradation, and ocean acidification on marine ecosystems, and (ii) examples of

*Corresponding author: kperez3@sjny.edu

© CAB International 2025. *One Health Concepts and the Aquatic Ecosystem*
(eds L.D. Urdes *et al.*)
DOI: 10.1079/9781800623248.0002

activities and their progress in mitigating these impacts.

2.2 Land-Based Pollution and Activities Aimed at Reducing Marine Pollution

2.2.1 Plastic debris

Plastics have been a staple in human society for several decades and are now a ubiquitous pollutant in marine environments (Derraik, 2002; Thushari and Senevirathna, 2020). While largely generated on land, the majority of this waste stays in coastal habitats (Onink et al., 2021). Plastic waste can range over a wide size spectrum, from megaplastics (>50 cm), macroplastics (>5 mm), and microplastics (<500 μm) to nanoplastics (<1 μm; Thushari and Senevirathna, 2020; Horton, 2022). Large plastic debris can accumulate and physically damage marine environments, disrupting the functioning of habitats (Le et al., 2024). Examples of direct impacts to marine organisms can also occur through entanglements or gastrointestinal issues after ingestion (Santos et al., 2021; Le et al., 2024). Smaller plastic particles, like microplastics and nanoplastics, are also a major cause for concern. These smaller particles pose threats to a broad spectrum of organ systems in marine organisms, including creating gastrointestinal blockages, affecting the immune system, reproductive system, nervous system, and endocrine system, and altering physiological processes (Vo and Pham, 2021; Zaki and Aris, 2022). Microplastic and nanoplastic pollution is particularly relevant when considering the importance of seafood to human populations (e.g. Danopoulos et al., 2020) and the potential for these contaminants to adsorb other toxicants and be transferred trophically (Parolini et al., 2023; Multisanti et al., 2025). Due to the broad spectrum of impacts to a vast array of organisms, including humans, unified and transdisciplinary approaches and policies are needed (Prata et al., 2021).

The severity of plastic pollution has spurred international efforts to address this issue. In 2022, the United Nations Environment Programme (UNEP) signed an agreement to develop a legally binding global plan to combat plastic pollution (UNEP, 2022a). This effort aimed to make progress toward a unified approach to reduce plastics (March et al., 2022). Some strategies to reduce plastic pollutants include formal policies to reduce plastic usage and promote formal and informal recycling, and the development of technologies to control plastics in solid wastes and in wastewaters (Nikiema and Asiedu, 2022; Velis et al., 2022). However, it is important to note that relying on current technology alone will likely not solve the plastics pollution issue (Nikiema and Asiedu, 2022). Ultimately, the policy details and magnitude of the global response will determine if humanity is able to reduce plastic pollution (Borrelle et al., 2020).

2.2.2 Eutrophication

Eutrophication is a persistent and menacing threat to the health of coastal, estuarine, and marine ecosystems worldwide (Rockström et al., 2024). Nutrients are released into coastal environments largely through human activities, including agriculture, wastewater treatment, septic systems, industry, and urban and suburban runoff (Dai et al., 2023). The over-enrichment of nutrients in coastal ecosystems has impacted marine organisms and ecosystem functioning, and often results in economic consequences (Horta et al., 2021). Eutrophication leads to a variety of impairments, including hypoxic zones (Dai et al., 2023), harmful algal blooms (Heisler et al., 2008), and coastal acidification (Wallace et al., 2014).

While nitrogen is the primary nutrient of concern for eutrophication in marine environments, a variety of other elements contribute (Malone and Newton, 2020; Kelly et al., 2021). For example, phosphorus, iron, and cobalt may serve as limiting or co-limiting nutrients for photoautotrophs in these systems (Browning and Moore, 2023). This can complicate policies and mitigation strategies to combat eutrophication.

Global efforts to combat eutrophication are further challenged by a heterogeneous landscape of water quality monitoring and data quality. Policies and strategies to combat eutrophication are typically found in regions where water quality monitoring is more robust and the effects of eutrophication have been well documented, while less attention has been given and less progress made in data-poor areas (Maúre *et al.*, 2021). Fortunately, progress has been made at the international level to produce awareness and frameworks for nutrient reduction strategies (UNEP, 2019). UNEP Resolution 4/14 was the strongest international nitrogen reduction commitment to date, with >80 countries taking national actions. The International Nitrogen Initiative is another example of a global effort with an ambitious goal, to halve the amount of nitrogen waste in the world by 2030 (Sutton *et al.*, 2021). Given the momentum, it is likely that reducing nitrogen waste will remain a priority at the international level, although global efforts have still not completely mitigated all of the negative externalities (Dai *et al.*, 2023).

2.2.2.1   *Total maximum daily loads*

In the United States, regulations focused on water pollution originated in 1948. It was not until the Clean Water Act (CWA) of 1972 that amendments were made related to sewage, nutrient discharge, and other water quality factors. The Environmental Protection Agency (EPA) is tasked to enforce the CWA (Russo, 2002), but there is still much more work to be done related to compliance throughout the USA (Mueller and Gasteyer, 2021). One of the strategies according to Section 303(d) of the CWA is the application of total maximum daily loads (TMDLs). The EPA mandates that states develop TMDLs for their approval, for pollutants like nutrients and others. The results of the usage of TMDLs will be highlighted below for the largest two estuaries of the USA, the Chesapeake Bay (CB) and Long Island Sound (LIS).

In the CB, nutrient management has involved a multi-stakeholder adaptive approach for the last four decades (Tango and Batiuk, 2016). The watershed represents six states and a variety of urban, suburban, rural farmland, and industrial zones. The results of the implementation of TMDLs have shown that CB's total nitrogen and total phosphorus levels are mostly improving, with reductions of 19% and 2.5%, respectively, from 1985 to 2017 (Zhang *et al.*, 2023). These results are promising, but the reduction of nutrient loads has not always

translated to better water quality (Murphy *et al.*, 2021).

Efforts to reduce nutrients in LIS have been the focus of New York and Connecticut since 1998. In addition, these states work in partnership with other stakeholders, including scientists, conservation organizations, and industry, in the LIS Study (Long Island Sound Study, n.d.). This program has been evaluating water quality in LIS for decades and is a leading source of information for the public. The results of the TMDLs for nutrients in LIS are striking, as not only has water quality generally improved in multiple areas, but bottom-water dissolved oxygen has also improved dramatically in areas prone to summer hypoxic conditions (Whitney and Vlahos, 2021).

2.2.2.2   *Harvesting aquatic organisms*

The harvesting of marine macroalgae and filter-feeding shellfish is a promising strategy to reduce nutrient loads in coastal waters (Racine *et al.*, 2021; Cakmak *et al.*, 2022). For example, areas around kelp farms have large decreases in nutrient concentrations in their vicinity, suggesting great opportunities for nutrient sequestration (Xu *et al.*, 2023). Macroalgae harvesting and/or farming activities would not only reduce nutrient levels in the water but also provide useful products for biomedical applications, feed, fertilizers, and other pharmaceutical applications (García-Poza *et al.*, 2022).

2.2.2.3   *Coastal habitat restoration*

Coastal wetlands, like salt marshes, are one of the most valuable habitats on earth, providing a myriad of ecosystem services to humans (Costanza *et al.*, 2021). Salt marshes intercept land-derived nutrients and sequester them into their biomass before they are released into the surrounding coastal waters (Valiela *et al.*, 2000). However, too much nutrient loading to these habitats can cause negative outcomes and overall habitat degradation (Turner *et al.*, 2009).

Salt marshes experience a variety of direct and indirect impacts at both the global level (e.g. atmospheric warming and sea level rise) and the local level (e.g. eutrophication, dredging, canal cutting, leveeing) (Kennish, 2001). The loss of coastal wetlands, like salt marshes, can have broader ecosystem impacts to food webs (Valiela et al., 2004). For example, Jordan et al. (2009) found that relatively small losses in this habitat could have far-reaching and long-term impacts on local fisheries. Unsurprisingly, salt marsh restoration science has a wealth of information available and strategies for successful implementation (Billah et al., 2022).

2.3 Sustainable Management and Protection of Marine and Coastal Ecosystems

2.3.1 Benefits and threats to marine ecosystems

The integrity of marine and coastal ecosystems is essential for a healthy world. These systems have major roles in regulating the climate and atmosphere. They are critical habitats for a vast array of animals, plants, and other life forms. The world's fisheries and their prominent role in human nutrition depend heavily on healthy coastal and marine environments (Sandifer and Sutton, 2014).

Human pressures on oceans, estuaries, and shores are strong and increasing. Because of CO_2 emissions from human activities, water temperatures are rising while oceans and coastal waters are becoming more acidic. Industrial, commercial, and residential infrastructure development in coastal areas can degrade or destroy habitats such as salt marshes, mangrove forests, seagrass beds, and coral reefs that are essential to fish, shellfish, and birds (Jordan and Benson, 2015). Petroleum and mineral extraction can do great damage, as in the case of the 2010 DeepWater Horizon oil well blowout off the Louisiana coast; ocean and coastal ecosystems, fisheries, and human health and welfare were affected over a wide area (Beyer et al., 2016).

Deep-sea mining is a developing enterprise with potential for damaging large areas of sensitive sea-bottom habitats along with their diverse, but poorly known, inhabitants, and for degrading ocean waters with suspended sediment and turbidity (Sharma, 2015).

To the extent that marine and coastal fisheries are not sustainable, they deplete populations of targeted species, leading to negative consequences for humans, economies, and ecosystems. Globally, a large proportion of fish stocks is fully or overexploited and under increasing pressure to supply the growing human population (Gebremedhin et al., 2021; FAO, 2024).

As described in Section 2.2, discharges of contaminated water from terrestrial sources threaten coastal waters with chemical, organic, and nutrient pollution, harmful algal blooms, and pathogens. Ocean acidification, covered in detail in Section 2.4, is also a growing global concern.

2.3.2 Conservation

The most direct and efficient approach to sustainable management of natural resources is to conserve and protect healthy habitats and biotic populations. In recent years, many marine protected areas (MPAs) have been established, with variable levels of protection, including complete exclusion of fishing and other exploitive uses (Sala et al., 2018; Maestro et al., 2019). The effectiveness of MPAs varies depending on size, location, level of protection, and enforcement, but larger and better-managed MPAs generally support healthier and more productive ecosystems than the alternatives (Graham et al., 2014). Scientifically based, ecosystem-oriented fishery management is another form of conservation that has emerged in recent decades (Fogarty, 2014; see Section 2.5). The sustainability of some fisheries is dependent on habitat conservation, where fishery managers have little or no influence (Jordan and Benson, 2013). Decisions that affect critical coastal habitats are typically made locally, with little or no consideration for cumulative effects. In this case as in others, conservation depends on sound governance (Jordan and Benson, 2013).

2.3.3 Restoration

Many marine and coastal ecosystems have been degraded by pollution, excessive fishing pressure, habitat destruction, or a combination of stresses. In these cases, the route to sustainability must be through restoration programs, which are far more costly and uncertain than conserving and protecting resources before they are degraded or depleted. Major sources of uncertainty in large-scale restoration efforts include inadequate funding and gaps in scientific knowledge (Waltham *et al.*, 2020).

In the USA, Long Island Sound, Chesapeake Bay, and Tampa Bay (TB) have been involved in major, long-term restoration programs. Each of these estuaries is subject to multiple stresses, but all have been concerned principally with nutrient (nitrogen and phosphorus) over-enrichment and attendant eutrophication. Deep water in western LIS and central CB experiences hypoxia and anoxia (depletion and absence of dissolved oxygen, respectively) seasonally. These conditions are symptoms of excess nutrient inputs stimulating algae blooms, high bacterial counts, increased turbidity, and proliferation of aquatic pathogens (Wolfe *et al.*, 2024). Management strategies in these systems will be further explored in Section 2.5. In TB, major symptoms were increasing turbidity and loss of seagrass. Despite decades of effort, LIS and CB have not fully recovered nor met all the goals established for their restoration (Carey, 2021). Tampa Bay is a different story. Nutrient management, largely through improved wastewater treatment, resulted in complete restoration of seagrass within 20 years after ecological decline was first observed. The general health of this coastal ecosystem has been maintained throughout a period of rapid population growth in the watershed. The different outcomes for the three systems are largely related to scale and complexity. Compared to western LIS and CB, TB is a compact system with low inflow and strong oceanic influence (Jordan and Benson, 2013).

Coastal ecosystem restoration programs are often directed toward specific areas or biomes, including seagrass beds, oyster reefs, coral reefs, salt marshes, or mangrove forests. The highest unit area costs were for coral reefs and seagrass beds; mangrove restoration was the least costly (Bayraktarov *et al.*, 2015). The technology of coastal habitat restoration has advanced greatly over the past few decades. For example, coral culture and transplantation have succeeded in restoration of coral reefs in some areas, although many of these projects are small in scale, poorly designed, or poorly documented (Boström-Einarsson *et al.*, 2020).

2.3.4 Resilience

The resilience of an ecosystem is its resistance to deleterious change in response to stress and disturbance. Any functioning ecosystem is in a metastable state; thus, resilience can be thought of as the magnitude of stress (e.g. climate change, pollution, resource exploitation) required to drive the system to collapse to a permanently altered, less desirable state (Cumming and Peterson, 2017). 'Less desirable' for the purposes of this chapter means less productive of ecosystem services and less supportive of human health.

What makes a marine ecosystem more or less resilient? There is no straightforward answer. Dozens of properties could influence the ability of a system to rebound or collapse under a given amount and type of pressure (Dakos and Kéfi, 2022). In practical terms, ecosystem resilience may depend more on controlling external pressures than on managing the properties of the ecosystem.

In a different context, resilience is a property of coastal communities that can be protected or enhanced by protection and restoration of marine ecosystems. Coral reefs, mangrove forests, and oyster reefs, for example, buffer shorelines against waves and storm surge, thus protecting people and property (Rolfing *et al.*, 2022; Dong and Lin, 2025).

Clearly, it is prudent to conserve and protect marine habitats and resources wherever possible. Where it is too late to conserve, then restoration is necessary. In either case, success depends on sustainable, adaptive management supported by long-term monitoring and effective, cooperative governance (Thom *et al.*, 2016).

2.4 Ocean Acidification and One Health

2.4.1 Changing chemistry in the oceans due to increasing carbon dioxide

The focus of this section is on understanding the chemistry and generalized impacts of ocean acidification, as well as strategies to minimize and address the effects of ocean acidification. Global climate change is causing a large number of impacts in the marine environment, including increasing average temperature, increasing storm activity, and increasing ocean acidification. Rising CO_2 levels are associated with lowered ocean pH (ocean acidification or OA). The ocean is one of the most important sinks for absorbing CO_2 and can buffer some of the harmful effects of increasing concentrations of greenhouse gases. When CO_2 is absorbed into the ocean, it reacts with water to form carbonic acid (H_2CO_3) (NOAA, 2025). Carbonic acid then dissociates into bicarbonate (HCO_3^-) and hydrogen ions (H^+). The extra hydrogen ions will consume carbonate to produce more bicarbonate, resulting in less carbonate available for calcifying organisms. The increasing hydrogen ion concentration in the water reduces the pH in the water. Importantly, pH is measured on the log scale, so a change from 8.1 to 8.0 can have a substantial impact.

Interestingly, acidification can also occur in coastal estuaries and bays; a number of coastal sites around Long Island were commonly found to have a pH ~7.4 in summer and fall (Wallace et al., 2014). This coastal OA is derived from eutrophication of these bays and estuarine regions, resulting in high algal biomass and subsequent breakdown by microbes, which causes hypoxia and increased carbon dioxide as a by-product of bacterial respiration (Wallace et al., 2014).

2.4.2 Impacts of OA—a focus on early life stages of shelled molluscs and fish

Impacts of OA and coastal acidification are widespread and vary based on the type of organism. Because the focus of other authors has been reviewing impacts across a wide range of organisms (i.e. Kroeker et al., 2013), here we first focus on describing the impacts on shelled molluscs, including bivalves and gastropods, due to their high diversity as well as economic and ecological value (Gazeau et al., 2013). Shelled molluscs begin to form their shell, composed of up to 99% calcium carbonate, in the form of aragonite and/or calcite, during the larval stage, and then continue to develop throughout their maturation. A review of the scientific literature on the impacts of OA on shell formation in molluscs

found that shell growth decreased (Gazeau et al., 2013). Interestingly, these authors noted that some studies reported that although overall shell area was not reduced following exposure to OA, the thickness of the shell was reduced, which would likely have subsequent impacts on species survival following predation exposure. Additionally, Gazeau et al. (2013) found that research studies had also identified exposure to OA causing a number of other impacts that were commonly either negative or neutral, including survival, clearance rates, and reduced immune responses. Of the traits examined, changes in respiration rates were highly variable with some studies reporting positive impacts from OA exposure, others reporting neutral impacts, and still others reporting negative impacts (Gazeau et al., 2013).

Early life stages are thought to be particularly important to the study of impacts from OA exposure as survival rates during the early life of invertebrate molluscs and fish are commonly very low. Thus, understanding the factors that influence the survivors is very valuable to the overall population. Of the studies reviewed by Gazeau et al. (2013) that studied the impacts of OA on embryonic and larval shelled molluscs, the effects were largely negative, particularly with regards to survival, size, development rate, shell structure, and metamorphosis. Of the studies reviewed, 64% of the articles measured the impacts of OA as a single stressor, while 27% explored the combined impacts of OA and temperature. Six percent or fewer of the studies measured the impacts of other stressors in combination with OA (Gazeau et al., 2013). Exposing larval stages of three shellfish species to OA showed interesting species-specific impacts, with larval hard clams (Mercenaria mercenaria) and bay scallops (Argopecten irradians) having substantially reduced survival compared to Eastern oysters (Crassostrea virginica) following exposure to increased acidity (Talmage and Gobler, 2009). All three species examined during their early life stages (Talmage and Gobler, 2009) showed reduced growth and developmental rates. Larval hard clams (M. mercenaria) and bay scallops (A. irradians) additionally had reductions in shell development, shell thickness, and lipid accumulation when exposed to low pH relative to current conditions and pre-industrial levels (Talmage and Gobler, 2010).

To increase the ecological realism of OA experiments, there is a need for more studies that include the effects of multiple stressors, instead of exploring the impact of a single factor. This approach allows for the identification of potential synergistic effects, where the combined impact of multiple factors is greater than the sum of the individual factors. These impacts may be common and may be particularly relevant for fish and invertebrates during their early life stages (Baumann, 2019). For example, an additive negative or synergistic negative impact was found with reduced survival of early life stages of two silverside species (*Menidia sp.*) when fish were exposed to low pH and low dissolved oxygen in combination compared to fish exposed to a single stressor or control conditions (DePasquale *et al.*, 2015). This research demonstrates that the effects of dual stressors can be complex and further study is needed on multi-stressors and also on the adaptive potential of marine species (Browman, 2017).

Further research is also warranted on a range of life stages, species, and traits of interest. For example, Clements *et al.* (2022) reported a 'decline effect' after reviewing the scientific literature on OA impacts on fish behavior, and concluded that although initial studies in the field reported significant impacts of OA on fish behavior, more recently conducted studies have not been successful in replicating those results (Clements *et al.*, 2022). This suggests that OA may not strongly impact behavioral traits (Clements *et al.*, 2022).

2.4.3   Strategies to limit impacts of OA and coastal OA

In light of the widespread impacts of OA and the open questions regarding the impacts of OA in combination with other factors, what are currently the ideal mitigation strategies to lessen these impacts? Due to their calcium carbonate skeleton, corals are also a group considered to be at high risk from future OA. Albright and Cooley (2019) described promising methods of coral reef restoration as well as strategies to reduce impacts of OA on corals. Importantly, this area of mitigation and restoration efforts requires further research (Albright and Cooley, 2019).

Following degradation, coral reefs may be rebuilt using passive or active methods. For example, Albright and Cooley (2019) describe passive restoration that focuses on removal of stressors (e.g. overfishing, poor water quality from nutrient loading) and maintaining high population size and genetic diversity to maximize the evolutionary potential. Another example of passive restoration is holistic spatial planning to account for resilient species and habitats and maintaining gene flow corridors that allow for migration and genetic exchange between MPAs (Albright and Cooley, 2019). These authors also describe active restoration of coral habitats, including development of artificial reefs and 'coral gardening,' where small corals are grown in a nursery habitat and then placed in degraded reefs (Albright and Cooley, 2019). These restoration practices tend to be smaller in scale and can be very costly. Another example of active restoration is 'assisted evolution,' which needs more research but includes using selective breeding for a desired trait, epigenetic acclimatization, and the addition of beneficial microbes or algae to increase the reliance of corals to OA. Epigenetic acclimatization leading to higher resistance to pH within a generation or lasting across multiple generations is a potential restoration strategy that requires more research (Albright and Cooley, 2019).

Several promising mitigation strategies are also discussed in Albright and Cooley (2019). These methods would reduce the amount of CO_2 either by maximizing sinks or through the addition of neutralizing chemicals. One method of maximizing sinks is through phytoremediation using seagrass beds and seaweeds (e.g. kelps), which absorb carbon dioxide while going through photosynthesis. Importantly, carbon storage efficiency varies with species of seagrass and habitat type (Lavery *et al.*, 2013). Supplementation with neutralizing chemicals like limestone or olivine could help buffer the harmful effects by neutralizing the acidity (Albright and Cooley, 2019). In addition, particularly for coastal regions, where coastal OA is occurring, reducing eutrophication is valuable as a mitigation effort to reduce increased carbon dioxide levels (see Section 2.2.2).

2.5   Sustainable Fishing in Global Oceans

2.5.1   State of fisheries and the need for more sustainable fishing

Fisheries have supported human societies for centuries. These industries continue to provide not only safe and nutritious food but also ample economic support to coastal communities (FAO, 2024). In the past 40 years, however, there has been increased concern about the longevity and health of global fisheries (Froese et al., 2012; FAO, 2024). Global fisheries catches have largely stagnated (FAO, 2024), illegal, unreported, and unregulated (IUU) catches are considerable (Watson and Tidd, 2018), and groups like large predatory fish have seen rapid declines in their global biomass (Christensen et al., 2014). These trends have supported the urgency for development of the UN's Sustainable Development Goal (SDG) 14, aimed at ending overfishing and promoting sustainable harvesting strategies. This section will review the progress in making global fisheries more sustainable from a One Health perspective. It will review the status and efforts for combating IUU fisheries and explore the growing need for and acceptance of ecosystem-based fisheries management (EBFM) approaches. Examples of EBFM applications will be provided for a variety of species groups.

2.5.2   Confronting illegal, unreported, and unregulated fisheries

The impact of IUU fishing remains an ever present global phenomenon, spanning all regions and fishery types (Temple et al., 2022). IUU fishing has considerable economic and ecological effects, making it a large concern for global fisheries (FAO, 2024). Although a plethora of legally binding (e.g. the United Nations Convention on the Law of the Sea (UNCLOS), the FAO's 'Agreement to Promote Compliance with International Conservation and Management Measures by Fishing Vessels on the High Seas,' and others) and voluntary (e.g. the FAO's 'International Plan of Action to Prevent, Deter and Eliminate Illegal, Unreported and Unregulated Fishing') regulatory instruments

have existed for decades to combat IUU fishing, progress is slow (Temple et al., 2022). For example, Shen and Huang (2021) found that the lack of progress in combating IUU fishing in China was largely due to ineffective policy implementation, lack of regulation, and ineffective supervision. More broadly, countries with effective fisheries management, regulations, and robust surveillance and patrols experience less IUU fishing (Petrossian, 2015). This is particularly important given the strong evidence of IUU fishing being linked with criminal organizations and governmental corruption (Liddick, 2014). Cabral et al. (2018) found that combating IUU fishing produces positive externalities, including increased catches and profits at the country, regional, and global levels. Ultimately, fisheries stakeholders in areas affected by IUU fishing should work collaboratively with criminologists to address this issue (Petrossian, 2015).

2.5.3   The need for ecosystem-based fisheries management and moving beyond single-species maximum sustainable yield approaches

While global efforts to improve fisheries have been a priority for several years, there is variable and uneven progress to meet the 2030 SDG goals at the country level (FAO, 2024). Data deficiencies for many of the world's assessed and unassessed fisheries also remain a major impediment to sustainable fisheries for this SDG (Ovando et al., 2021). Although the issues impeding sustainable global fisheries have been widely acknowledged, there are a variety of perspectives as to the solutions (Browman et al., 2004; Pikitch et al., 2004).

Traditional fisheries management approaches are based on a single-species perspective, which evaluates the biological characteristics (i.e. biomass, growth rates, reproduction, recruitment, etc.) and the mortality terms (i.e. natural and fisheries mortality terms) of the species being targeted by the fishery (Jennings et al., 2009). Based on this information, fisheries scientists calculate the maximum sustainable yield (MSY), which is the biomass that could be extracted while still ensuring that the population growth rate be at its highest. This is typically at a biomass of 50% of the estimated carrying capacity (i.e. K/2) of the population. The use of traditional MSY-based fisheries management is effective in certain cases, when considering a single-species perspective and in areas that have good regulations (Hilborn and Ovando, 2014). However, the health and longevity of fish stocks managed from a single-species perspective has very limited success globally (Link, 2010) and is not a panacea for all types of fisheries (Pikitch *et al.*, 2004; Karnauskas *et al.*, 2021).

By definition, MSY is 'sustainable' in the context of the targeted species (Mace, 2001). However, no targeted species, or the fishery targeting them, exists in a vacuum where they do not interact with other components of the ecosystem (Link, 2010; Karnauskas *et al.*, 2021). Applying a single-species approach is therefore risky in coastal and marine environments, where all species, targeted or not, are integrated into food webs and several fisheries gear types have the ability to cause negative externalities (e.g. habitat degradation, bycatch, etc.) (e.g. Link, 2010; Karnauskas *et al.*, 2021). Appraisal of important ecosystem-based considerations for the management of single species typically leads to more conservative quotas than single-species MSY estimates alone would, thus improving the likelihood of satisfactory management outcomes for the ecosystem and its integrity (Link, 2010; Karnauskas *et al.*, 2021).

2.5.4 Considerations and application of EBFM for marine species

As global fisheries target a diversity of ocean life, it is unrealistic and harmful to assume that a singular management approach, like a single-species MSY approach, would lead to sustainable fisheries outcomes (Pikitch *et al.*, 2004; Link, 2010). This is also due to the fact that catches often exceed the estimates derived from single-species management approaches (Mace, 2001). While all species would benefit by being managed with an ecosystem-based approach, the importance of EBFM is exemplified when considering species with unique characteristics.

Ecosystem engineers are organisms that create or ameliorate their surrounding habitat (Wright and Jones, 2006). These species, and

the broader ecosystems they inhabit, benefit from EBFM as opposed to single-species focused management. For example, reef building oysters provide a variety of ecosystem services, ranging from erosion control and water filtration to habitat, in addition to their support of fisheries (Freitag *et al.*, 2018). The overharvesting and collapse of oyster reefs due largely to traditional single-species management has created significant turmoil in many estuaries, such as decreased water quality, loss of nursery habitat, fisheries losses, and other issues (Kirby, 2004). In many coastal areas, efforts to restore oyster reefs are a high priority in order to try to restore these fisheries and habitats (e.g. Schulte and Burke, 2014; Gobler *et al.*, 2022).

Long-lived marine species have a variety of attributes which make them particularly vulnerable to overfishing (Heppell *et al.*, 2005). This includes their high migration, schooling and reproductive aggregating behaviors, and relatively late age of reproductive maturity (Schindler *et al.*, 2002; Heppell *et al.*, 2005). Tuna species are an example of a fast growing and highly fecund fish where EBFM has allowed for improvements in their assessments but less progress on bycatch and other effects of the ecosystem (Juan-Jordá *et al.*, 2018). These species are vulnerable due to their high market value (FAO, 2024) and often predictable migration and spawning aggregation behavior (Fromentin and Lopuszanski, 2014; Hernández *et al.*, 2019). In contrast, sharks are large predators with slow growth rates and low fecundity, making their populations more vulnerable to bycatch (Schindler *et al.*, 2002). It is not surprising, therefore, that calls for more holistic and ecosystem-based approaches to shark management have been recommended (Booth *et al.*, 2019). EBFM for both tuna and shark species could be better facilitated by greater observer monitoring, particularly in regions where fisheries regulations need to be more closely scrutinized (Gilman *et al.*, 2017).

At intermediate trophic levels, forage species are relatively small (typically ≤30 cm), short-lived, and schooling species that include fish (e.g. anchovies, sardines, menhaden, etc.) and invertebrates (e.g. krill, cephalopods, etc.) (Rountos, 2016). Forage species occupy a critical ecological role in transferring energy from the base of the food web to upper trophic-level predators (Pikitch *et al.*, 2014). In addition to serving as important prey for larger fish, seabirds, and marine mammals, these species also make significant economic contributions, both directly and indirectly, to global fisheries (Pikitch *et al.*, 2014; Ruzicka *et al.*, 2024). Traditional management approaches have largely failed in the sustainable management of forage fisheries, as they do not consider their relatively short life span, 'boom and bust' population cycles, and vulnerability to overexploitation due to their schooling behavior (Pikitch *et al.*, 2012; Peck *et al.*, 2024). Based on these attributes, and the broad overlap in the trophic levels that forage fisheries and marine predators target (Rountos *et al.*, 2015), EBFM approaches for forage species are imperative (Pikitch *et al.*, 2012; Siple *et al.*, 2019.). This has been further supported through management strategy evaluation (MSE), which has assessed the fishing of forage species using traditional MSY approaches, as opposed to more conservative EBFM-derived catch quotas. These studies have revealed trade-offs in these approaches and the increased risk of forage fishery collapse when using traditional constant harvest rates at MSY (Pikitch *et al.*, 2012; Siple *et al.*, 2019). Ultimately, knowledge gaps still exist, and the use of MSE to assess harvest control rules will likely drive the management of forage species toward more sustainable outcomes (Peck *et al.*, 2024; Rooper *et al.*, 2024).

2.6 Conservation of Coastal and Marine Areas: A Focus on National and International Laws and Available Scientific Information

2.6.1 International overview

The laws, policies, practices, and entities involved in marine conservation form complex webs at scales ranging from local to global. At the grandest scale, UNCLOS went into effect in 1994; presently it has 169 signatory countries plus the European Union (EU). The treaty has not been ratified by the US Senate. The broad policy statement of UNCLOS is 'the area of the sea-bed and ocean floor and the subsoil thereof, beyond the limits of national jurisdiction, as well as its resources, are the common heritage of mankind, the exploration and exploitation of which shall be carried out for the benefit of mankind as a whole, irrespective of the geographical location of States' (United Nations Convention on the Law of the Sea, n.d.).

The EU enacted legislation in 2008, the Marine Strategy Framework Directive (European Commission, EU Marine Strategy Framework Directive, n.d.), that required member states to develop strategies for all marine ecosystems within their jurisdictions (Baltic, Black, and Mediterranean Seas, and the north-east Atlantic Coast) to achieve 'good' status, based on several indicators, by 2020. These qualitative indicators are:

1. Biodiversity is maintained.
2. Non-indigenous species do not adversely alter ecosystems.
3. Populations of commercial fish and shellfish species are healthy.
4. Food webs ensure long-term abundance and reproduction of species.
5. Eutrophication is reduced.
6. Sea floor integrity ensures the proper functioning of ecosystems.
7. Permanent alteration of hydrographical conditions does not adversely affect ecosystems.
8. Concentrations of contaminants give no pollution effects.
9. Contaminants in seafood are at safe levels.
10. Marine litter does not cause harm.
11. Introduction of energy (including underwater noise) does not adversely affect the ecosystem.

The EU's Integrated Maritime Policy, with the purpose of better coordination across policy areas, has five cross-cutting topics: blue growth (i.e. optimizing maritime economic benefits while maintaining a sound environment), maritime data and knowledge, maritime spatial planning, integrated maritime surveillance (i.e. monitoring and data collection), and sea basin strategies for each of the five major basins in the EU (European Commission, Integrated Maritime Policy, n.d.).

The goal of marine (or maritime) spatial planning (MSP) is to organize spatially and temporally the many and various activities in the marine environment, including fishing, aquaculture, energy production, mining, shipping, etc., so that they are sustainable and do not conflict. A majority of maritime nations are engaged in various stages of MSP (Santos *et al.*, 2019). In 2014, the EU adopted a directive establishing a Framework for Maritime Spatial Planning, requiring member states to complete their plans by 2021 (EU Framework for Maritime Spatial Planning, n.d.).

2.6.2 Marine conservation at the national level

There are more than 200 countries and dependencies with marine coastlines, so for the purposes of this chapter, we focus on a few prominent examples to illustrate similarities and differences in how nations approach marine conservation. It is probably a safe assumption that qualitative policy goals for marine conservation do not differ greatly among maritime nations, but the means and processes of implementation might vary significantly. Below, we briefly summarize the approaches taken by three large countries with expansive maritime zones: Australia, China, and the USA.

2.6.2.1 Australia

Australia is completely surrounded by the ocean, which is of vital importance to the nation's identity and economy. The country is in the late stages of completing its Sustainable Ocean Plan, in draft as of March 2025 (Commonwealth of Australia, 2024). One of the seven focus areas of the plan is 'Protect and Restore,' matching the theme of this chapter. 'The ocean is healthy; species and ecosystems are increasingly protected, resilient and recovering; and key threats are being addressed effectively' (Commonwealth of Australia, 2024, p. 19) is the proposed outcome. Australia has adopted the '30 x 30' goal of the Kunming-Montreal Global Biodiversity Framework (UNEP, 2022b) by setting national targets to protect 30% of its lands and waters by 2030, including 30% of its marine and coastal waters. The country has one of the world's largest distributions of MPAs, covering 48% of their Exclusive Economic Zone; 22% of this area is in no-take zones. The greatest

concern for the health of Australia's marine habitats is climate change, followed by ocean pollution, especially plastics.

2.6.2.2 China

The mainland coast of China extends 18,400 km; islands add another 13,600 km to the total coastline (Wang and Aubrey, 1987). The country appears to have a strong, sophisticated approach to protecting and restoring marine ecosystems. An excerpt from a government white paper states: 'China adopts a holistic approach to protecting the marine eco-environment, attaching equal importance to development, protection, pollution prevention and control, and restoration. It has improved the management of the marine eco-environment through land-sea coordination, coordinating protection work for rivers and seas, mountain and sea areas, onshore and offshore areas, and upstream and downstream river basins. China has established a mechanism for collaborative protection, governance, supervision, and law enforcement across different regions and departments, in order to create a comprehensive system for governing coastal areas, river basins, and sea areas' (The State Council Information Office of the People's Republic of China, 2025). Even if this policy is more aspirational than operational, it expresses well the philosophy needed for sound management of marine and coastal systems.

The country has developed a multi-tiered classification system for coastal and marine ecosystems, applied a 'red line system' to protect sensitive areas, and has an expansive array of MPAs. China recognizes two types of MPAs: Marine Nature Reserves, which are no-take, fully protected MPAs, and Special Marine Protected Areas, which may have multiple uses while protecting biodiversity and habitat values. Aquatic Germplasm Reserves protect habitats for reproduction and migration of important fish species but do not qualify as MPAs. Overall, about 13% of China's marine systems are protected in these zones. Protection is skewed toward shallow-water habitats, with more attention needed for offshore, deepwater areas (Bohorquez *et al.*, 2021).

2.6.2.3 United States

In the USA, a wide range of laws, policies, and agencies have roles in marine conservation, as summarized in Jordan and Benson (2013). The history of marine environmental policy in the USA goes back to 1871, with the creation of a federal Fish Commission to study declines in important fisheries. Its successor is today's National Marine Fisheries Service (NMFS), a part of the Department of Commerce (US Department of Commerce, National Oceanic and Atmospheric Administration, n.d.). The NMFS administers the Magnuson–Stevens Fishery Conservation and Management Act, which was enhanced by the Sustainable Fisheries Act of 1996 and a 2007 reauthorization (NOAA, 2007). Management of fisheries in US federal waters (from state boundaries to the 200 nautical mile (370.4 km) limit of the exclusive economic zone—EEZ) is through eight regional fishery management councils. Voting council members include state fishery management officials and representatives from commercial, recreational, and charter fisheries; others (e.g. scientists) are appointed by state governors with the approval of the Secretary of Commerce. The MSA is a comprehensive body of law designed to, *inter alia*, prevent overfishing, minimize bycatch and bycatch mortality, identify essential fish habitat areas, protect deep-sea corals, and provide for sustainable and equitable fisheries.

The Coastal Zone Management Act of 1972 (CZMA) established the world's first comprehensive set of policies and programs intended to conserve and protect coastal areas, including the Estuarine Research Reserve System (Office for Coastal Management, n.d.). The CZMA provided planning assistance to states, each of which developed its own coastal management program consistent with overall federal policy. The national program continues to support state programs with grants for research and management (Botero *et al.*, 2023).

The US National Ocean Policy was established by an Executive Order in 2010. It requires coastal and marine spatial planning, implemented through several state and regional programs (Halpern *et al.*, 2012). In this way, it serves as a modern extension and enhancement of the CZMA.

2.6.3 Marine conservation at sub-national scales

States, provinces, municipalities, regional consortia, and non-governmental organizations all have roles in protecting and restoring the marine environment, mostly in the coastal zone and connecting watersheds. In the USA, coastal states manage marine fisheries within their boundaries through regional interstate fishery management commissions (see e.g. the Atlantic States Marine Fisheries Commission, n.d.), similar to the fishery management councils that manage offshore fisheries. The Coastal Zone Management programs are operated by the states with federal assistance. Major programs to protect and restore coastal ecosystems involve multi-jurisdictional compacts, such as the Chesapeake Bay Program (six states and the District of Columbia; Chesapeake Bay Program, n.d.).

2.6.4 Scientific information

Over the past several decades, the amount of scientific data and information relevant to marine conservation has increased enormously. The more that is known about these ecosystems and their functioning, the more likely it is that they can be sustained through sound management. Nevertheless, current knowledge is not perfect and is especially deficient for deep ocean realms, where data collection is difficult. It is encouraging that the UN is now conducting global ocean assessments (Evans *et al.*, 2021).

Monitoring is a crucial element of conservation, restoration, and sustainability. Well-designed, long-term monitoring supports: (i) tracking the status and trends of ecosystems and their components; (ii) modeling to predict outcomes of management measures and environmental change; and (iii) adaptive management, which builds on experience and knowledge to improve the ways in which we attempt to build a more sustainable environment (Ringold *et al.*, 1996).

2.7   Development of Marine Technology to Improve Conservation Efforts

2.7.1   Introduction on emerging technologies relevant to marine conservation

The Intergovernmental Oceanographic Commission (IOC) branch of the United Nations Educational, Scientific and Cultural Organization (UNESCO) is part of the United Nations Decade of Ocean Science for Sustainable Development from 2021 to 2030 (Intergovernmental Oceanographic Commission, n.d.). This branch of the UN promotes the development and transfer of marine technology, which is particularly important for developing nations. As many technologies are relevant to aiding marine conservation initiatives, here we focus on several technologies as case studies that include unmanned vehicles like drones, underwater acoustic devices, emerging survey technologies, aquaculture advances, and artificial intelligence.

2.7.2   Unmanned aerial vehicles and underwater listening devices

The establishment of MPAs is a conservation strategy that has been reported to provide many benefits (Sections 2.3 and 2.6). One of the challenges associated with MPAs lies with the ability of developing countries to effectively manage and enforce the regulations across the entire protected region with ongoing surveillance to prevent illegal fishing within the MPA. One possible marine technology that may improve the ability to effectively monitor protected areas is the use of drones (Reis-Filho *et al.*, 2022). Jiménez López and Mulero-Pázmány (2019) identified 99 reports in the literature of studies evaluating the use of drones as a surveillance tool in management of MPAs. Drones provide a cost-efficient manner of recording pictures or videos to monitor species and volume of catch

and track illegal activities. Although Jiménez López and Mulero-Pázmány (2019) note that a number of challenges still remain, such as short flying time and the inability to fly during inclement weather, these authors suggest that drones show great promise as a conservation tool for MPAs. One example of an organization that focuses on increasing drone usage for conservation applications in developing nations is Conservation Drones (n.d.).

Stationary and mobile acoustic devices, as part of the Integrated Undersea Surveillance System (IUSS) developed by the United States Navy, have been used to track whale movement for a period of over a month by their unique songs and calls and can help determine species distribution and abundance (De Alessi, 2003). Additionally, hydrophones, part of the IUSS, can precisely identify vessel- and propellor-specific sound characteristics to determine the location of boats should they be fishing illegally in protected waters.

Marine technologies involving underwater listening devices also have applications to lessen harmful fishing practices, such as blast fishing. Blast fishing, where explosives are used, damages critical habitats and stuns fish that are then collected with nets or by divers (Braulik *et al.*, 2017). The entire coast of Tanzania was acoustically surveyed using hydrophones towed from a catamaran (Braulik *et al.*, 2017). Hydrophones were an effective tool to identify blast fishing in Tanzania, and 62% of the 318 blasts were found to be located near a specific region of the country. This data could be used to help identify ideal locations for placement of enforcement vessels.

2.7.3   Emerging survey technologies: BRUVs and eDNA

Accurate survey data is critical for successful conservation efforts. Due to advances in video resolution, baited remote underwater video systems (BRUVs) are an efficient and non-destructive method of surveying fish species (Harvey *et al.*, 2013). With this video system, the size of fish can subsequently be measured along with species abundance and diversity. Some limitations that were noted by Harvey *et al.* (2013)

were the lack of standard operating methodologies and that some fish species may be over- or under-estimated based on unique behavioral interactions (i.e. hiding) or attraction to the bait.

Marine environmental DNA (eDNA) is another non-invasive method that can be used to survey species diversity, where DNA is measured from a sample of water. Estimates of species diversity determined at four sites in New Zealand were compared for sampling using eDNA and sampling using BRUVs (Jeunen et al., 2020). Representatives from many more families were detected using the eDNA approach compared to the BRUVs technology (i.e. 56 vs 7), however, some fish that were known to be present from the video were not detected by particular eDNA assays, suggesting that some false negatives had occurred and that multiple eDNA assays should be run to reduce the rate of false negatives (Jeunen et al., 2020).

One of the strengths of eDNA technology is the ability to detect rare species as well as species that are highly camouflaged or have the behavioral tendency to remain hidden during video surveys. A total of 21 studies that had utilized eDNA approaches to survey the abundance of marine mammals was reviewed, and this technology was found to commonly identify more species than traditional survey techniques and to be efficient in detecting the presence of highly endangered species (Suarez-Bregua et al., 2022).

2.7.4 Artificial intelligence

Large monitoring datasets can have long time lags before enough data is processed to be useful for conservation managers. One possible way that technology can reduce this issue is by implementing artificial intelligence (AI) to speed up data processing (Ditria et al., 2022). For example, AI was found to be very accurate, matching expert readers up to 97% when analyzing coral abundance from images of coral reefs collected by digital photography (Gonzalez-Rivero et al., 2020). The potential benefit of incorporating AI into image processing is a significant reduction in processing time and overall total cost.

2.7.5 Satellite technologies

In many marine habitats, from coral reefs to the high seas, illegal fishing is an ongoing challenge for management. For coral reef conservation in particular, several technologies show great promise, such as satellite imagery (Madin et al., 2019). Madin et al. (2019) list a number of open access or low-cost emerging technologies that can help to protect reefs and other marine habitats, such as Global Fishing Watch (Global Fishing Watch, n.d.), which maps the activity and routes of boats from space to allow detection of illegal fishing. These authors also note that, although emerging technologies can be of value, effective training is needed. Fishing can be a challenge to regulate, particularly in open ocean regions that are outside the boundaries of the 200 nautical mile (370.4 km) EEZ (Dunn et al., 2018). Dunn et al. (2018) proposed using the automatic identification system (AIS), a system that is mandatory on many vessels (although specific requirements vary by country) to record and transmit GPS location and boat identity to AIS-specific satellites to help track and identify illegal fishing activity.

2.7.6 Aquaculture technology

Due to their high protein content, fish are extremely valuable as a source of food. Improvements that have been developed over time to increase gear efficiency, such as sonar technology, have made it easier to overexploit wild fish stocks (Jennings et al., 2009). The majority of fisheries have not been sustainable even in historical times when the human population, and thus demand, was much smaller than it is now (Pauly et al., 2003). North Atlantic cod populations have been decimated to a point where fishing of this species was banned in 1992, and this collapse is thought to be due to overfishing and does not appear to be the result of poor recruitment (Myers et al., 1997). Thus, to meet the high demand for fish protein, aquaculture production, or rearing fish and shellfish in pens or ponds, has been rapidly increasing in many parts of the world, whereas wild fishery collections are static or are decreasing (Muir, 2005). In fact, the proportion of fish and shellfish protein that originates from aquaculture is expected to double in volume by 2050 (Stentiford *et al.*, 2020). To help meet this enhanced aquaculture production in a sustainable manner, Stentiford *et al.* (2020) proposed One Health-focused strategies including human health, organism health, and environmental health with relevant success metrics.

There are three different types of aquaculture practices (Muir, 2005). The first, intensive aquaculture, is the most costly, as the individuals reared are completely reliant on artificial feeding. For carnivorous species that require fish in their diet, aquaculture is an increasing strain on wild fisheries as these species may require a biomass of prey (commonly small planktivorous fish, such as anchoveta) that is five times greater than their generated production biomass (Naylor *et al.*, 2000). Semi-intensive aquaculture is where food is only supplemented for reared individuals, whereas extensive aquaculture, being the most cost-efficient practice, is where reared individuals are not fed at all but are able to obtain nourishment naturally through their environment (Muir, 2005). Many bivalve molluscs, filter-feeding fish, and marine macroalgae lie in the extensive category as they feed on phytoplankton drifting through the hatchery area or are photosynthetic and do not rely on outside nourishment (FAO, 2024).

What strategies are there for improving aquaculture efficiency for the future? There have been several decades of extensive research conducted on how to make aquaculture feeding more cost-efficient and less reliant on wild fish stocks. This has been a challenge because fish are naturally rich in a wide range of essential fatty acids, which can be crucial to rapidly growing larval fish for cell membrane structure and function and cardiovascular function (Sargent *et al.*, 1999). Fortunately, the aquaculture industry has become more efficient in reducing the amount of marine resources used for fish food (Naylor *et al.*, 2021). Although continued research is warranted, substitution of fish meal and fish oil with alternative options, indicated by a lower fish-in to fish-out ratio, is becoming increasingly possible and may reduce exploitation of wild fish for aquaculture feed (Naylor

et al., 2009; Ma and Hu, 2024). Utilizing life cycle assessment (LCA) to understand environmental impacts can help design aquaculture systems to minimize harmful effects from nutrient pollution, aquaculture feed, and high energy usage (Henriksson *et al.*, 2012). Additionally, increased incorporation of multi-trophic aquaculture, where rearing of multiple species that each fill a trophic niche is implemented, could decrease pollution and reduce energy expenditure (Bostock *et al.*, 2010). Similarly, the release of pollution from aquaculture into surrounding environments could be decreased by increased incorporation of recirculating systems to filter and reuse the water (Naylor *et al.*, 2021). The size of pens and ponds and the volume of water exchange are some of the limits of aquaculture. One potential solution is moving aquaculture offshore, particularly for larger carnivorous species like tuna, salmon, and cobia, which would increase water exchange rates and reduce pollution impacts in nearshore regions (Bostock *et al.*, 2010). Although extensive aquaculture, specifically macroalgae and mollusc production, has increased over time, this sector has more opportunity for continued growth as these species are extractive and thus more efficient to raise due to having no reliance on fish meal from wild fish stocks (Naylor *et al.*, 2021).

In conclusion, the future of healthy and sustainable oceans relies on an understanding of how humans, plants and animals, and the abiotic world interact and are affected by global stressors. The intersection of humans, animals, and the integrity of aquatic ecosystems through the One Health lens reveals a plethora of human-derived impacts, persistent challenges, and fruitful and constructive ways forward. As humans are now a major driving force in the changes observed on our planet (Lewis and Maslin, 2015), there is much work to be done to find effective solutions (Rockström *et al.*, 2024). This chapter has demonstrated that there exists a growing inventory of scientifically justified solutions to improve the condition of our global oceans. Scientifically based policies which consider socio-economic aspects and are derived from the efforts of all stakeholders should be the most effective. Global, regional, national, and local policies and regulations have the potential to continue to make progress for our planet, through the reduction of human-derived wastes, overconsumption, bycatch, and ecosystem effects. Ultimately, the key to long-lasting solutions for the sustainability of our oceans lies, first, in a paradigm shift in the day-to-day living expectations and behavior of our conspecifics (Antal and Hukkinen, 2010), and second, in recognition, acceptance, and constructive engagement by the governmental and commercial sectors (FAO, 2024).

References

Albright, R. and Cooley, S. (2019) A review of interventions proposed to abate impacts of ocean acidification on coral reefs. *Regional Studies in Marine Science* 29, 100612.

Antal, M. and Hukkinen, J.I. (2010) The art of the cognitive war to save the planet. *Ecological Economics* 69, 937–943.

Atlantic States Marine Fisheries Commission (n.d.) Atlantic States Marine Fisheries Commission. Available at: https://asmfc.org (accessed 16 June 2025).

Baumann, H. (2019) Experimental assessments of marine species sensitivities to ocean acidification and co-stressors: How far have we come? *Canadian Journal of Zoology* 97, 399–408.

Bayraktarov, E., Saunders, M.I., Abdullah, S., Mills, M., Beher, J. *et al.* (2015) The cost and feasibility of marine coastal restoration. *Ecological Applications* 26, 1055–1074. DOI: 10.1890/15-1077.

Beyer, J., Trannum, H.C., Bakke, T., Hodson, P.V. and Collier, T.K. (2016) Environmental effects of the Deepwater Horizon oil spill: A review. *Marine Pollution Bulletin* 110, 28–51. DOI: 10.1016/j.marpolbul.2016.06.027.

Billah, M.M., Bhuiyan, M.K.A., Islam, M.A., Das, J. and Hoque, A.R. (2022) Salt marsh restoration: An overview of techniques and success indicators. *Environmental Science and Pollution Research* 29, 15347–15363.

Bohorquez, J.J., Xue, G., Frankstone, T., Grima, M.M., Kleinhaus, K. *et al.* (2021) China's little-known efforts to protect its marine ecosystems safeguard some habitats but omit others. *Science Advances* 7, 46. DOI: 10.1126/sciadv.abj1569.

Booth, H., Squires, D. and Milner-Gulland, E.J. (2019) The neglected complexities of shark fisheries, and priorities for holistic risk-based management. *Ocean & Coastal Management* 182, 104994.

Borrelle, S.B., Ringma, J., Law, K.L., Monnahan, C.C., Lebreton, L. *et al.* (2020) Predicted growth in plastic waste exceeds efforts to mitigate plastic pollution. *Science* 369(6510), 1515–1518.

Bostock, J., McAndrew, B., Richards, R., Jauncey, K., Telfer, T. *et al.* (2010) Aquaculture: Global status and trends. *Philosophical Transactions of the Royal Society B: Biological Sciences* 365(1554), 2897–2912.

Boström-Einarsson, L., Babcock, R.C., Bayraktarov, E., Ceccarelli, D., Cook, N. *et al.* (2020) Coral restoration – a systematic review of current methods, successes, failures and future directions. *PLoS ONE* 15(1), e0226631. DOI: 10.1371/journal.pone.0226631.

Botero, C.M., Milanes, C.B. and Robledo, S. (2023) 50 years of the Coastal Zone Management Act: The bibliometric influence of the first coastal management law on the world. *Marine Policy* 150, 105548. DOI: 10.1016/j.marpol.2023.105548.

Braulik, G., Wittich, A., Macaulay, J., Kasuga, M., Gordon, J. *et al.* (2017) Acoustic monitoring to document the spatial distribution and hotspots of blast fishing in Tanzania. *Marine Pollution Bulletin* 125(1–2), 360–366.

Browman, H.I. (2017) Towards a broader perspective on ocean acidification research. *ICES Journal of Marine Science* 74, 889–894.

Browman, H.I., Stergiou, K.I., Cury, P.M., Hilborn, R., Jennings, S. *et al.* (2004) Perspectives on ecosystem-based approaches to the management of marine resources. *Marine Ecology Progress Series* 274, 269–303.

Browning, T.J. and Moore, C.M. (2023) Global analysis of ocean phytoplankton nutrient limitation reveals high prevalence of co-limitation. *Nature Communications* 14(1), 5014.

Cabral, R.B., Mayorga, J., Clemence, M., Lynham, J., Koeshendrajana, S. *et al.* (2018) Rapid and lasting gains from solving illegal fishing. *Nature Ecology & Evolution* 2, 650–658.

Cakmak, E.K., Hartl, M., Kisser, J. and Cetecioglu, Z. (2022) Phosphorus mining from eutrophic marine environment towards a blue economy: The role of bio-based applications. *Water Research* 219, 118505.

Carey, J. (2021) The complex case of Chesapeake bay restoration. *Proceedings of the National Academy of Sciences* 118(25), 1–5. DOI: 1073/pnas.2108734118.

Chesapeake Bay Program (n.d.) Our regional partnership guides the restoration and protection of the nation's largest estuary. Available at: https://www.chesapeakebay.net (accessed 16 June 2025).

Christensen, V., Coll, M., Piroddi, C., Steenbeek, J., Buszowski, J. *et al.* (2014) A century of fish biomass decline in the ocean. *Marine Ecology Progress Series* 512, 155–166.

Clements, J.C., Sundin, J., Clark, T.D. and Jutfelt, F. (2022) Meta-analysis reveals an extreme "decline effect" in the impacts of ocean acidification on fish behavior. *PLOS Biology* 20(2), e3001511.

Commonwealth of Australia (2024) *Australia's Draft Sustainable Ocean Plan: Navigating a Course to 2040.* Commonwealth of Australia, Canberra, Australia. Available at: https://www.dcceew.gov.au/sites/default/files/documents/draft-sustainable-ocean-plan.pdf (accessed 12 June 2025).

Conservation Drones (n.d.) A drone-based population survey of Delacour's langur (*Trachypithecus delacouri*) in the karst forests of northern Vietnam. Available at: conservationdrones.org (accessed 20 February 2025).

Costanza, R., Anderson, S.J., Sutton, P., Mulder, K., Mulder, O. *et al.* (2021) The global value of coastal wetlands for storm protection. *Global Environmental Change* 70, 102328.

Cumming, G.S. and Peterson, G.D. (2017) Unifying research on social–ecological resilience and collapse. *Trends in Ecology & Evolution* 32(9), 695–713. DOI: 10.1016/j.tree.2017.06.014.

Dai, M., Zhao, Y., Chai, F., Chen, M., Chen, N. *et al.* (2023) Persistent eutrophication and hypoxia in the coastal ocean. *Cambridge Prisms: Coastal Futures* 1, e19.

Dakos, V. and Kéfi, S. (2022) Ecological resilience: What to measure and how. *Environmental Research Letters* 17, 043003. DOI: 10.1088/1748-9326/ac5767.

Danopoulos, E., Jenner, L.C., Twiddy, M. and Rotchell, J.M. (2020) Microplastic contamination of seafood intended for human consumption: A systematic review and meta-analysis. *Environmental Health Perspectives* 128(12), 126002. DOI: 10.1289/EHP7171.

De Alessi, M. (2003) Technology, marine conservation, and fisheries management. In: Foldvary, F.E. and Klein, D.B. (eds) *The Half-Life of Policy Rationales: How New Technology Affects Old Policy Issues.* New York University Press, New York, pp. 21–37.

DePasquale, E., Baumann, H. and Gobler, C.J. (2015) Vulnerability of early life stage Northwest Atlantic forage fish to ocean acidification and low oxygen. *Marine Ecology Progress Series* 523, 145–156.

Derraik, J.G. (2002) The pollution of the marine environment by plastic debris: A review. *Marine Pollution Bulletin* 44, 842–852.

Ditria, E.M., Buelow, C.A., Gonzalez-Rivero, M. and Connolly, R.M. (2022) Artificial intelligence and automated monitoring for assisting conservation of marine ecosystems: A perspective. *Frontiers in Marine Science* 9, 918104.

Dong, W.W. and Lin, G. (2025) Comparison of coastal resilience policies: A perspective on effective global governance strategies. *E3S Web of Conferences* 617, 01019. DOI: 10.1051/e3sconf/202561701019.

Dunn, D.C., Jablonicky, C., Crespo, G.O., McCauley, D.J., Kroodsma, D.A. *et al.* (2018) Empowering high seas governance with satellite vessel tracking data. *Fish and Fisheries* 19, 729–739.

EU Framework for Maritime Spatial Planning (n.d.) EUR-Lex. Available at: https://eur-lex.europa.eu/legal-content/EN/TXT/?uri=celex%3A32014L0089 (accessed 16 June 2025).

European Commission, EU Marine Strategy Framework Directive (n.d.) Research and innovation. Available at: https://research-and-innovation.ec.europa.eu/research-area/environment/oceans-and-seas/eu-marine-strategy-framework-directive_en (accessed 16 June 2025).

European Commission, Integrated Maritime Policy (n.d.) Research and innovation. Available at: https://research-and-innovation.ec.europa.eu/research-area/environment/oceans-and-seas/integrated-maritime-policy_en (accessed 16 June 2025).

Evans, K., Zielinski, T., Chiba, S., Garcia-Soto, C., Ojaveer, H. *et al.* (2021) Transferring complex scientific knowledge to useable products for society: The role of the global integrated ocean assessment and challenges in the effective delivery of ocean knowledge. *Frontiers in Environmental Science* 9, 626532. DOI: 10.3389/fenvs.2021.626532.

FAO (2024) *The State of World Fisheries and Aquaculture 2024: Blue Transformation in Action.* FAO, Rome.

Fogarty, M.J. (2014) The art of ecosystem-based fishery management. *Canadian Journal of Fisheries and Aquatic Sciences* 71, 479–490. DOI: 10.1139/cjfas-2013-0203.

Freitag, A., Vogt, B. and Hartley, T. (2018) Ecosystem-based fisheries management in the Chesapeake: Developing functional indicators. *Coastal Management* 46(3), 127–147.

Froese, R., Zeller, D., Kleisner, K. and Pauly, D. (2012) What catch data can tell us about the status of global fisheries. *Marine Biology* 159, 1283–1292.

Fromentin, J.M. and Lopuszanski, D. (2014) Migration, residency, and homing of bluefin tuna in the western Mediterranean Sea. *ICES Journal of Marine Science* 71, 510–518.

García-Poza, S., Pacheco, D., Cotas, J., Marques, J.C., Pereira, L. *et al.* (2022) Marine macroalgae as a feasible and complete resource to address and promote Sustainable Development Goals (SDGs). *Integrated Environmental Assessment and Management* 18, 1148–1161.

Gazeau, F., Parker, L.M., Comeau, S., Gattuso, J.P., O'Connor, W.A. *et al.* (2013) Impacts of ocean acidification on marine shelled molluscs. *Marine Biology* 160, 2207–2245.

Gebremedhin, S., Bruneel, S., Getahun, A., Anteneh, W. and Goethals, P. (2021) Scientific methods to understand fish population dynamics and support sustainable fisheries management. *Water* 13(4), 574. DOI: 10.3390/w13040574.

Gilman, E., Weijerman, M. and Suuronen, P. (2017) Ecological data from observer programmes underpin ecosystem-based fisheries management. *ICES Journal of Marine Science* 74, 1481–1495.

Gissi, E., Manea, E., Mazaris, A.D., Fraschetti, S., Almpanidou, V. *et al.* (2021) A review of the combined effects of climate change and other local human stressors on the marine environment. *Science of the Total Environment* 755, 142564.

Global Fishing Watch (n.d.) Revolutionizing Ocean Monitoring and Analysis. Available at: https://globalfis hingwatch.org/ (accessed 18 February 2025).

Gobler, C.J., Doall, M.H., Peterson, B.J., Young, C.S., DeLaney, F. *et al.* (2022) Rebuilding a collapsed bivalve population, restoring seagrass meadows, and eradicating harmful algal blooms in a temperate lagoon using spawner sanctuaries. *Frontiers in Marine Science* 9, 911731.

Gonzalez-Rivero, M., Beijbom, O., Rodriguez-Ramirez, A., Bryant, D.E., Ganase, A. *et al.* (2020) Monitoring of coral reefs using artificial intelligence: A feasible and cost-effective approach. *Remote Sensing* 12(3), 489.

Graham, J.E., Stuart-Smith, R.D., Willis, T.J., Kininmonth, S., Baker, S.C. *et al.* (2014) Global conservation outcomes depend on marine protected areas with five key features. *Nature* 506, 216–220.

Halpern, B.S., Diamond, J., Gaines, S., Gelcich, S., Gleason, M. *et al.* (2012) Near-term priorities for the science, policy and practice of Coastal and Marine Spatial Planning (CMSP). *Marine Policy* 36, 198–205.

He, Q. and Silliman, B.R. (2019) Climate change, human impacts, and coastal ecosystems in the Anthropocene. *Current Biology* 29(19), R1021–R1035.

Harvey, E., McLean, D., Frusher, S., Haywood, M., Newman, S.J. *et al.* (2013) *The Use of BRUVs as A Tool for Assessing Marine Fisheries and Ecosystems: A Review of the Hurdles and Potential.* Project no.2010/002. University of Western Australia, Crawley, Australia. Available at: https://www.frdc.com .au/sites/default/files/products/2010-002-DLD.pdf (accessed 9 June 2025).

Heinze, C., Blenckner, T., Martins, H., Rusiecka, D., Döscher, R. *et al.* (2021) The quiet crossing of ocean tipping points. *Proceedings of the National Academy of Sciences* 118(9), e2008478118.

Heisler, J., Glibert, P.M., Burkholder, J.M., Anderson, D.M., Cochlan, W. *et al.* (2008) Eutrophication and harmful algal blooms: A scientific consensus. *Harmful Algae* 8(1), 3–13.

Henriksson, P.J., Guinée, J.B., Kleijn, R. and de Snoo, G.R. (2012) Life cycle assessment of aquaculture systems — a review of methodologies. *The International Journal of Life Cycle Assessment* 17, 304–313.

Heppell, S.S., Heppell, S.A., Read, A.J. and Crowder, L.B. (2005) Effects of fishing on long-lived marine organisms. In: Norse, E.A. and Crowder, L.B. (eds) *Marine Conservation Biology: The Science of Maintaining the Sea's Biodiversity*. Island Press, Washington, DC, pp. 211–231.

Hernández, C.M., Witting, J., Willis, C., Thorrold, S.R., Llopiz, J.K. *et al.* (2019) Evidence and patterns of tuna spawning inside a large no-take Marine Protected Area. *Scientific Reports* 9(1), 10772.

Hilborn, R. and Ovando, D. (2014) Reflections on the success of traditional fisheries management. *ICES Journal of Marine Science* 71, 1040–1046.

Horta, P.A., Rörig, L.R., Costa, G.B., Baruffi, J.B., Bastos, E. *et al.* (2021) Marine eutrophication: Overview from now to the future. In: Häder, D.-P., Helbling, E.W. and Villafañe, V.E. (eds) *Anthropogenic Pollution of Aquatic Ecosystems*. Springer, Cham, Switzerland, pp. 157–180.

Horton, A.A. (2022) Plastic pollution: When do we know enough. *Journal of Hazardous Materials* 422, 126885.

Intergovernmental Oceanographic Commission (IOC) (n.d.) Ocean Science for Sustainable Development. United Nations Education, Scientific and Cultural Organization (UNESCO). Available at: https://www .ioc.unesco.org/en (accessed 18 February 2025).

Jennings, S., Kaiser, M. and Reynolds, J.D. (2009) *Marine Fisheries Ecology*. Wiley, Oxford, UK.

Jeunen, G.-J., Urban, L., Lewis, R., Knapp, M., Lamare, M. *et al.* (2020) Marine environmental DNA (eDNA) for biodiversity assessments: A one-to-one comparison between eDNA and baited remote underwater video (BRUV) surveys. *Authorea Preprints* 486941.

Jiménez López, J. and Mulero-Pázmány, M. (2019) Drones for conservation in protected areas: Present and future. *Drones* 3(1), 10.

Jordan, S.J. and Benson, W.H. (2013) Governance and the Gulf of Mexico coast: How are current policies contributing to sustainability? *Sustainability* 5, 4688–4705. DOI: 10.3390/su5114688.

Jordan, S.J. and Benson, W.H. (2015) Sustainable watersheds: Integrating ecosystem services and public health. *Environmental Health Insights* 9(Suppl. 2), 1–7. DOI: 10.4137/EHI.S19586.

Jordan, S.J., Smith, L.M. and Nestlerode, J.A. (2009) Cumulative effects of coastal habitat alterations on fishery resources: Toward prediction at regional scales. *Ecology and Society* 14(1).

Juan-Jordá, M.J., Murua, H., Arrizabalaga, H., Dulvy, N.K. and Restrepo, V. (2018) Report card on ecosystem-based fisheries management in tuna regional fisheries management organizations. *Fish and Fisheries* 19(2), 321–339.

Karnauskas, M., Walter III, J.F., Kelble, C.R., McPherson, M. and Sagarese, S.R. (2021) To EBFM or not to EBFM? That is not the question. *Fish and Fisheries* 22(3), 646–651.

Kelly, N.E., Guijarro-Sabaniel, J. and Zimmerman, R. (2021) Anthropogenic nitrogen loading and risk of eutrophication in the coastal zone of Atlantic Canada. *Estuarine, Coastal and Shelf Science* 263, 107630.

Kennish, M.J. (2001) Coastal salt marsh systems in the US: A review of anthropogenic impacts. *Journal of Coastal Research* 17, 731–748.

Kirby, M.X. (2004) Fishing down the coast: Historical expansion and collapse of oyster fisheries along continental margins. *Proceedings of the National Academy of Sciences* 101, 13096–13099.

Kroeker, K.J., Kordas, R.L., Crim, R., Hendriks, I.E., Ramajo, L. *et al.* (2013) Impacts of ocean acidification on marine organisms: Quantifying sensitivities and interaction with warming. *Global Change Biology* 19, 1884–1896.

Lavery, P.S., Mateo, M.-A., Serrano, O. and Rozaimi, M. (2013) Variability in the carbon storage of seagrass habitats and its implications for global estimates of blue carbon ecosystem service. *PLoS ONE* 8(9), e73748. DOI: 10.1371/journal.pone.0073748.

Le, V.G., Nguyen, H.L., Nguyen, M.K., Lin, C., Hung, N.T.Q. *et al.* (2024) Marine macro-litter sources and ecological impact: A review. *Environmental Chemistry Letters* 22(3), 1257–1273.

Lewis, S.L. and Maslin, M.A. (2015) Defining the anthropocene. *Nature* 519(7542), 171–180.

Liddick, D. (2014) The dimensions of a transnational crime problem: The case of IUU fishing. *Trends in Organized Crime* 17, 290–312.

Link, J. (2010) *Ecosystem-Based Fisheries Management: Confronting Tradeoffs*. Cambridge University Press, Cambridge, UK.

Long Island Sound Study (LISS) (n.d.) Clean Waters and Healthy Watersheds. Available at: https://longisl andsoundstudy.net/ (accessed 18 February 2025).

Ma, M. and Hu, Q. (2024) Microalgae as feed sources and feed additives for sustainable aquaculture: Prospects and challenges. *Reviews in Aquaculture* 16, 818–835.

Mace (2001) A new role for MSY in single-species and ecosystem approaches to fisheries stock assessment and management. *Fish and Fisheries* 2(1), 2–32.

Madin, E.M., Darling, E.S. and Hardt, M.J. (2019) Emerging technologies and coral reef conservation: Opportunities, challenges, and moving forward. *Frontiers in Marine Science* 6, 727.

Maestro, M., Pérez-Cayeira, M.L., Chica-Ruiz, J.A. and Reyes, H. (2019) Marine protected areas in the 21st century: Current situation and trends. *Ocean and Coastal Management* 171, 28–36.

Malone, T.C. and Newton, A. (2020) The globalization of cultural eutrophication in the coastal ocean: Causes and consequences. *Frontiers in Marine Science* 7, 670.

March, A., Roberts, K.P. and Fletcher, S. (2022) A new treaty process offers hope to end plastic pollution. *Nature Reviews Earth & Environment* 3(11), 726–727.

Maúre, E.D.R., Terauchi, G., Ishizaka, J., Clinton, N. and DeWitt, M. (2021) Globally consistent assessment of coastal eutrophication. *Nature Communications* 12(1), 6142.

Mueller, J.T. and Gasteyer, S. (2021) The widespread and unjust drinking water and clean water crisis in the United States. *Nature Communications* 12(1), 3544.

Muir, J. (2005) Managing to harvest? Perspectives on the potential of aquaculture. *Philosophical Transactions of the Royal Society B* 360, 191–218.

Multisanti, C.R., Ferrara, S., Piccione, G. and Faggio, C. (2025) Plastics and their derivatives are impacting animal ecophysiology: A review. *Comparative Biochemistry and Physiology Part C: Toxicology & Pharmacology* 291, 110149.

Murphy, R.R., Keisman, J., Harcum, J., Karrh, R.R., Lane, M. *et al.* (2021) Nutrient improvements in Chesapeake Bay: Direct effect of load reductions and implications for coastal management. *Environmental Science & Technology* 56, 260–270.

Myers, R.A., Hutchings, J.A. and Barrowman, N.J. (1997) Why do fish stocks collapse? The example of cod in Atlantic Canada. *Ecological Applications* 7(1), 91–106.

Naylor, R.L., Goldburg, J., Primavera, J.H., Kautsky, N., Beveridge, C.M. *et al.* (2000) Effect of aquaculture on world fish supplies. *Nature* 405, 1017–1024.

Naylor, R.L., Hardy, R.W., Bureau, D.P., Chiu, A., Elliott, M. *et al.* (2009) Feeding aquaculture in an era of finite resources. *Proceedings of the National Academy of Sciences* 106, 15103–15110.

Naylor, R.L., Hardy, R.W., Buschmann, A.H., Bush, S.R., Cao, L. *et al.* (2021) A 20-year retrospective review of global aquaculture. *Nature* 591(7851), 551–563.

Nikiema, J. and Asiedu, Z. (2022) A review of the cost and effectiveness of solutions to address plastic pollution. *Environmental Science and Pollution Research* 29, 24547–24573.

NOAA (2025) Ocean Acidification. Available at: https://www.noaa.gov/education/resource-collections/ocean-coasts/ocean-acidification (accessed 4 February 2025).

NOAA (2007) Magnuson-Stevens Fishery Conservation and Management Act. Available at: https://media.fisheries.noaa.gov/dam-migration/msa-amended-2007.pdf (accessed 27 July 2025).

Office for Coastal Management, National Oceanic and Atmospheric Administration (n.d.) Coastal Zone Management Act. Available at: https://coast.noaa.gov/czm/act/ (accessed March 2025).

Onink, V., Jongedijk, C.E., Hoffman, M.J., van Sebille, E. and Laufkötter, C. (2021) Global simulations of marine plastic transport show plastic trapping in coastal zones. *Environmental Research Letters* 16(6), 064053.

Ovando, D., Hilborn, R., Monnahan, C., Rudd, M., Sharma, R. *et al.* (2021) Improving estimates of the state of global fisheries depends on better data. *Fish and Fisheries* 22, 1377–1391.

Parolini, M., Stucchi, M., Ambrosini, R. and Romano, A. (2023) A global perspective on microplastic bioaccumulation in marine organisms. *Ecological Indicators* 149, 110179.

Pauly, D., Alder, J., Bennett, E., Christensen, V., Tyedmers, P. *et al.* (2003) The future for fisheries. *Science* 302(5649), 1359–1361.

Peck, M.A., Catalán, I.A., Garrido, S., Rykaczewski, R.R., Asch, R.G. *et al.* (2024) Small pelagic fish: New frontiers in ecological research. *Marine Ecology Progress Series* 741, 1–6.

Petrossian, G.A. (2015) Preventing illegal, unreported and unregulated (IUU) fishing: A situational approach. *Biological Conservation* 189, 39–48.

Pikitch, E.K., Santora, C., Babcock, E.A., Bakun, A., Bonfil, R. *et al.* (2004) Ecosystem-based fishery management. *Science* 305(5682), 346–347.

Pikitch, E.K., Boersma, P.D., Boyd, I.L., Conover, D.O., Cury, P. *et al.* (2012) *Little Fish, Big Impact: Managing a Crucial Link in Ocean Food Webs.* Lenfest Ocean Program, Washington, DC.

Pikitch, E.K., Rountos, K.J., Essington, T.E., Santora, C., Pauly, D. *et al.* (2014) The global contribution of forage fish to marine fisheries and ecosystems. *Fish and Fisheries* 15, 43–64.

Prata, J.C., de Costa, J.P., Lopes, I., Andrady, A.L., Duarte, A.C. *et al.* (2021) A one health perspective of the impacts of microplastics on animal, human and environmental health. *Science of the Total Environment* 777, 146094.

Racine, P., Marley, A., Froehlich, H.E., Gaines, S.D., Ladner, I. *et al.* (2021) A case for seaweed aquaculture inclusion in US nutrient pollution management. *Marine Policy* 129, 104506.

Reimann, L., Vafeidis, A.T. and Honsel, L.E. (2023) Population development as a driver of coastal risk: Current trends and future pathways. *Cambridge Prisms: Coastal Futures* 1, e14.

Reis-Filho, J.A., Joyeux, J.C., Pimentel, C.R., Teixeira, J.B., Macieira, R. *et al.* (2022) The challenges and opportunities of using small drones to monitor fishing activities in a marine protected area. *Fisheries Management and Ecology* 29(5), 745–752.

Ringold, P.L., Alegria, J., Czaplewski, R.L., Mulder, B.S., Tolle, T. *et al.* (1996) Adaptive monitoring design for ecosystem management. *Ecological Applications* 6(3), 745–747.

Rockström, J., Donges, J.F., Fetzer, I., Martin, M.A., Wang-Erlandsson, L. *et al.* (2024) Planetary Boundaries guide humanity's future on earth. *Nature Reviews Earth & Environment* 5(11), 773–788.

Rolfing, L., Celliers, L. and Abson, D.J. (2022) Resilience and coastal governance: Knowledge and navigation between stability and transformation. *Ecology and Society* 27(2), 40. Available at: https://www.ecologyandsociety.org/vol27/iss2/art40/

Rooper, C.N., Boldt, J.L., Uriarte, A., Hansen, C., Ward, T. *et al.* (2024) Small pelagic fish: New frontiers in science and sustainable management. *Canadian Journal of Fisheries and Aquatic Sciences* 81, 984–989.

Rountos, K.J. (2016) Defining forage species to prevent a management dilemma. *Fisheries* 41(1), 16–17.

Rountos, K.J., Frisk, M.G. and Pikitch, E.K. (2015) Are we catching what they eat? Moving beyond trends in the mean trophic level of catch. *Fisheries* 40(8), 376–385.

Russo, R.C. (2002) Development of marine water quality criteria for the USA. *Marine Pollution Bulletin* 45(1–12), 84–91.

Ruzicka, J., Chiaverano, L., Coll, M., Garrido, S., Tam, J. *et al.* (2024) The role of small pelagic fish in diverse ecosystems: Knowledge gleaned from food-web models. *Marine Ecology Progress Series* 741, 7–27. DOI: 10.3354/meps14513.

Sala, E., Lubchenco, J., Grorud-Colvert, K., Novelli, C., Roberts, C. *et al.* (2018) Assessing real progress towards effective ocean protection. *Marine Policy* 91, 11–13.

Sandifer, P.A. and Sutton, A.E. (2014) Connecting stressors, ocean ecosystem services, and human health. *Natural Resources Forum* 38(3), 157–167. DOI: 10.1111/1477-8947.12047.

Santos, C.F., Ehler, C.N., Agardy, T., Andrade, F., Orbach, M.K. *et al.* (2019) Marine spatial planning. In: Sheppard, C. (ed.) *World Seas: An Environmental Evaluation*. Academic Press, Cambridge, Massachusetts, pp. 571–592.

Santos, R.G., Machovsky-Capuska, G.E. and Andrades, R. (2021) Plastic ingestion as an evolutionary trap: Toward a holistic understanding. *Science* 373(6550), 56–60.

Sargent, J., McEvoy, L., Estevez, A., Bell, G., Bell, M. *et al.* (1999) Lipid nutrition of marine fish during early development: Current status and future directions. *Aquaculture* 179(1–4), 217–229.

Schindler, D.E., Essington, T.E., Kitchell, J.F., Boggs, C. and Hilborn, R. (2002) Sharks and tunas: Fisheries impacts on predators with contrasting life histories. *Ecological Applications* 12, 735–748.

Schulte, D.M. and Burke, R.P. (2014) Recruitment enhancement as an indicator of oyster restoration success in Chesapeake Bay. *Ecological Restoration* 32(4), 434–440.

Sharma, R. (2015) Environmental issues of deep-sea mining. *Procedia Earth and Planetary Science* 11, 204–211.

Shen, H. and Huang, S. (2021) China's policies and practice on combating IUU in distant water fisheries. *Aquaculture and Fisheries* 6(1), 27–34.

Siple, M.C., Essington, T.E. and E. Plagányi, É. (2019) Forage fish fisheries management requires a tailored approach to balance trade-offs. *Fish and Fisheries* 20(1), 110–124.

Stentiford, G.D., Bateman, I.J., Hinchliffe, S.J., Bass, D., Hartnell, R. *et al.* (2020) Sustainable aquaculture through the One Health lens. *Nature Food* 1(8), 468–474.

Suarez-Bregua, P., Álvarez-González, M., Parsons, K.M., Rotllant, J., Pierce, G.J. *et al.* (2022) Environmental DNA (eDNA) for monitoring marine mammals: Challenges and opportunities. *Frontiers in Marine Science* 9, 987774.

Sutton, M.A., Howard, C.M., Kanter, D.R., Lassaletta, L., Móring, A. *et al.* (2021) The nitrogen decade: Mobilizing global action on nitrogen to 2030 and beyond. *One Earth* 4(1), 10–14.

Talmage, S.C. and Gobler, C.J. (2009) The effects of elevated carbon dioxide concentrations on the metamorphosis, size, and survival of larval hard clams (*Mercenaria mercenaria*), bay scallops (*Argopecten irradians*), and Eastern oysters (*Crassostrea virginica*). *Limnology and Oceanography* 54, 2072–2080.

Talmage, S.C. and Gobler, C.J. (2010) Effects of past, present, and future ocean carbon dioxide concentrations on the growth and survival of larval shellfish. *Proceedings of the National Academy of Sciences* 107, 17246–17251.

Tango, P.J. and Batiuk, R.A. (2016) Chesapeake Bay recovery and factors affecting trends: Long-term monitoring, indicators, and insights. *Regional Studies in Marine Science* 4, 12–20.

Temple, A.J., Skerritt, D.J., Howarth, P.E., Pearce, J. and Mangi, S.C. (2022) Illegal, unregulated and unreported fishing impacts: A systematic review of evidence and proposed future agenda. *Marine Policy* 139, 105033.

The State Council Information Office of the People's Republic of China (2025) Marine Eco-Environmental Protection in China. Available at: http://english.scio.gov.cn/whitepapers/2024-07/11/content_11730 2527.htm (accessed 16 June 2025).

Thom, R., St. Clair, T., Burns, R. and Anderson, M. (2016) Adaptive management of large aquatic ecosystem recovery programs in the United States. *Journal of Environmental Management* 183(2), 424–430.

Thushari, G.G.N. and Senevirathna, J.D.M. (2020) Plastic pollution in the marine environment. *Heliyon* 6(8).

Turner, R.E., Howes, B.L., Teal, J.M., Milan, C.S., Swenson, E.M. *et al.* (2009) Salt marshes and eutrophication: An unsustainable outcome. *Limnology and Oceanography* 54, 1634–1642.

UNEP (2019) Resolution Adopted by the United Nations Environment Assembly on 15 March 2019. 4/14. Sustainable Nitrogen Management. Available at: https://wedocs.unep.org/bitstream/handle/20.500. 11822/28478/English.pdf?sequence=3&isAllowed=y (accessed 9 June 2025).

UNEP (2022a) Resolution Adopted by the United Nations Environment Assembly on 2 March 2022. 5/14. End Plastic Pollution: Towards an International Legally Binding Instrument. Available at: https:// digitallibrary.un.org/record/3999257?ln=en&v=pdf (accessed 16 June 2025).

UNEP (2022b) Convention on Biological Diversity. Available at: https://www.cbd.int/doc/decisions/cop-15/cop-15-dec-04-en.pdf (accessed 16 June 2025).

United Nations Convention on the Law of the Sea (n.d.) International maritime law. Available at: https://treaties.un.org/doc/publication/CTC/Ch_XXI_6_english_p.pdf (accessed 16 June 2025).

US Department of Commerce, National Oceanic and Atmospheric Administration, National Marine Fisheries Service (n.d.) Magnuson-Stevens Fishery Conservation and Management Act. Available at: https://www.fisheries.noaa.gov/s3//dam-migration/msa-amended-2007.pdf (accessed 16 June 2025).

Valiela, I., Cole, M.L., Mcclelland, J., Hauxwell, J., Cebrian, J. *et al.* (2000) Role of salt marshes as part of coastal landscapes. In: Weinstein, M.P. and Krieger, D.A. (eds) *Concepts and Controversies in Tidal Marsh Ecology*. Kluwer, Dordrecht, Netherlands, pp. 23–36.

Valiela, I., Rutecki, D. and Fox, S. (2004) Salt marshes: Biological controls of food webs in a diminishing environment. *Journal of Experimental Marine Biology and Ecology* 300(1–2), 131–159.

Velis, C.A., Hardesty, B.D., Cottom, J.W. and Wilcox, C. (2022) Enabling the informal recycling sector to prevent plastic pollution and deliver an inclusive circular economy. *Environmental Science & Policy* 138, 20–25.

Vo, H.C. and Pham, M.H. (2021) Ecotoxicological effects of microplastics on aquatic organisms: A review. *Environmental Science and Pollution Research* 28, 44716–44725.

Wallace, R.B., Baumann, H., Grear, J.S., Aller, R.C. and Gobler, C.J. (2014) Coastal ocean acidification: The other eutrophication problem. *Estuarine, Coastal and Shelf Science* 148, 1–13.

Waltham, N.J., Elliott, M., Lee, S.Y., Lovelock, C., Duarte, C.M. *et al.* (2020) UN decade on ecosystem restoration 2021–2030—what chance for success in restoring coastal ecosystems. *Frontiers in Marine Science* 7, 1–5. DOI: 10.3389/fmars.2020.00071.

Wang, Y. and Aubrey, G.B. (1987) The characteristics of the China coastline. *Continental Shelf Research* 7(4), 329–349.

Watson, R.A. and Tidd, A. (2018) Mapping nearly a century and a half of global marine fishing: 1869–2015. *Marine Policy* 93, 171–177.

Whitney, M.M. and Vlahos, P. (2021) Reducing hypoxia in an urban estuary despite climate warming. *Environmental Science & Technology* 55(2), 941–951.

Wolfe, D.A., Monahan, R., Stacey, P.F., Farrow, D.R.G. and Robertson, A. (2024) Environmental quality of long island sound: Assessment and management issues. *Estuaries* 14(3), 224–236.

Wright, J.P. and Jones, C.G. (2006) The concept of organisms as ecosystem engineers ten years on: Progress, limitations, and challenges. *BioScience* 56(3), 203–209.

Xu, S., Yu, Z., Zhou, Y., Yue, S., Liang, J. *et al.* (2023) The potential for large-scale kelp aquaculture to counteract marine eutrophication by nutrient removal. *Marine Pollution Bulletin* 187, 114513.

Zaki, M.R.M. and Aris, A.Z. (2022) An overview of the effects of *nanoplastics* on marine organisms. *Science of the Total Environment* 831, 154757.

Zhang, Q., Blomquist, J.D., Fanelli, R.M., Keisman, J.L., Moyer, D.L. *et al.* (2023) Progress in reducing nutrient and sediment loads to Chesapeake Bay: Three decades of monitoring data and implications for restoring complex ecosystems. *Wiley Interdisciplinary Reviews: Water* 10(5), e1671.

3 Biodiversity Conservation

Devon R. Dublin*

Ocean Policy Research Institute,Sasakawa Peace Foundation, 1-15-16 Toranomon, Minato-ku, Tokyo 105-8524, Japan

Abstract

The chapter begins by defining and describing the concept of biodiversity conservation, followed by a section on key components for the administration of natural assets and ecological diversity. The biodiversity of various life forms that exist on earth, including animals, plants, microorganisms, and their ecosystems, with a focus on the aquatic ecosystem, is also presented, with a discussion of the four main levels of biodiversity: species, genetic, ecosystem, and global biodiversity.

3.1 Concepts and Definitions

Biodiversity can be described as the variety of different life forms on earth, including plants, animals, microorganisms, the genes they possess, and the ecosystem which they form (Rawat and Agarwal, 2015). Conservation science is a new and evolving discipline (Pimm, 2021). Biodiversity is very beneficial to the existence of man (Fig. 3.1) and therefore its conservation is invaluable (Adom *et al.*, 2019). Biodiversity conservation is therefore geared toward saving life on earth in all its forms and ensuring that the natural ecosystems that they make up are functioning and remain healthy (Rawat and Agarwal, 2015).

3.2 Levels of Biodiversity

The Convention on Biological Diversity, a landmark treaty which was signed by more than 150 nations on June 5, 1992, at the United Nations Conference on Environment and Development, held in Rio de Janeiro, states that: 'Biological diversity' means the variability among living organisms from all sources including, inter alia, terrestrial, marine and other aquatic ecosystems and the ecological complexes of which they are part; this includes diversity within species, between species and of ecosystems' (Secretariatfor the Convention on Biological Diversity, 2011, p. 4). It is broadly accepted that there are three levels of biodiversity: genetic, species/organismal, and ecosystem/ecological. A fourth level is recognized as global/cultural biodiversity (Table 3.1).

3.2.1 Genetic diversity

Genetic diversity refers to the variety of genes found within a given species or population. Studies are done to classify an individual or population in comparison to other individuals or populations (Mukhopadhyay and Bhattacharjee, 2016). It encompasses the components of genetic coding, including nucleotides, genes

*Corresponding author: devdub02@gmail.com

© CAB International 2025. *One Health Concepts and the Aquatic Ecosystem*
(eds L.D. Urdes *et al.*)
DOI: 10.1079/9781800623248.0003

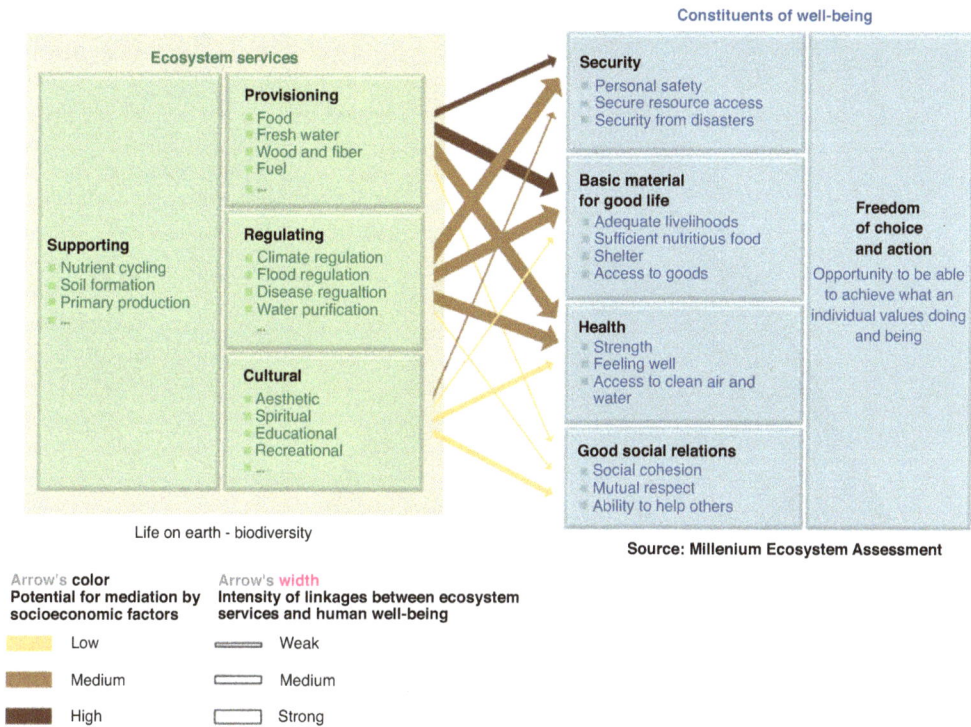

Fig. 3.1. Relationships between ecosystem services and human well-being. From Millennium Ecosystem Assessment, 2005.

Table 3.1. The levels of organization and components of biodiversity. Modified from Yankelevich, 2007.

Ecological diversity	Genetic diversity	Organismal diversity	Cultural diversity
Biomes	Populations	Kingdoms	Human Interactions at all levels
Bioregions	Individuals	Phyla	
Landscapes	Chromosomes	Families	
Ecosystems	Genes	Genera	
Habitats	Nucleotides	Species	
Niche		Subspecies	
Populations		Populations	
		Individuals	

(fundamental units of all biological variations), and chromosomes that structure organisms (Gaston, 2010). The number of possible combinations of genes and of the molecules making up genes is immense, corresponding to a number much larger than the individuals making up a species.

This variety of genetic material within species has resulted in distinct species evolving through natural selection. Species that inhabit large areas and interbreed throughout the whole area have a high rate of gene flow when compared to those living in small or isolated areas. Additionally, behavioral traits can influence the geographical distribution of genetic characteristics within a given species. This is easily demonstrated through North American eels, which inhabit streams along 4000 km of coastline but migrate to the Sargasso Sea to reproduce as one massive population, resulting in no geographic differentiation. Salmon, on the other hand, which breed in different streams

Fig. 3.2. The hierarchy of biological classification.

but spend most of their life at sea, develop into distinct, localized populations (Southeast Asian Fisheries Development Center, 1994).

3.2.2   Species/organismal diversity

The study of species diversity gives ecologists insight into the stability of communities, defined as the ability of a system to recover to an equilibrium state after disturbance (Hamilton, 2005). Species diversity encompasses the full taxonomic hierarchy and its components (Fig. 3.2). This level of organization is often expressed as species richness, which corresponds to the number of various species or the range of different types of species in a given ecosystem or environment.

3.2.3   Ecological diversity

Ecological diversity is the variety of ecosystems in an area. The wide range of terrestrial and aquatic environments on earth has been classified into several ecosystems (coral reefs, freshwater, grasslands, mangroves, peatlands, tropical rain forests, and wetlands). The measurement of ecological diversity to understand the processes that sustain biodiversity is essential for any study of ecosystem health and its viability (Daly *et al.*, 2018).

3.3   Biodiversity Conservation

The planetary boundaries, nine processes that regulate the stability and resilience of the earth system, were identified by a group of 28 internationally renowned scientists. They suggest that our developmental trend is not sustainable and therefore change is needed (Fig. 3.3). Biodiversity loss degrades ecosystem functionality and, if left unchecked, can lead to mass extinctions and even total ecosystem collapse (Gaston, 2010). As demonstrated by the study on planetary boundaries, biodiversity represented as biosphere integrity has gone way beyond safe levels (Steffen *et al.*, 2015). Other processes, such as ocean acidification, climate change, and land system change, have direct

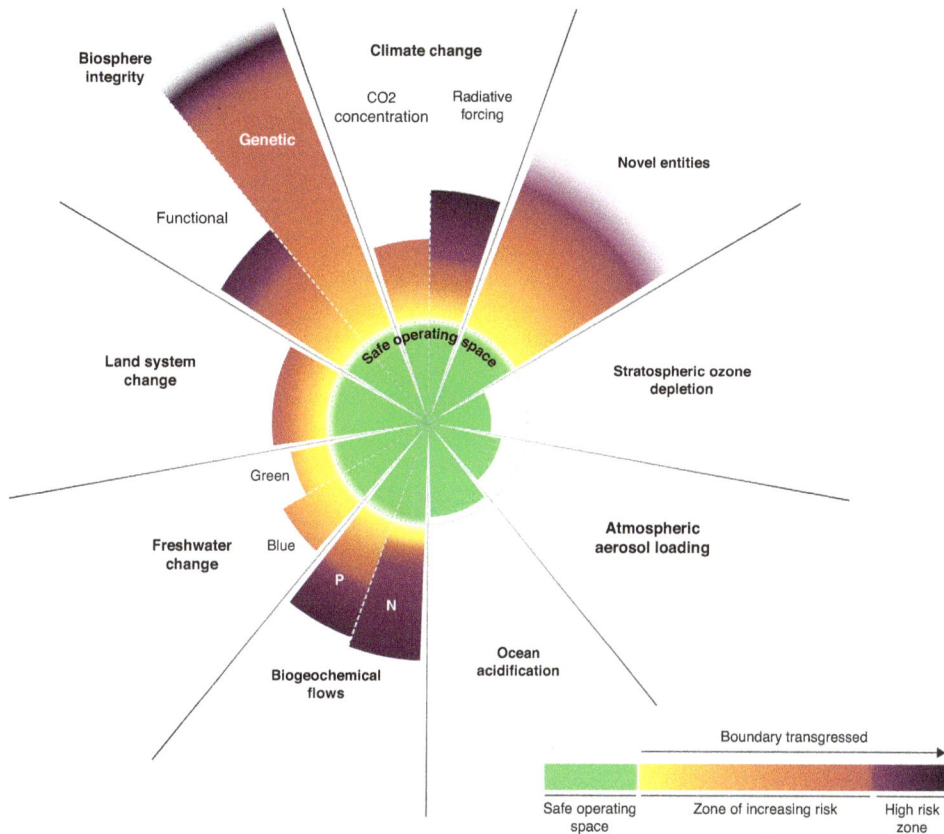

Fig. 3.3. The planetary boundaries (Reproduced from Richardson *et al.*, *Science Advances*, doi:10.1126/sciadv.adh2458 2023, AAAS).

impacts on biodiversity loss, such as slowed growth of coral skeletons, powerful natural disasters, and loss of habitats, respectively.

This has occurred largely due to human actions, primarily since the earth entered the Anthropocene epoch in which humans are the dominant driver of change on the planet (Rockström *et al.*, 2009). The term the 'Evil Quartet' was coined in reference to the four main causes of biodiversity loss. These are habitat loss and fragmentation, overexploitation, alien species invasions, and co-extinctions.

Biodiversity conservation is therefore premised on the fact that humanity needs to adopt several activities and actions to avoid the risk of deleterious or possible catastrophic environmental change at the global level. In other words, it fits into the safe operating space as defined by the planetary boundaries

framework (Steffen *et al.*, 2015). The question will be asked as to why biodiversity conservation is necessary. There are three points of view. The first is the narrowly utilitarian argument, which is premised on the fact that humans derive innumerable direct economic benefits from nature, such as food, firewood, fiber, construction material, industrial products, and medicine (Provisioning in Fig. 3.1). The second is the broad utilitarian argument, which deals with the role biodiversity plays in many ecosystem services that nature provides, such as pollination and photosynthesis (Supporting and Regulating in Fig. 3.1). The third is the ethical argument that we owe a debt to millions of plants, animals, and microbe species with whom we share this planet. The idea is that, philosophically or spiritually, every species has an intrinsic value, even if it may not presently

Most effective

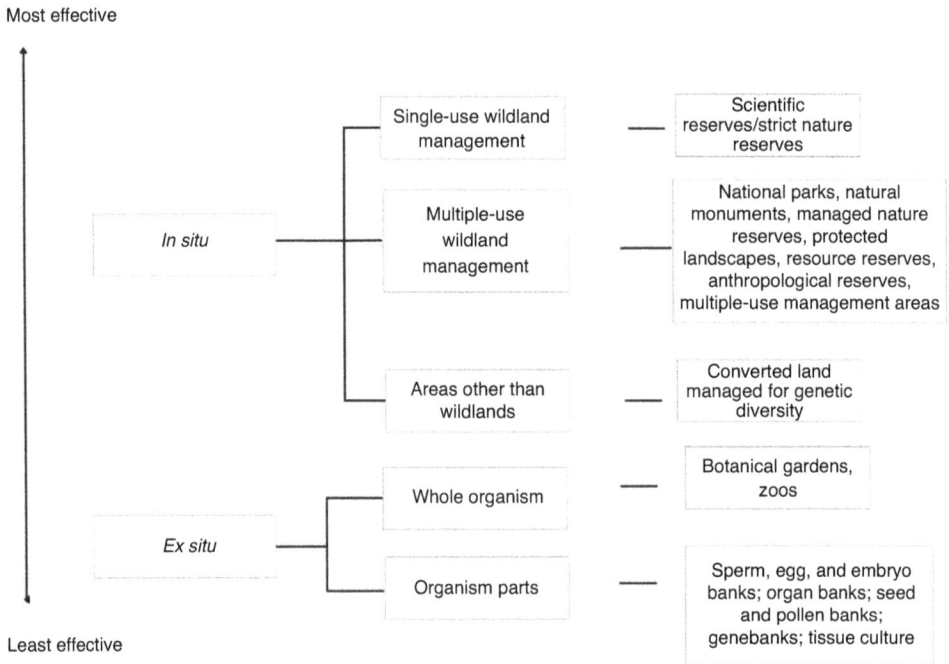

Least effective

Fig. 3.4. The roles of *in situ* and *ex situ* conservation. Modified from Whitmore, 1990. Figure used with permission from Oxford Publishing Ltd.

have any economic value to us (Cultural in Fig. 3.1).

There are two types of Biodiversity conservation, namely: *in situ* (where species are conserved in their own ecosystem) and *ex situ* (where species are conserved outside their habitat). *In situ* and *ex situ* conservation can be subdivided as shown in Fig. 3.4.

The choice of the type of conservation is dependent on circumstances on the ground. *In situ* is the main way of conserving a species, where we conserve and protect the whole ecosystem, thus its biodiversity at all levels is protected even if a specific species is targeted. However, when the case arises that an animal or plant is endangered or threatened and needs urgent measures to save it from extinction, then *ex situ* conservation is the desirable approach. *Ex situ* conservation refers to efforts conducted outside the natural habitat of a species, such as at laboratories, zoos, and aquariums. There is also the view that *in situ* and *ex situ* conservation are mutually

reinforcing and complementary approaches which, when combined, help to achieve the most stable and cost-effective conservation effort for a given gene pool under locally prevailing conditions (Zegeye, 2016). We can therefore conclude that biodiversity conservation requires a multidisciplinary approach and must be continuous in nature.

3.3.1 Conservation versus preservation

Preservation is equated with the authoritarian protection of ecosystems by the state, and conservation with community-based natural resource management and wise use (Table 3.2). The preservationist view removes humans from the ecosystem rather than seeing us as part of nature, which is an anthropocentric view. Preservation allows us to preserve something for future use while ignoring the current needs of existing humans.

Table 3.2. Preservation versus conservation. Modified from van der Ploeg *et al.*, 2011. Table used with permission of Sage.

	Preservation	Conservation
Policy Tool	Protected areas	Sustainable use
	'Fines and fences'	'Use it or lose it'
Philosophy	Intrinstic values	Utilitarian values
Rural Communities	Destructive	'Stewards of the environment'
	Ignoranat	Traditional ecological knowledge
	Irrational	Marginalized,e.g.alitarian
Nature	Pristine wilderness	Human-dominated landscapes
Wildlife and People	Conflict	Coexistence
Governance	Authoritarian	Participatory
	Centralized ('top-down')	Devolved ('bottom-up')
	Technocratic	People-oriented
	Protectionist	Comanagement
	'Fortress conservation'	'Community-based natural resource management'

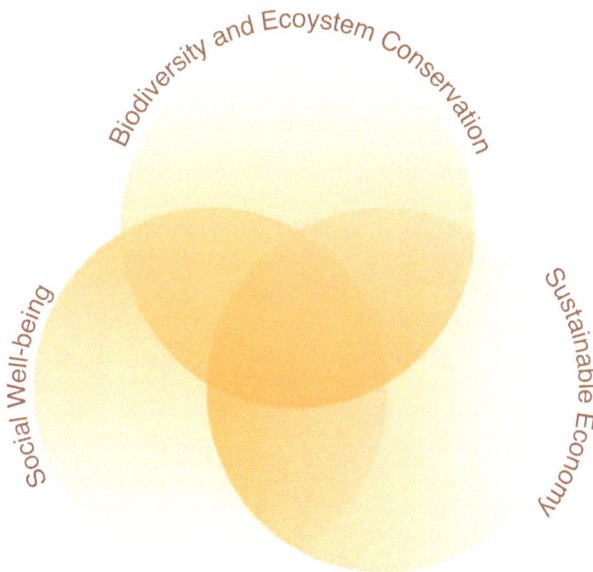

Fig. 3.5. The concept of ecosystem-based management. Modified from Laurila-Pant *et al.*, 2015. Figure used with permission of Elsevier.

3.3.2 Key components for administration of ecological diversity

Biodiversity is globally recognized as a cornerstone of healthy ecosystems, with biodiversity conservation being an important aspect of the administration of ecological diversity (Laurila-Pant *et al.*, 2015). Ecosystem-based management is premised on a comprehensive decision-making process which harmonizes social, economic, and ecological elements (Fig. 3.5).

To achieve effective ecosystem-based management, three important aspects need to be considered: (i) the ultimate objectives of management actions, (ii) ecosystem boundaries, and (iii) what the management actions are directed to (Pirot *et al.*, 2000). When it comes to the objective, the purpose of ecosystem management is to use ecosystems in such a way that we ensure that their goods and services are available on a sustainable basis. In this regard, it must be noted that biodiversity and ecosystem service conservation represented by the Sustainable Development Goals (SDGs) 14: Life below water ('Conserve and sustainably use the oceans, seas and marine resources for sustainable development') and 15: Life on land ('Protect, restore and promote sustainable use of terrestrial ecosystems, sustainably manage forests, combat desertification, and halt and reverse land degradation and halt biodiversity loss') (United Nations General Assembly, 2015) and their contribution to ecosystem services and human well-being underpins the achievement of all other goals (Fig. 3.6). Any area that is to be managed would have a geographic boundary and would need to factor in how different societies interact with the systems within which they live and the jurisdiction that governs the area being managed. Since ecosystem-based management should focus on the role of people and their interactions with other components of the system, as an ecosystem component, human actions should be the focus of ecosystem management.

An ecosystem approach therefore aims to balance the needs of humans and ecosystems in a harmonious relationship at all levels. The basic concepts are as follows.

- All components of the ecosystem are interdependent.
- Due to the fact that ecosystems are dynamic and complex in nature, a flexible adaptable approach should be taken.

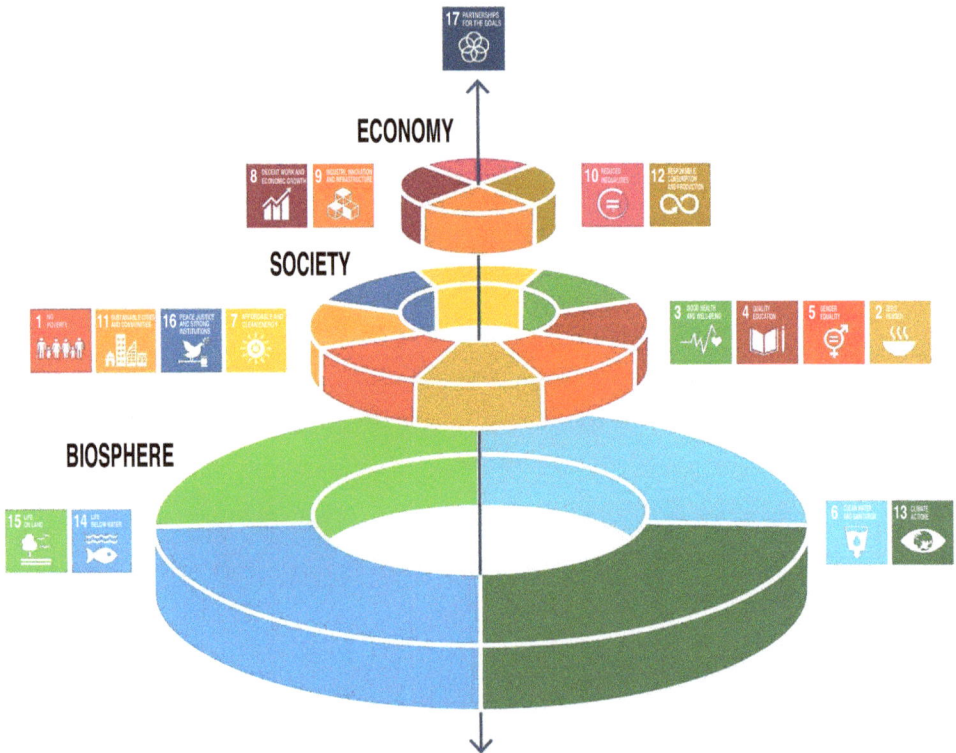

Fig. 3.6. The biosphere SDGs underpin all other SDGs. Courtesy of Azote for Stockholm Resilience Centre, Stockholm University CC BY-ND 3.0.

- Scientific, social, and financial concerns must be included.

3.4 The Institutional Framework on Biodiversity Conservation and Management

The institutional framework on biodiversity conservation and management includes a variety of organizations operating on different spatial scales (Wells, 1998). These include community groups, local governments, national governments, non-governmental organizations (NGOs), the private sector, and international organizations.

Global conventions are useful in minimizing negative transboundary ecosystem issues when two or more countries share a common ecosystem. Oceans are a very good example of this. These conventions can provide a useful framework for implementing joint agreements for ecosystem-based management. These include:

- the Convention on Wetlands (Ramsar, 1971);
- the World Heritage Convention (Paris, 1972);
- the Convention on Migratory Species (Bonn, 1979);
- the Law of the Sea (Montego Bay, 1982);
- the Convention on Biological Diversity (Rio de Janeiro, 1992); and
- the Convention to Combat Desertification (Paris, 1994).

3.4.1 Convention on Biological Diversity

Perhaps the most important convention, which is instructive for effective ecosystem-based management and why it should be conducted, is the Convention on Biological Diversity (CBD), which was opened for signature at the Earth Summit in Rio de Janeiro on June 5, 1992, and entered into force on December 29, 1993. The Convention has three main goals:

1. the conservation of biological diversity (or biodiversity);

2. the sustainable use of its components; and
3. the fair and equitable sharing of benefits arising from genetic resources.

Since the Convention is legally binding, countries that join are obliged to implement its provisions, and this is generally manifested in National Biodiversity Strategies and Action Plans (NBSAPs), which are contained in Article 6 of the Convention. Although NBSAPs are government-approved biodiversity documents, they are not binding legal texts. Nevertheless, an NBSAP legal framework will be made up of legislation, regulations, and policy; legally defined institutional arrangements; and regulatory mechanisms (including compliance and enforcement) (United Nations Environment Programme, 2018). It is important to note that the CBD came into being because of an existing common interest in the coordinated management of domestic natural resources, rather than that of a common resource.

3.5 The Aquatic Ecosystem

Approximately 75% of the earth's surface is covered by water, forming various types of aquatic ecosystems which act as carbon sinks and regulate global temperature, helping to mitigate climate change. Aquatic ecosystems are classified as freshwater ecosystems and marine ecosystems. These ecosystems provide a home to approximately 230,000 identified species, including phytoplankton, zooplankton, aquatic plants, insects, fish, birds, mammals, and other organisms (Fujikura et al., 2010). They provide enormous economic and aesthetic value to human civilization and welfare through tangible and intangible goods and services (Table 3.3). This includes aquatic resources for food, medicines, recreation, fishing, transportation, mental health, and tourism. As shown in Table 3.4, there are several types of these ecosystems. Freshwater ecosystems are generally classified into lentic ecosystems, which are standing or still water bodies, and lotic ecosystems, referring to flowing water bodies. Marine ecosystems can be classified as offshore and onshore, the latter of which are subjected to the action of oceanic waves and tides.

Table 3.3. Ecosystem structure, functions, and benefits. Modified from Pirot *et al.*, 2000.

Biome	Functions and services provided by ecosystems	Common ecosystem benefits and attributes
Forests	• Micro climate stabilization • Carbon uptake and storage • Soil and watershed protection • Energy storage	• Carbon dioxide removal • Fuel products • Timber products • Non-timber products • Wildlife resources • Biodiversity conservation
Wetlands	• Groundwater recharge and discharge • Flood control • Water quality and quantity • Water purification • Sediment/toxicant/nutrient retention	• Medicinal and biomedical products • Water supply • Pollution clean-up • Fish nurseries and fisheries products • Forage products • Agricultural products • Transport
Mangroves	• Storm protection • Provision and renewal of nutrients • Sediment accumulation	• Aesthetic and recreational values • Historical and cultural values • Fish nurseries and fisheries products
Coral reefs	• coastal protection • Sand protection	• Construction material • Genetic resources • Global heritage • Education and scientific interest
Oceans	• Global climate regulation	• Fisheries products

Table 3.4. Types of aquatic ecosystems. Table author's own.

Freshwater		Marine	
Lentic	Lotic	Onshore	Offshore
Lakes	Rivers	Salt marshes	Oceans
Ponds	Streams	Wetlands (along shores and river mouths)	Estuaries
Wetlands	Springs	Mangrove forests	Coral reefs
Swamps	Creeks	Foreshores (muddy, sandy, shingle, rocky)	Tidal inlets
Bogs	Tributaries		Lagoons
Reservoirs	Brooks		
Groundwater			

Oceans have a great capacity to transport heat from the earth, and the ocean ecosystem absorbs carbon in the form of a 'carbon sink.' Additionally, marine phytoplanktons process around 50 billion tonnes of carbon annually (Irfan and Alatawi, 2019). Many concerns have arisen due to our dependence on aquatic resources. Such concerns include overharvesting, invasive species, pollution, habitat loss, and alteration and diversion of water bodies. It is therefore necessary to implement conservation measures to protect and conserve these aquatic ecosystems.

The nature of these conservation measures should be based on an analysis of the interaction between human activities, water resources, and the aquatic environment. A ridge-to-reef (landscape–seascape) approach, which consists of integrated management of both freshwater bodies and coastal areas, is a

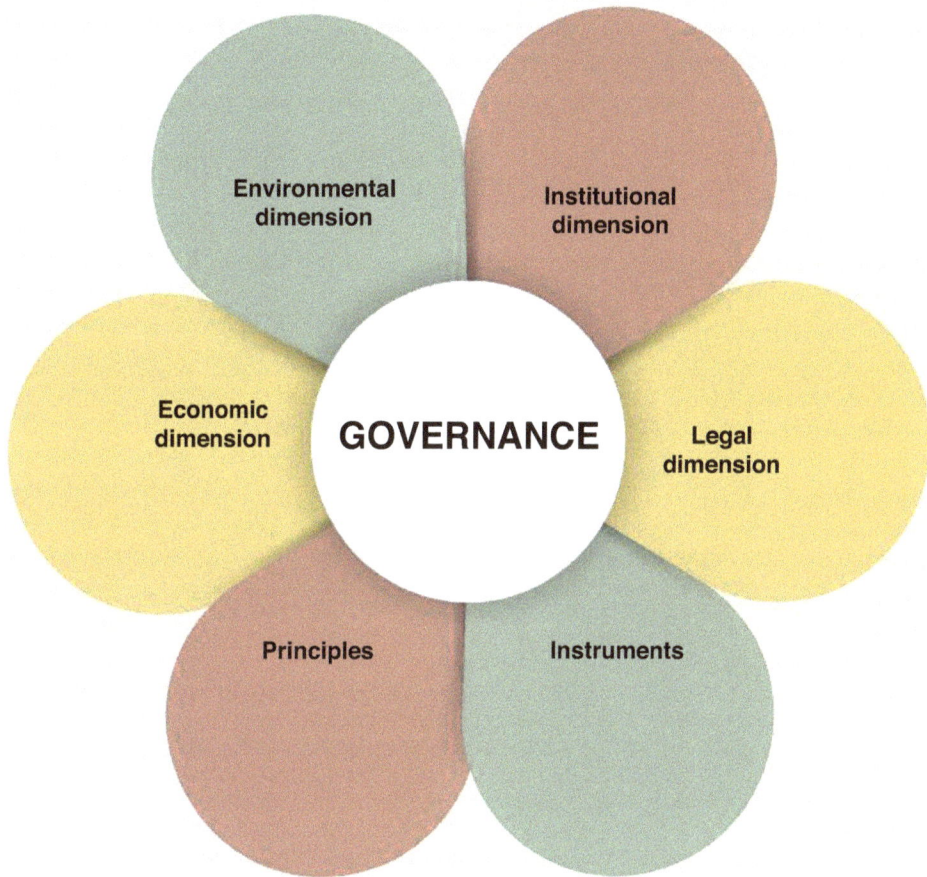

Fig. 3.7. Governance of aquatic ecosystems. Modified after Brachet *et al.*, 2015.

good example of this. This type of conservation requires a cross-cutting approach that is multidisciplinary and multisectoral in nature (Fig. 3.7). This is because expertise and skill in various disciplines are needed in all stages, including planning, designing, constructing, and monitoring. Since the quality of aquatic ecosystems is reliant on good stewardship in territorial areas, activities such as agriculture, forestry, and infrastructural development must be factored in, hence the reason for a territorial approach.

Governance of aquatic ecosystems should be done to ensure that the provisions, regulation, and cultural services that we obtain from them can be sustained—in other words, maintaining healthy aquatic ecosystems where the aquatic environment has the capacity to sustain its ecological structure, processes, functions, and resilience well within the range of natural variability that it possesses. Hence, good governance is closely tied to sustainable development, ecosystem conservation, and the capacity for facing climate change.

References

Adom, D., Umachandran, K., Ziarati, P., Sawicka, B. and Sekyere, P. (2019) The concept of biodiversity and its relevance to mankind: A short review. *Journal of Agriculture and Sustainability* 12, 219–231.

Brachet, C., Magnier, J., Valensuela, D., Petit, K., Fribourg-Blanc, B. *et al.* (2015) *The Handbook for Management and Restoration of Aquatic Ecosystems in River and Lake Basins.* International Network of Basin Organizations (INBO) and Global Water Partnership (GWP), Paris.

Daly, A.J., Baetens, J.M. and De Baets, B. (2018) Ecological diversity: Measuring the unmeasurable. *Mathematics* 6(7), 119. DOI: 10.3390/math6070119.

Fujikura, K., Lindsay, D., Kitazato, H., Nishida, S. and Shirayama, Y. (2010) Marine biodiversity in Japanese waters. *PLoS ONE* 5(8), e11836. DOI: 10.1371/journal.pone.0011836.

Gaston, K.J. (2010) Biodiversity. In: Sodhi, N.S. and Ehrlich, P.R. (eds) *Conservation Biology for All.* Oxford University Press, Oxford, UK, pp. 27–44.

Hamilton, A.J. (2005) Species diversity or biodiversity? *Journal of Environmental Management* 75, 89–92.

Irfan, S. and Alatawi, A.M.M. (2019) Aquatic ecosystem and biodiversity: A review. *Open Journal of Ecology* 9, 1–13. DOI: 10.4236/oje.2019.91001.

Laurila-Pant, M., Lehikoinen, A., Uusitalo, L. and Venesjärvi, R. (2015) How to value biodiversity in environmental management? *Ecological Indicators* 55, 1–11. DOI: 10.1016/j.ecolind.2015.02.034.

Millennium Ecosystem Assessment (2005) *Ecosystems and Human Well-being: Synthesis.* Island Press, Washington, DC.

Mukhopadhyay, T. and Bhattacharjee, S. (2016) Genetic diversity: Importance and measurements. In: Mir, A.H. and Bhat, N.H. (eds) *Conserving Biological Diversity: A Multiscaled Approach.* Research India Publications, New Delhi, pp. 251–295.

Pimm, S.I. (2021) What is biodiversity conservation? *Ambio* 50, 976–980. DOI: 10.1007/s13280-020-01399-5.

Pirot, J.-Y., Meynell, P.J. and Elder, D. (2000) *Ecosystem management: Lessons from around the world. A guide for development and conservation practitioners.* IUCN, UK.

Rawat, U.S. and Agarwal, N.K. (2015) Biodiversity: Concept, threats and conservation. *Environment Conservation Journal* 16(3), 19–28.

Richardson, K., Steffen, W., Lucht, W., Bendtsen, J., Cornell, S. *et al.* (2023) Earth beyond six of nine planetary boundaries. *Science Advances* 9, 37. DOI: 10.1126/sciadv.adh2458.

Rockström, J., Steffen, W., Noone, K., Persson, A., Chapin, III. *et al.* (2009) Planetary boundaries: Exploring the safe operating space for humanity. *Ecology and Society* 14(2), 32. Available at: http://www.ecologyandsociety.org/vol14/iss2/art32/

Secretariat for the Convention on Biological Diversity (2011) *Convention on Biological Diversity: Text and Annexes.* Montreal, Canada.

Southeast Asian Fisheries Development Center, Aquaculture Department (1994) Genetic, species, and ecosystem diversity. *Aqua Farm News* 12(3), 2–3.

Steffen, W., Richardson, K., Rockström, J., Cornell, S.E., Fetzer, I. *et al.* (2015) Planetary boundaries: Guiding human development on a changing planet. *Science* 347, 6223. DOI: 10.1126/science.1259855.

United Nations Environment Programme (2018) *Law and National Biodiversity Strategies and Action Plans.* United Nations Environment Programme, Nairobi, Kenya. Available at: https://wedocs.unep.org/bitstream/handle/20.500.11822/25655/LawBiodiversity_Strategies.pdf?sequence=1&isAllowed=y

United Nations General Assembly (2015) *Transforming Our World: The 2030 Agenda for Sustainable Development.* Available at: https://docs.un.org/en/A/RES/70/1

Van der Ploeg, J., Araño, R.R. and Van Weerd, M. (2011) What local people think about crocodiles: Challenging environmental policy narratives in the Philippines. *The Journal of Environment & Development* 20(3), 303–328.

Wells, M.P. (1998) Institutions and incentives for biodiversity conservation. *Biodiversity and Conservation* 7, 815–835. DOI: 10.1023/A:1008896620848.

Whitmore, T.C. (1990) *An Introduction to Tropical Rain Forests.* Clarendon Press, Oxford, UK.

Yankelevich, S.N. (2007) What do we mean by biodiversity? *Ludus Vitalis* XV(28), 45–68.

Zegeye, H. (2016) In situ and ex situ conservation: Complementary approaches for maintaining biodiversity. *International Journal of Research in Environmental Studies* 4, 1–12.

4 Pollution and Mitigation of Environmental Risks

Devon R. Dublin*

Ocean Policy Research Institute, Sasakawa Peace Foundation, 1-15-16 Toranomon, Minato-ku, Tokyo 105-8524, Japan

Abstract

The chapter begins by looking at the concept behind pollutants and mitigation of environmental risks. This is followed by a section that discusses sources of contaminants (including antimicrobial contaminants and heavy metals) and their impact on environmental components (water, soil, air), with a focus on the aquatic ecosystem as an interface where the three health components (human, animal, and environment) meet. It also describes innovative biotechnologies for environmental pollutants/hazards mitigation and sustainable technologies in use to mitigate/adapt to climate change.

4.1 The Concept Behind Pollution and Pollutants

Pollution can be defined as the negative/undesirable changes in our surroundings that are due to the presence or addition of contaminants in amounts that have harmful effects on plants, animals, and human beings. These contaminants are called pollutants. Pollutants can be classified depending on various perspectives, as shown in Table 4.1.

4.2 Types of Pollution

Environmental pollution occurs in different forms: air, water, soil, land, radioactive, noise, heat/thermal, and light. Every form of pollution has two sources of occurrence: (i) the point sources, which are easy to identify, monitor, and control (e.g. oil refineries), and (ii) the non-point sources, which cannot be easily identified and are hard to control (e.g. acid rain).

4.2.1 Air pollution

Clean air is essential for sustaining life on earth, and this is threatened by air pollution. The earliest recorded major air pollution disaster was the 1952 Great Smog of London, which resulted in more than 4000 deaths due to the accumulation of air pollutants over the city lasting for five days. With the industrial revolution and the development of factories and transportation systems, air pollution has become a serious environmental issue.

Air pollution may be defined as the presence of any solid, liquid, or gaseous substance in sufficient quantities in the air as to be directly or indirectly injurious to humans, animals, plants, or property or that interferes with the normal

*Corresponding author: devdub02@gmail.com

© CAB International 2025. *One Health Concepts and the Aquatic Ecosystem*
(eds L.D. Urdes *et al.*)
DOI: 10.1079/9781800623248.0004

Table 4.1. Classification of pollutants. Table author's own.

Perspective	Types	Definition	Example	Additional notes
Persistency	Persistent pollutants	Pollutants that remain consistent in the environment for a long period of time without any change in their original form.	Nuclear wastes	
	Non-persistent pollutants	Pollutants that break down into simple form.	Organic waste	If the process of breaking down includes living things, then they are referred to as biodegradable pollutants.
Consistency	Primary pollutants	Pollutants that remain in the form in which they were added to the environment.	Plastic	
	Secondary pollutants	Pollutants formed due to interaction of primary pollutants among themselves.	Nitrogen dioxide (NO_2)	
Existence in nature	Quantitative pollutants	Substances that are already in the environment but become pollutants when their concentration level passes the threshold limit.	CO_2	
	Qualitative pollutants	Substances that are not normally present in the environment and are added by human beings.	Herbicides	
Origin	Man-made pollutants	Pollutants produced because of human activities.	Vehicular emissions	
	Natural pollutants	Pollutants produced because of natural phenomena.	Volcanic eruptions	
Nature of disposal	Biodegradable pollutants	Pollutants decomposed by natural processes.	Sewage	
	Non-biodegradable pollutants	Pollutants which don't decompose naturally or decompose slowly.	Aluminum cans	

environmental processes. Although these are examined separately in this chapter, it includes noise and radioactive radiation. Air pollution is one of the few forms that have regional and global reaches since contaminants can travel long distances depending on their properties and the climatic conditions (Fig. 4.1). An example of this is uncontrolled forest fires, mainly in Indonesia, which result in lengthy and severe episodes of air pollution affecting several countries in South-east Asia.

There are diverse types of air pollutants coming from various sources (Table 4.2). The air may get polluted by natural causes such as volcanoes, which release ash, dust, and sulfur into the atmosphere. Unlike pollutants from

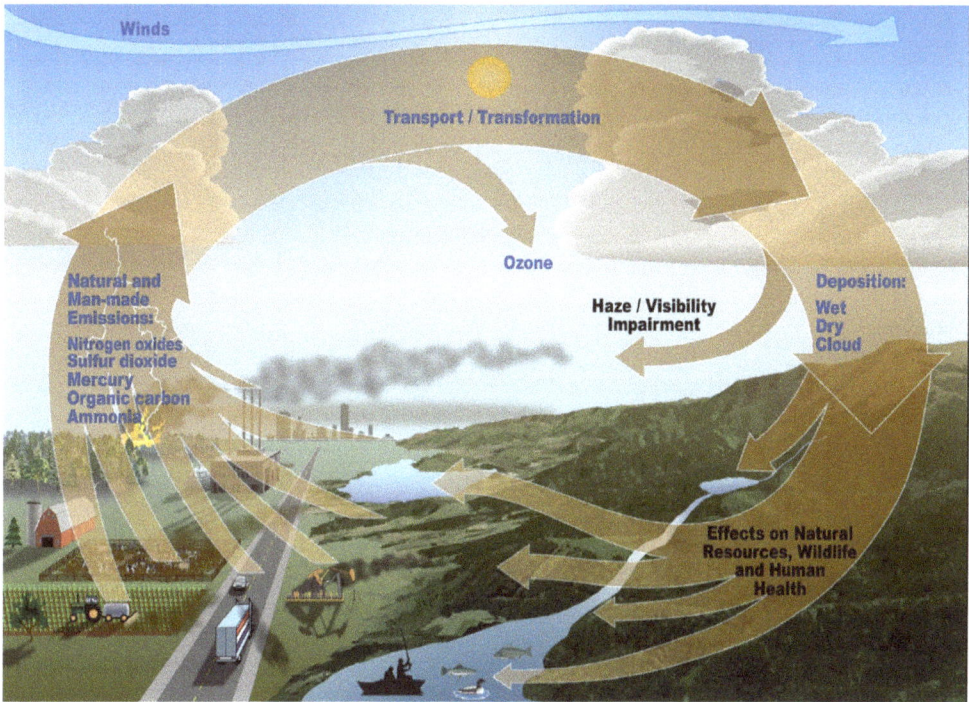

Fig. 4.1. Transmission of air pollutants. From Sharma, 2020. https://naturalhealthcourses.com/2018/11/all-we-need-is-the-air-that-we-breathe-and-a-little-less-pollution/snhs-air-quality-3/

human activity, naturally occurring pollutants do not remain in the atmosphere for a long time and thus do not lead to permanent atmospheric change.

4.2.1.1 Impact on environmental components

The impacts on environmental components are manifested through the carbon cycle, the sulfur cycle, the nitrogen cycle, and the fluoride cycle.

4.2.1.1.1 THE CARBON CYCLE. The carbon cycle has a large effect on the function and well-being of the earth since it plays a key role in regulating its climate by controlling the concentration of carbon dioxide (CO_2) in the atmosphere. When various activities occur, such as the burning of biomass, coal, and fossil fuels to generate heat and electricity, carbon is released to the atmosphere and oceans in the form of CO_2 and carbonate (Fig. 4.2). The rate of the constant release of CO_2 into the atmosphere far outweighs the

relatively slow reaction and removal rates of its concentration, thus leading to global warming.

Carbon is transported within the ocean by three mechanisms: (i) the 'solubility pump' (a physico-chemical process that transports carbon as dissolved inorganic carbon (DIC) from the ocean's surface to its interior), (ii) the 'biological pump' (biologically driven sequestration of carbon from the atmosphere and land runoff to the ocean interior and seafloor sediments), and (iii) the 'marine carbonate pump' (generated by the formation of calcareous shells of certain oceanic microorganisms in the surface ocean, which, after sinking to depth, are remineralized) (Ciais *et al.*, 2013).

4.2.1.1.2 THE SULFUR CYCLE. The sulfur cycle is an important part of global geochemical cycles where sulfur moves between rocks, waterways, and living systems. Air pollution occurs due to the production of sulfur from fossil fuel and ore smelting and this sulfur passes through the cycle

Table 4.2. Types of air pollutants and their sources. Table author's own.

Types	Examples	Sources	Additional notes
Carbon monoxide	–	Burning of natural gas, coal, or wood; vehicular exhausts	Colorless, odorless, and toxic.
Sulfur oxides	SO_2, H_2S	Produced when sulfur-containing fossil fuels are burned	
Nitrogen oxides		Vehicular exhausts	Involved in the production of secondary air pollutants such as ozone.
Hydrocarbons	Benzene, ethylene	Evaporation from fuel supplies; remnants of fuel that did not burn completely	Group of compounds consisting of carbon and hydrogen atoms.
Lead	Tetraethyl lead (TEL)	Vehicular exhausts from leaded petrol	
Particulates	Aerosol	Sprays from pressurized cans	General term for particles suspended in air.
	Mist	Small droplets of water suspended in cold air	Aerosol consisting of liquid droplets.
	Dust	Dust storms	Aerosol consisting of solid particles that are blown into the air or are produced from larger particles by grinding them down.
	Smoke	Cigarette smoke; burning garbage; forest fires	Aerosol consisting of solid particles, or a mixture of solid and liquid particles produced by chemical reaction.
	Fumes (zinc, lead)	Sprays from pressurized cans; aluminum refineries, steel plants	Generally, means the same as smoke but often applies specifically to aerosols produced by condensation of hot vapors of metals.
	Plumes	Factories; volcanic eruptions	Geometrical shape or form of the smoke coming out of a chimney.
	Fog	Sprays from pressurized cans	Aerosol consisting of water droplets.
	Smog	Vehicular exhausts; coal burning	Term used to describe a mixture of smoke and fog.

subject to redox reactions with mainly oxygen (O), carbon (C), and iron (Fe), which lead to a tight coupling of the global cycles for these elements (Fig. 4.3) (Schoonen, 2016).

Some of the SO_2 is converted to sulfate, and this is a principal player in acid deposition resulting in low pH of freshwater bodies and soils. With the lower pH in freshwater bodies, the viability of some plant and aquatic species is affected.

4.2.1.1.3 THE NITROGEN CYCLE. The largest reservoir of nitrogen is found in the atmosphere mainly in the form of nitrogen gas (N_2) (Fig. 4.4).

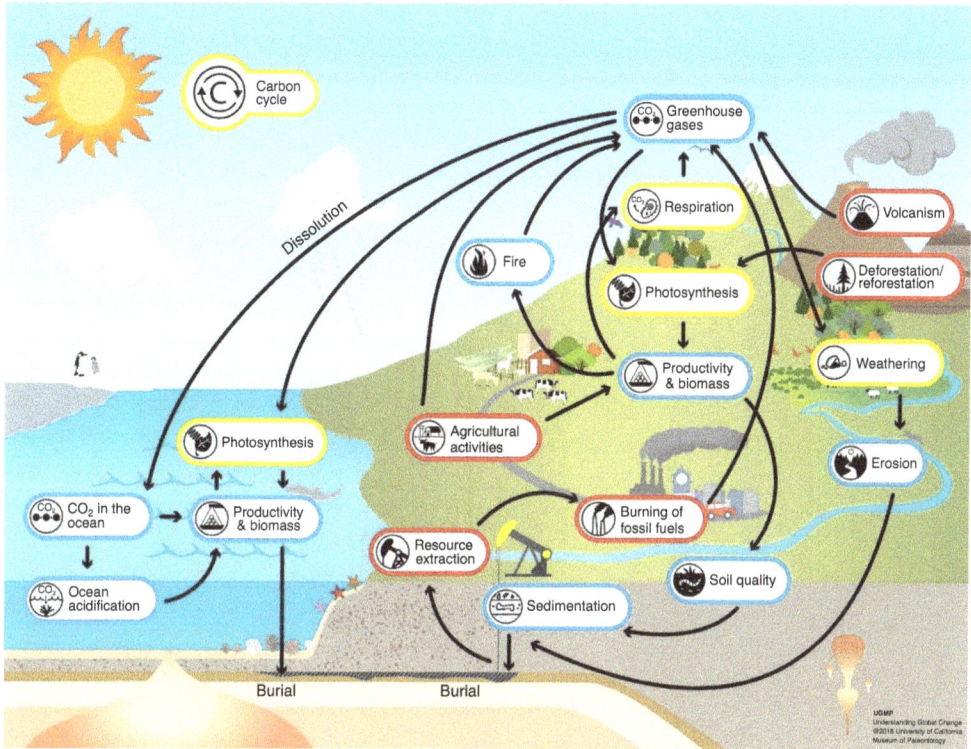

Fig. 4.2. Carbon cycle system. From University of California Regents (2023a).

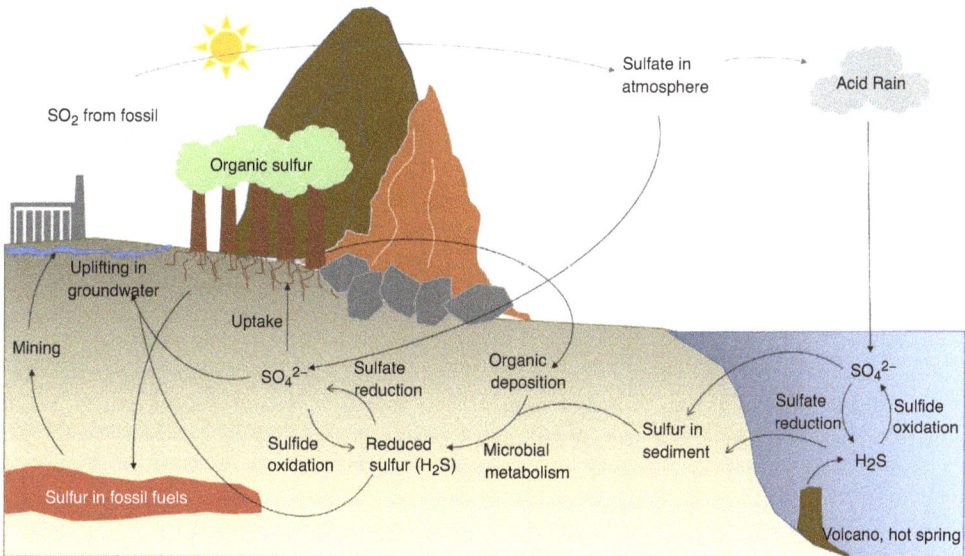

Fig. 4.3. Global sulfur cycle. From Fan *et al.*, 2023.

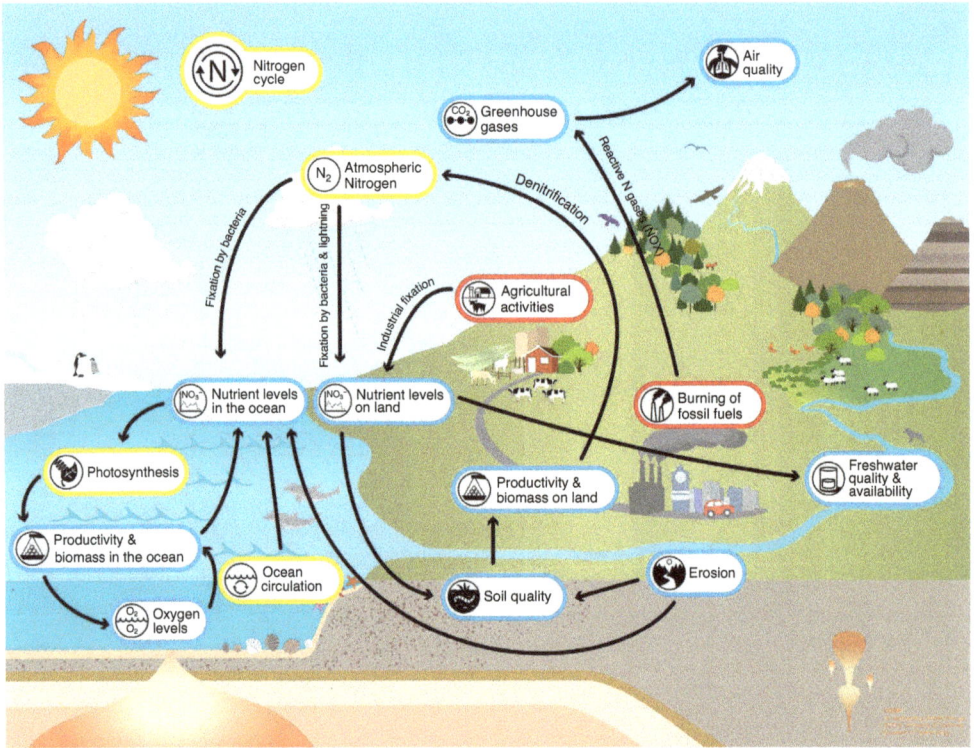

Fig. 4.4. Nitrogen cycle. From University of California Regents, 2023b.

Nitrogen compounds are formed during combustion, with 90–95% of the nitrogen oxides generated in the form of nitric oxide (NO). In the atmosphere, NO is converted photochemically into nitrogen dioxide (NO_2).

Nitrogen oxides react in the atmosphere with the ozone (O_3) to make nitric acid, which is one of the components of acid rain, which damages soil and forest richness by destroying communities of organisms. In this way, the burning of fossil fuels by humans contributes to acid rain and to the destruction of the ozone layer in the atmosphere, leading to global warming.

4.2.1.1.4 THE FLUORIDE CYCLE. The dominant natural sources of fluoride in the atmosphere are emissions from volcanic eruptions, eolian mobilization of surface deposits, production of sea salt aerosols, and natural biomass burning (Schlesinger *et al.*, 2020). Fluoride moves through the atmosphere and into a food chain, illustrated through an air–water interaction

(Fig. 4.5). This occurs when fluoride that is released into the air is deposited and accumulated in vegetation, which, when consumed in sufficient amounts, can cause damage to the teeth and bone structure of the animals.

4.2.1.2 Impact on aquatic ecosystems

Aquatic ecosystems are often impacted by atmospheric deposition of both nutrients and toxic chemicals, contributing to acidification of lakes and eutrophication of estuaries and coastal waters (Fig. 4.6). The effects of nutrient deposition on food web structure and ecological function influence how other toxic substances are processed by the ecosystem, how they bioaccumulate, and how they impact fish, wildlife, and humans (Swackhamer *et al.*, 2004; Lovett *et al.*, 2009).

As shown in Table 4.3, the effects of air pollution on the environment include global warming and acid rain, with eutrophication having a more direct effect on aquatic

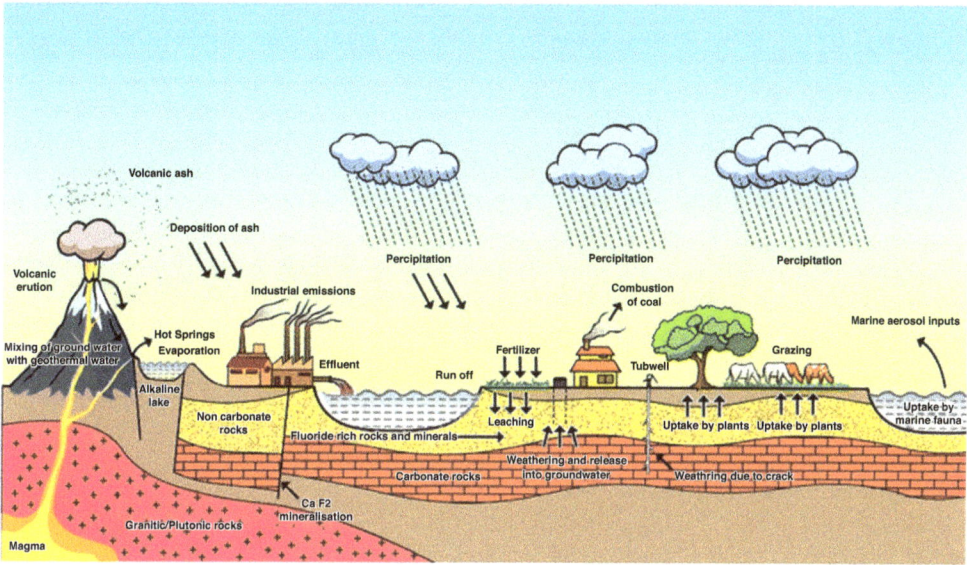

Fig. 4.5. Fluorine hydrogeochemical cycle. From Mukherjee and Singh, 2018. Figure used with permission from Springer Nature.

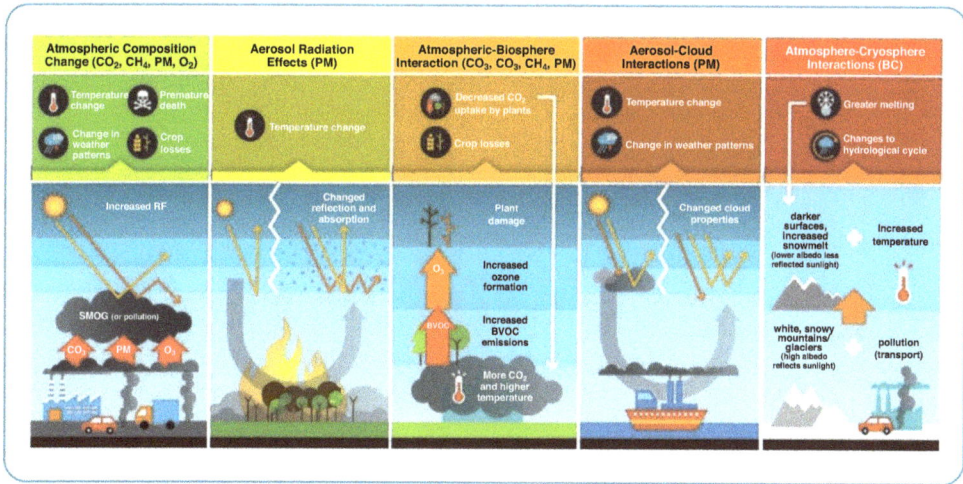

BVOC=Biogenic Volatile Organic Compounds, RF=radiofrequency

Fig. 4.6. The interactions between air pollution and climate change. From von Schneidemesser *et al.*, 2015.

ecosystems, while respiratory and heart problems are the main ways in which human health is affected (Vallero, 2008; Sneh, 2018).

Air pollution causes damage to plants, resulting in a reduction in their ecosystem functions, which affects the well-being of humans. When plant color is changed due to air pollution, it affects the plant–insect relationship, interfering with pollination, which poses a human food security problem. Air pollution also causes suppression of the immune systems of wild animals,

Table 4.3. Effects of air pollution on human and environmental health. Table author's own.

Pollutants	Effects on human health	Effects on environmental health
Carbon monoxide	Poisoning; severe headache; irritation to mucous membrane; unconsciousness; low birth weight; death	Contributes indirectly to climate change; weak direct effect on climate
Sulfur oxides	Suffocation; throat and eye irritation; respiratory diseases; allergies; cough; asthma; bronchitis; emphysema; chronic and acute respiratory disease	Crop destruction; reduced yield; chlorosis; necrosis
Nitrogen oxides	Irritation; bronchitis; oedema of lungs; asthma; bronchitis; heart problems	Damage to foliage; reduced crop yields
Hydrocarbons	Kidney problems; irritation in eyes, nose, and throat; asthma; hypertension; carcinogenic effects on lungs	Climate change; reduced photosynthetic ability of plants; ozone depletion
Lead	Degradation of renal function; impairment of hemoglobin synthesis; alteration of the nervous system	Biodiversity losses; changes in community composition; decreased growth and reproductive rates in plants and animals; neurological effects in vertebrates
Aerosol	Damage to lung tissue leading to lung diseases	Changes the size or lifetime of water droplets inside clouds; direct scattering and absorption of incoming solar radiation and trapping of outgoing long-wave radiation
Mist	Irritation of airways; asthma; allergies;	–
Dust	Respiratory diseases; silicosis; wheezing; asthma; chronic bronchitis and chronic obstructive pulmonary disease (COPD); cardiovascular disease; death	Reduced crop yields by burying seedlings; causing loss of plant tissue; reducing photosynthetic activity; increasing soil erosion
Smoke	Eye irritations; nose irritations; throat irritations, nausea	Soil contamination; groundwater contamination
Fumes (zinc, lead)	Damage to lungs, brain, nervous system, and other organs	Climate change; reduced soil and water quality
Plumes	Respiratory effects; eye symptoms; skin irritations	Damage to vegetation
Fog	Chills; irritation causing coughs and sniffles	Reduced agricultural productivity
Smog	Lung irritation; asthma; bronchitis; eye problems	Crop destruction

increasing the spread of zoonotic diseases. Air pollution causes accumulation of metals in soil where soil microorganisms make them bioavailable, affecting human health when the animals in which the metals are accumulated are consumed (World Health Organization, 2022).

4.2.2 Land pollution

Land is the most basic production factor for socio-economic development. This development over the years has resulted in pollution due to the way in which modern agriculture, mining, urbanization, and industrialization have evolved

Table 4.4. Causes and effects of land pollution. Table author's own.

Causes	Effects	Additional notes
Deforestation	Soil erosion, desertification, shifting habitat	Misuse of land
Agriculture	Improper waste disposal, groundwater poisoning, water nutrient enrichment, soil pollution	
Industrialization	Improper waste disposal, groundwater poisoning	
Mining	Improper waste disposal, removal of topsoil, groundwater poisoning	
Landfill	Improper waste disposal, groundwater poisoning, air pollution	
Sewage management	Improper waste disposal	Untreated human waste
Construction activities	Improper waste disposal	
Nuclear waste management	Improper waste disposal, groundwater poisoning, soil pollution	
Overpopulation	Improper waste disposal	
Urbanization	Improper waste disposal	

(Gao *et al.*, 2022). The causes and effects of land pollution are intricately linked and are reflected in Table 4.4.

Land pollution may be defined as solid or liquid waste materials that, when deposited on land, further degrade and deteriorate the quality and the productive capacity of the land surface (Savaşan, 2017). This is almost always due to direct and indirect actions of humans.

4.2.2.1 Impact on aquatic ecosystem

Aquatic ecosystems are generally impacted through runoff, where effluent from farms, industrial sites, and landfills are released directly or indirectly into waterways. In the case of farms, only a small portion of the nutrients end up benefiting the crops, while the remainder, which ends up in the waterways, results in eutrophication, where the water is depleted of oxygen, negatively affecting aquatic life forms. Overpopulation and urbanization result in greater need for larger and/or more landfill sites due to the increase in waste generation. While leaching into waterways does occur, there is a growing issue of indiscriminate dumping directly into waters, resulting in marine plastic pollution and toxic chemicals being released. These pollutants are consumed by the aquatic animals and plants, making their way into the ecosystem. This is referred to as biomagnification and poses a serious threat to the ecology.

4.2.3 Light pollution

Light pollution, also known as photopollution, may be defined as the alteration of the natural quantity of light in the night environment due mainly to the introduction of excess artificial light by humans. When compared to other forms of pollution, light pollution goes unnoticed, tends to get very little attention, or is simply viewed as a nuisance or something that has purely aesthetic effects. However, there is evidence that it negatively affects the environment.

Sources of light pollution include street lighting, commercial lights, urban areas, yard lights, park lights, billboards, road lights, and decorative outdoor lights (Lechner and Arns, 2013).

4.2.3.1 Impact on aquatic ecosystem

The effect of light on organisms has been known by man for centuries and has been documented with regard to the phases of the moon. With this traditional knowledge, artisanal and industrial fisheries have used light to attract fish to their nets. More recently, this knowledge is being deployed by natural resource managers, where light is set out to attract larval fish to coral reefs to boost fish stocks (Depledge *et al.*, 2010; Rajkhowa, 2012). Submerged light increases swimming depth and reduces fish density of Atlantic salmon in production cages. It is also used to postpone sexual maturation and increase growth. On halibut farms, artificial light is used to influence swimming depth and swimming activity, allowing halibut to swim less and grow more (Rajkhowa, 2012).

The natural environment is closely linked to the cycles of night and day, so the alteration of darkness into light affects these rhythms in the ecosystem. Known and potential impacts include those on navigation, foraging, reproduction, predator–prey interactions, and communication (Longcore and Rich, 2004; Davies *et al.*, 2014).

While light is important for primary production by photoautotrophs, excessive irradiance can cause over-excitation of the photoactive centers and thus damage the photosynthetic machinery of primary producers. The free-swimming organisms of aquatic habitats need light for different purposes, and respond differently to any alterations in the natural patterns of light and dark (Khanduri and Saxena, 2020). When it comes to the aquatic ecosystem, sea turtles would be one of the best examples to illustrate the effects of light pollution. Female turtles avoid illuminated beaches for their nests, with the effect that the nests are concentrated on the less illuminated and shaded parts, which may not be an ideal nesting site (Kamrowski *et al.*, 2012; Rajkhowa, 2012). Sea turtles then lay their eggs on these dark beaches. When the hatchlings emerge, they gravitate toward the brightness of the sea. However, with the advent of beachfront development and its associated artificial lighting, the light from the sea is often eclipsed by these artificial lights from the beach, thus disorienting the hatchlings, leaving them vulnerable and unlikely to successfully reach the sea (Cohen and Robbins, 2012; Kamrowski *et al.*, 2012).

Artificial light at night changes the daytime behavior of certain species of fish. Artificial lighting also disturbs the reproductive cycles of many animals, mainly amphibians such as frogs. Artificial lighting of the water surface affects the vertical migration of zooplankton such as *Daphnia* because they avoid feeding in the subsurface layers of the reservoir. This results in the growth of phytoplankton, leading to eutrophication (Ściężor, 2019).

The introduction of lighthouses in the 19th century and, more recently, that of offshore oil/gas platforms has resulted in marine bird fatalities. These vary from species to species, depending on the type of signal used (e.g., white lights, colored lights, flashing lights). The fatalities occur due to direct injury or death by heat, collision, or oil, but also indirectly by the trapping effect of the light that leads birds to circle around the light source, reducing their energy reserves and hampering their ability to reach the next shore or decreasing their ability to survive the winter or reproduce. Because there are fewer artificial light sources out at sea in comparison to terrestrial locations, the effect of a single artificial light is greater (Rajkhowa, 2012).

4.2.4   Noise pollution

Noise pollution may be defined as the elevation of natural ambient noise levels attributed to sound-generating human activities (which could be deliberate and wanted or unwanted) and may have detrimental consequences for both humans and animals. Some sources of sounds are music, sirens, seismic survey sounds, military sonar, traffic (motor vehicles, trains, aircraft, and sea/ocean vessels), generator noise, pile driving, explosives, construction sites, motorized recreation, air conditioners, cleaning machines, dredging, and pumping systems (Slabbekoorn, 2019).

4.2.4.1   *Impact on aquatic ecosystem*

The locations of noise pollution in the ocean tend to be along well-traveled paths in the sea and particularly encompass coastal and continental shelf waters. These sources of noise pollution have increased due to an intensification of commercial shipping, an expansion of geophysical exploration, and the advent of advanced warfare (Kunc *et al.*, 2016).

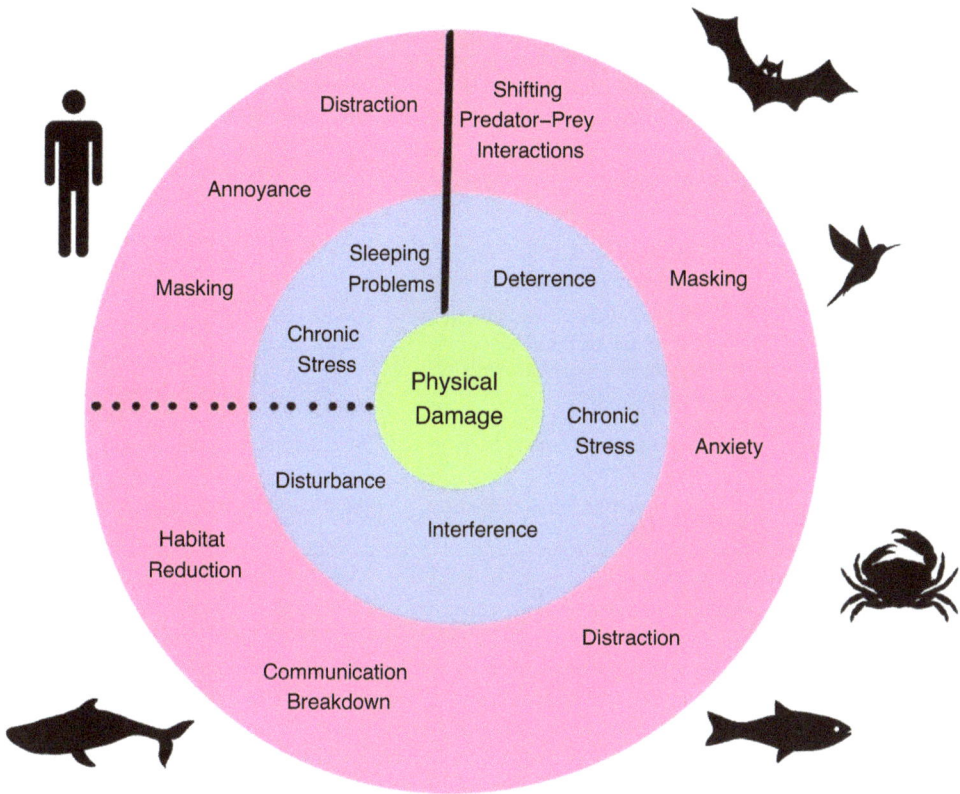

Fig. 4.7. Diverse effects of noise pollution. Modified from Slabbekoorn, 2019. Figure used with permission of Elsevier.

Noise can affect both the anatomy and the morphology of an organism by damaging single cells or organs and hampering hearing capabilities. It can also induce a variety of behavioral changes by affecting navigation, leading to strandings and abnormal schooling. It can cause problems with attracting mates and communicating with other members of their species, with prey location, and with threat/predator avoidance. This can result in difficulties locating suitable habitats and feeding and breeding grounds (Fig. 4.7) (Firestone and Jarvis, 2007; Slabbekoorn *et al.*, 2010; Kunc *et al.*, 2016; Weilgart, 2018; Slabbekoorn, 2019; Arctic Council's Working Group on Protection of the Arctic Marine Environment, 2021).

A review of 115 primary studies which encompassed various human-produced underwater noise sources, 66 species of fish, and 36 species of invertebrates showed that noise impacts include body malformations, higher egg or immature mortality, developmental delays, delays in metamorphosing and settling, and slower growth rates. Zooplankton also suffered high mortality in the presence of noise (Hazra, 2017).

4.2.5 Radioactive pollution

Radiation is energy traveling through space transported in the form of either electromagnetic waves or a stream of energetic particles in either non-ionizing (longer wavelength, from near ultraviolet rays to radio waves) or ionizing (high energy such as alpha (α), beta (β), and gamma (γ) radiations) forms. Radioactive pollution may be defined as the emission of radioactive radiation at a level where it causes harmful effects

Table 4.5. Types of soil pollutants and their sources. Table author's own.

Types	Examples	Sources
Industrial wastes	Coal ash, mud, pulp, paper	Chemical industries, sugar factories, tanneries, textile mills, steel industries, distilleries, pulp and paper mills, oil refineries, petroleum industries, thermal power plants, atomic power plants
Agricultural wastes	Fertilizers, pesticides, insecticides, weedicides, pharmaceuticals	Farms
Urban wastes	Plastics, glass, metallic cans, fibers, paper, street sweepings, leaves, rubble, pharmaceuticals	Commercial centers, domestic areas/households
Radioactive materials	Uranium	Accidents, fallout from nuclear bombs
Biological agents	Biological organisms from human and animal excreta	Faulty sanitation, disposal of wastewater, septic tank leachate
Mining	Tail-end seepages, mercury, cyanide	Mining pits, abandoned mines

to living beings. Sources of radioactive pollution are spent fuel from nuclear power plants; radioactive waste from medical institutions, industrial operations, and research activities; nuclear explorations; nuclear weapon tests; and nuclear and radiation accidents (Rahman *et al.*, 2014; Hazra, 2017).

4.2.5.1 Impact on aquatic ecosystem

Radioactive contamination and radiation exposure occurs when radioactive materials are released into the environment as a result of an accident, an event in nature, or an act of war or terrorism, contaminating the surrounding areas (Rahman *et al.*, 2014; Hazra, 2017). When living things are exposed to this contamination in the environment, radiation sickness may occur, which differs in intensity and the types of symptoms present depending on whether it is acute, chronic, moderate, or severe. Radiation sickness affects humans and animals, with symptoms including nausea, weakness, hair loss, skin burns, diminished organ function, lowering of the white blood cell count, bacterial infections, vomiting, loss of appetite, reddening of the skin, diarrhea, fatigue, fever, abdominal pains, sterility, internal bleeding, infection, shock, convulsions, coma, and ultimately death. When exposed to ionizing radiation, living tissue can be damaged when it absorbs energy. When

the body attempts to repair the damage, mistakes may occur in the natural repair process, leading to the development of cancerous cells (Rahman *et al.*, 2014).

Pollution of the aquatic and marine environment can occur as a result of accidents, as in the case of the Fukushima Daiichi Nuclear Power Plant accident in March 2011. This occurred as a result of an earthquake and tsunami on March 11, 2011, with river catchments draining radioactive materials (Evrard *et al.*, 2021). Radioactive substances can be transported by ocean currents and spread over large distances, thus becoming diluted in large bodies of water. As a result of this, along with the fact that the potassium content of the sea causes the animals to absorb less radioactive caesium from the environment, radioactive contamination in marine fish and shellfish tends to be extremely low (Norwegian Radiation and Nuclear Safety Authority, 2023).

4.2.6 Soil pollution

Soil pollution may be defined as the entry of elements that change the composition and organism of the soil, reducing its fertility, increasing its vulnerability to drought, and making it unsuitable for agriculture and/or having adverse

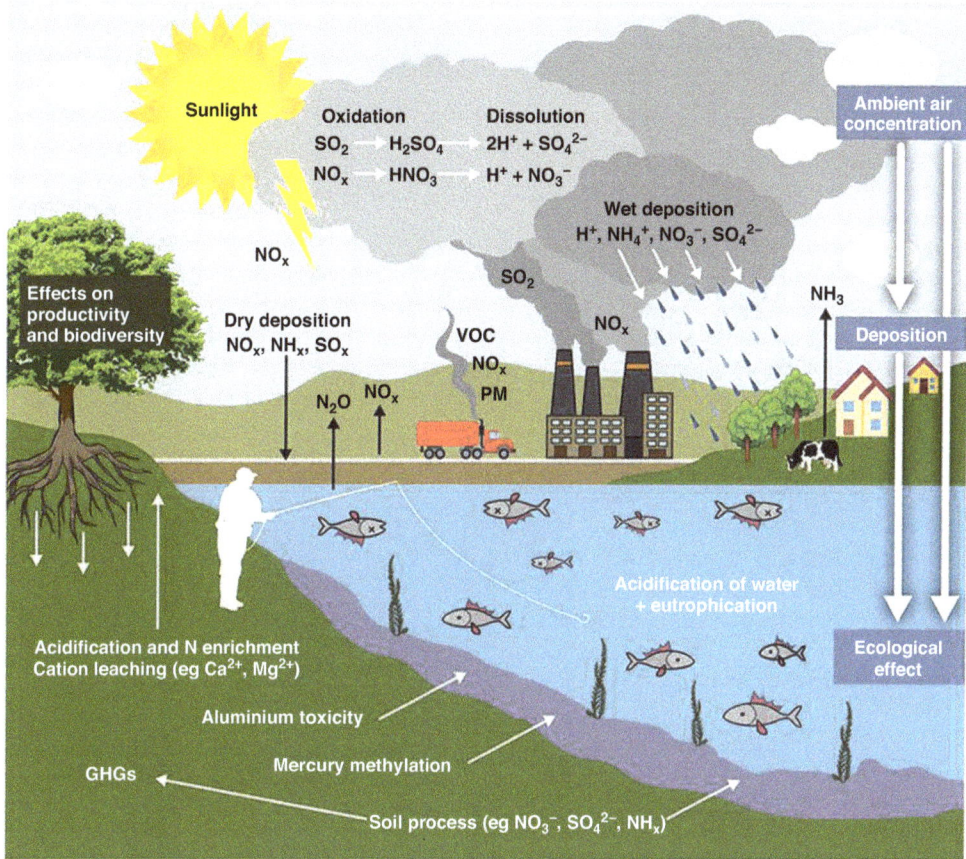

Fig. 4.8. Process of atmospheric fluxes of contaminants resulting in soil pollution. From FAO and UNEP, 2021.

effects on any non-targeted organism (Eugenio et al., 2018; Hassan Al-Taai, 2021). As shown in Table 4.5, soil pollution has diverse sources, which could be natural or anthropogenic (Eugenio et al., 2018).

Industrial activities and transportation (automobiles and trains) release pollutants to the atmosphere which enter the soil through acid rain or atmospheric deposition (Fig. 4.8). Mining activities usually result in soils heavily contaminated with heavy metals that persist long after these activities are over. Waste and sewage generation is increasing due to population growth, leading to soil pollution by landfill sites through leaching and by incineration sites through ash fallout. Farming results in soil pollution due to the use of pesticides and fertilizers (Eugenio et al., 2018).

4.2.6.1 Impact on aquatic ecosystem

Land-based activities are mainly responsible for marine pollution, represented primarily through plastic and nutrient contamination. Nevertheless, contaminants in aquatic ecosystems originating from soil pollution are diverse in nature. The impact of soil pollution on the aquatic ecosystem is primarily through increased sediment runoff due to precipitation, floods, snowmelt, or irrigation resulting in the degradation of inland and coastal waters with contaminants ultimately reaching the seas and oceans (Mateo-Sagasta et al., 2017; FAO and UNEP, 2021). In the case of agriculture, the main impact on the aquatic ecosystem is that of eutrophication due to over-enrichment of the soil. This is not limited to crop farming but includes piggeries, dairies, meat and vegetable processing plants, fertilizer factories, and tanneries (Ansari and Gill, 2014). Soil particles cause water turbidity, while plants growing in contaminated waters act as a pathway for contaminants to be further dispersed. Harmful algal blooms resulting from excessive nutrients entering the waterways cause dead zones, which are areas with reduced levels of dissolved oxygen. This happens when the algae growth dies and sinks to the floor of the water body and the bacteria involved in its decomposition consumes significant amounts of oxygen in the process (FAO and UNEP, 2021).

4.2.7   Thermal pollution

Thermal pollution can be defined as the increase in the average temperature of natural waterways due to heated wastewater discharges produced as an unavoidable by-product of industrial and electric power plants (Geurdes, 2023).

4.2.7.1   Impact on aquatic ecosystem

Warm water holds less dissolved oxygen than cold water. Thermal pollution resulting in low levels of dissolved oxygen causes aquatic organisms to be more vulnerable to disease, parasites, and toxic chemicals. It also affects their reproductive cycles, digestion rates, and respiration rates. Additionally, when a power plant starts or shuts down for scheduled maintenance or needed repair, fish and other organisms adapted to the temperature range produced by the heated water discharge can die from stress and thermal shock due to the abrupt temperature change. Increase in temperature may change the inland waterway's capacity to assimilate wastes, due to changes in algal composition because of increased metabolism, or lead to algal blooms due to accelerated growth, ultimately causing eutrophication (Geurdes, 2023).

4.2.8   Water pollution

Water pollution can be defined as the contamination of streams, lakes, seas, underground water, or oceans by substances that are harmful for living beings and are transported into the water cycle (Owa, 2014). The types and examples of water pollution are the same as those for soil pollution but are categorized as point or non-point sources (Table 4.6). Point-source water pollution refers to contaminants that enter a waterway from a single, well-defined, identifiable source, such as a pipe or ditch, while non-point-source pollution refers to diffuse contamination that is scattered or spread over large areas and occurs when rainfall, snowfall, or irrigation runs over land or through the ground and picks up pollutants and deposits them in bodies of water (Kılıç, 2021)

4.2.8.1   Impact on aquatic ecosystem

The majority of aquatic organisms are very sensitive to variations in the environment, and the impacts of pollution are manifested in different ways, with the most drastic being death or migration (Bassem, 2020). The latter was demonstrated through studies in China which showed that organic pollutants affect macroinvertebrate taxa richness and composition of functional feeding groups, with scrapers, shredders, predators, and collector-filterers decreasing while collector-gatherers dominate, resulting in extremely non-uniform distribution of the functional feeding groups (Xu *et al.*, 2014).

Various harmful substances present in polluted water bodies, such as insecticides, pesticides, heavy metals, and crude oil, cause several effects on the aquatic ecosystem due to changes in the physico-chemical properties of the water. Aquatic plants are affected by poor plant growth, reduction in photosynthetic rate, phytotoxicity, and death. Agricultural runoff pollutes waterways with nitrogen- and phosphorus-rich fertilizers, paving the way to algal blooms, which lead to eutrophication. This results in oxygen depletion and the death of fish due to suffocation (Singh and Gupta, 2017; Malik *et al.*, 2020). Furthermore, pollution has non-fatal impacts on fish nutrition, migration, genetics, and morphology (Malik *et al.*, 2020).

While the true extent of ocean pollution on fish and other marine life is difficult to determine due to the large number of animals involved and the vastness of the ocean, there is enough evidence that demonstrates that it is taking its

Table 4.6. Types of water pollutants and their sources. Table author's own.

Types	Examples	Sources	
		Point	Non-point
Industrial wastes	Coal ash, mud, pulp, paper	Runoff from chemical industries, sugar factories, tanneries, textile mills, steel industries, distilleries, pulp and paper mills, oil refineries, petroleum industries, thermal power plants, atomic power plants	Runoff from construction sites
Agricultural wastes	Fertilizers, pesticides, insecticides, weedicides, pharmaceuticals	Runoff from farms, from animal feedlots	Runoff from agriculture, pasture, and range
Urban wastes	Plastics, glass, metallic cans, fibers, paper, street sweepings, leaves, rubbles, pharmaceuticals	Runoff from landfill/ waste disposal sites, storm sewer outfalls	Littering from commercial centers, domestic areas/ households, urban runoff (storm water washed off of parking lots, roads, and highways)
Radioactive materials	Uranium	Accidents, fallout from nuclear bombs	Leakage of radioactive materials
Biological agents	Biological organisms from human and animal excreta	Wastewater effluent, sewer outfalls	Septic tank leachate, runoff from failed septic systems
Mining	Tail-end seepages, mercury, cyanide	Accidents, leakage from mining pits	Abandoned mines

toll on species. This includes the discovery of man-made debris in the guts of marine species and the encounter of marine species entangled in man-made waste, which leads to injury, deformations, and restriction in movement, in turn leading to death by drowning, starvation, or predation (Sharma and Sharma, 2020). Another example is the flushing of cat feces into human septic systems, resulting in the release of *Toxoplasma gondii* into the oceans. *T. gondii* is known to cause mortality in several species of marine mammals (Díaz-Delgado *et al.*, 2020).

Pharmaceutical pollutants in aquatic ecosystems originating from both human and veterinary sources can have short-, medium-, and long-term impacts on the environment, wildlife, and humans due to their wide range of bioactivity (Rzymski *et al.*, 2017; Kayode-Afolayan *et al.*, 2022; Kock *et al.*, 2023). While pharmaceutical

pollutants are a growing cause for concern and considered as emerging contaminants, there is insufficient data to determine the exact extent of the effects of these pollutants in the environment, and most of what we know comes from research done in controlled environments (Gworek *et al.*, 2019; Ortúzar *et al.*, 2022). Among the impacts of concern are antimicrobial resistance on a global scale and the increased use of endocrine-disrupting compounds which are exacerbated by bioaccumulation in aquatic systems (Eapen *et al.*, 2024).

4.2.9 Biopollution

Biopollution can be defined as intentional or unintentional biological introduction by humans or the natural biological invasion of an alien species into an environment. Introduction by humans occurs due to escapes from aquaculture facilities, hatcheries, or research stations/laboratories (Occhipinti-Ambrogi, 2021). Marine debris is a well-documented source of biopollution where facultative species (those known to live on debris items) and obligate rafters (known to associate with floating debris for their entire life cycle) become introduced into new areas/habitats. Microorganisms are also capable of biopollution through the colonization of various surfaces of marine debris through which a microbial biofilm is created. Sessile organisms (those that do not move), such as bryozoans, barnacles, ascidians, hydroids, macroalgae, and some mollusks, also cause biopollution by clinging to marine vessels. Mobile organisms, also called hitchhikers, hangers-on, or aquatic rafters, such as arthropods, mollusks, and cnidarians, also cause biopollution by way of marine debris. This is also the case for terrestrial organisms such as ants (National Oceanic and Atmospheric Administration Marine Debris Program, 2017).

The effects of biopollution include biodiversity loss, food web alteration, physical habitat disruption, hybridization with native species, and the introduction of parasites and diseases uncommon in the habitat in question (Joshi, 2008; Occhipinti-Ambrogi, 2021).

4.3 Mitigation of Environmental Risks

Pollution mitigation is an important policy goal because of the interplay between pollution and ecosystems, health, biodiversity, climate change, and economic growth, as well as the speed at which effects become seen, felt, and quantified.

Environmental risks can be reduced through:

- regulatory measures such as international conventions;
- market-based measures such as subsidies and taxes;
- training and capacity building;
- public awareness building to change behaviors; and

- employing engineering controls and technology (United Nations Environment Programme, 2022).

4.3.1 Innovative biotechnologies for environmental pollution/hazards mitigation

There are two policy frameworks for biotechnology: one for environmental biotechnology and another for industrial biotechnology. The primary emphasis of environmental biotechnology is biotechnologies for environmental cleanup of contaminated soil and water. Industrial biotechnology is related to the development of biofuels and the production of biotechnology products such as biobased chemicals and bioplastics (Organisation for Economic Co-operation and Development, 2013). There are various factors that could be taken into account when selecting the appropriate method to be employed, as shown in Table 4.7.

Bioremediation is one of the most recent biotechnology measures; it includes a group of procedures that employ biological systems such as microorganisms to either clean up or repair polluted areas (Vishwakarma et al., 2020). Ex situ technology known as 'soil washing' is another; it uses two methods to remove pollutants from soil: chemical leaching using aqueous solutions and physical size separation (Fernández et al., 2024). Landfill is the least expensive solution for contaminated soil problems, in comparison with bioremediation and soil washing (Organisation for Economic Co-operation and Development, 2013).

The treatment of wastewater using conventional systems is unable to completely remove pharmaceuticals (Kock et al., 2023; Eapen et al., 2024). Currently, new and alternative add-on systems for traditional wastewater treatment plants capable of addressing emerging contaminants are being developed to mitigate and reduce the impacts and effects they have on the environment and animal and human health. These include the biological transformation of pharmaceutical pollutants as a green technology and the incorporation of bioremediation technologies (Ortúzar et al., 2022).

Table 4.7. Eco-efficiency of selected contaminated land remediation technologies. Modified from Organisation for Economic Co-operation and Development, 2013.

Remediation method	Positive factors	Negative factors
Reactive barrier	Removal of the barrier is not required	Long-term operating costs, applicable for some contaminants
Soil stabilization, isolation	Soil removal is not necessary; quick; economical	No removal of contaminants from environment; can be energy-intensive
Soil vapor extraction (SVE)	Generally cost-effective; high certainty in risk reduction	Suitable only for volatile contaminants; exhaust air requires treatment
Incineration (mobile)	Effective contaminant removal	Flue gas treatment needed; energy-intensive, requiring fuel
Composting	Low cost; treated soil may be used for landscaping; no emissions	Suitable only for some organic contaminants; can be long duration; depends on contaminant concentrations
Landfill	Effective control of risks; soil can be used in daily cover	Not treatment; unsuitable for reuse; becoming costly; inefficient use of landfill sites

4.3.2 Sustainable technologies in use to combat climate change

One of the main aspects that need to be tackled when it comes to climate change is the world's reliance on fossil fuels. This can be done effectively and efficiently with the employment of green technologies (sustainable and friendly), which will help to create new relationships between humans and nature (Wu and Strezov, 2023). Therefore, technologies are needed to facilitate energy and economic transition for sustainable development. These include offshore wind farms, the capture, storage, and utilization of carbon dioxide (CO_2), particularly the electrochemical reduction and bioconversion of CO_2, sustainable ammonia production through energy-saving innovations, and cellular agriculture (Coccia, 2023).

The use of industrial biotechnology is cross-cutting since it can be implemented in different industrial sectors, which would result in a reduction in greenhouse gas emissions. Some examples are:

- cold-water enzymes in detergents resulting in a reduction in temperature required for washing clothes;
- polylactic acid, or PLA, produced from corn starch;
- biotechnology processes to reduce energy consumption during the bleaching process in pulp and paper production;
- biobased substances used in building blocks;
- removing the malting process by use of an enzyme;
- enzymes formed by microbial fermentation for tanning; and
- biofuels (Organisation for Economic Co-operation and Development, 2011).

In summary, green technologies, which rely on non-toxic chemical processes, non-toxic end products, renewable energy sources, and environmental monitoring equipment, can lessen or offset the adverse effects of human activity on the environment, including the marine ecosystem (Rene *et al.*, 2021).

References

Ansari, A. and Gill, S. (2014) *Eutrophication: Causes, Consequences and Control*. Springer Science, Dordrecht, Netherlands.

Arctic Council's Working Group on Protection of the Arctic Marine Environment (2021) *Underwater Noise Pollution from Shipping in the Arctic*. Protection of the Arctic Marine Environment Working Group (PAME). Available at: https://oaarchive.arctic-council.org/server/api/core/bitstreams/c7d409c8-e56 7-471f-89af-b8dc14ec7ca0/content (accessed 11 June 2025).

Bassem, S. (2020) Water pollution and aquatic biodiversity. *Biodiversity International Journal* 4(1), 10–16. DOI: 10.15406/bij.2020.04.00159.

Ciais, P., Sabine, C., Bala, G., Bopp, L., Brovkin, V. *et al.* (2013) Carbon and other biogeochemical cycles. In: Stocker, T.F., Qin, D., Plattner, G.K., Tignor, M. and Allen, S.K. (eds) *Climate Change 2013: The Physical Science Basis. Contribution of Working Group I to the Fifth Assessment Report of the Intergovernmental Panel on Climate Change*. Cambridge University Press, Cambridge, United Kingdom.

Coccia, M. (2023) New directions of technologies pointing the way to a sustainable global society. *Sustainable Futures* 5, 100114.

Cohen, N. and Robbins, P. (2012) *Green Cities: An A-to-Z Guide – Light Pollution*. SAGE Publications, Thousand Oaks, California.

Davies, T., Duffy, J., Bennie, J. and Gaston, K. (2014) The nature, extent, and ecological implications of marine light pollution. *Frontiers in Ecology and the Environment* 12(6), 347–355.

Depledge, M., Godard-Codding, C. and Bowen, R. (2010) Light pollution in the sea. *Marine Pollution Bulletin* 60, 1383–1385.

Díaz-Delgado, J., Groch, K.R., Ramos, H.G.C., Colosio, A.C., Alves, B.F. *et al.* (2020) Fatal systemic toxoplasmosis by a novel non-archetypal *Toxoplasma gondii* in a Bryde's Whale (*Balaenoptera edeni*). *Frontiers in Marine Science* 7, 336. DOI: 10.3389/fmars.2020.00336.

Eapen, J.V., Thomas, S., Antony, S., George, P. and Antony, J. (2024) A review of the effects of pharmaceutical pollutants on humans and aquatic ecosystem. *Exploration of Drug Science* 2, 484–507. DOI: 10.37349/eds.2024.00058.

Eugenio, N., McLaughlin, M. and Pennock, D. (2018) *Soil Pollution: A Hidden Reality*. FAO, Rome.

Evrard, O., Chartin, C., Laceby, P., Onda, Y., Wakiyama, Y. *et al.* (2021) Radioactive dose rates and fallout radionuclide activities in sediment deposits along rivers draining the main Fukushima Plume, Japan. *PANGAEA*. DOI: 10.1594/PANGAEA.928594.

Fan, K., Wang, W., Xu, X., Yuan, Y., Ren, N. *et al.* (2023) Recent advances in biotechnologies for the treatment of environmental pollutants based on reactive sulfur species. *Antioxidants* 12(3), 767. DOI: 10.3390/antiox12030767.

FAO and UNEP (2021) *Global Assessment of Soil Pollution: Report*. FAO, Rome. DOI: 10.4060/cb4894en.

Fernández, M., Sánchez-Arguello, P. and García-Gómez, C. (2024) Soil pollution remediation. In: Wexler, P. (ed.) *Encyclopedia of Toxicology*, 4th edn. Academic Press, Cambridge, Massachusetts, pp. 631–645.

Firestone, J. and Jarvis, C. (2007) Response and responsibility: Regulating noise pollution in the marine environment. *Journal of International Wildlife Law and Policy* 10, 109–152.

Gao, L., Hu, T., Li, L., Zhou, M. and Zhu, B. (2022) Land pollution research: Progress, challenges, and prospects. *Environmental Research Communications* 4, 112001.

Geurdes, M. (2023) Thermal pollution and its impact on the environment. *Journal of Pollution Effects and Control* 11, 361.

Gworek, B., Kijeńska, M., Zaborowska, M., Wrzosek, J., Tokarz, L. *et al.* (2019) Pharmaceuticals in aquatic environment. Fate and behavior, ecotoxicology and risk assessment – A review. *Acta Poloniae Pharmaceutica – Drug Research* 76(3), 397–407. DOI: 10.32383/appdr/103368.

Hassan Al-Taai, S. (2021) Soil pollution - causes and effects. *IOP Conference Series: Earth and Environmental Science* 790, 012009.

Hazra, G. (2017) Radioactive pollution: An overview. *The Holistic Approach to Environment* 8(2), 48–65.

Joshi, S. (2008) Invasive species - ecological corrections for control of pollution. In: Sengupta, M. and Dalwani, R. (eds) *Proceedings of Taal 2007: The 12th World Lake Conference, Ministry of Environment and Forests, Government of India, 28 October–2 November, 2007*, Jaipur, Rajasthan, India, pp. 889–893.

Kamrowski, R., Limpus, C., Moloney, J. and Hamann, M. (2012) Coastal light pollution and marine turtles: Assessing the magnitude of the problem. *Endangered Species Research* 19, 85–98. DOI: 10.3354/esr00462.

Kayode-Afolayan, S.D., Ahuekwe, E.F. and Nwinyi, O.C. (2022) Impacts of pharmaceutical effluents on aquatic ecosystems. *Scientific African* 17, e01288. DOI: 10.1016/j.sciaf.2022.e01288.

Khanduri, M. and Saxena, A. (2020) Ecological light pollution: Consequences for the aquatic ecosystem. *International Journal of Fisheries and Aquatic Studies* 8(3), 01–05.

Kock, A., Glanville, H.C., Law, A.C., Stanton, T., Carter, L.J. *et al.* (2023) Emerging challenges of the impacts of pharmaceuticals on aquatic ecosystems: A diatom perspective. *Science of The Total Environment* 878, 162939. DOI: 10.1016/j.scitotenv.2023.162939.

Kunc, H., McLaughlin, K. and Schmidt, R. (2016) Aquatic noise pollution: Implications for individuals, populations, and ecosystems. *Proceedings of the Royal Society B* 283, 1836. DOI: 10.1098/rspb.2016.0839.

Kılıç, Z. (2021) Water pollution: Causes, negative effects and prevention methods. *Istanbul Sabahattin Zaim University Journal of the Institute of Science and Technology* 3(1), 129–132. DOI: 10.47769/izufbed.862679.

Lechner, S. and Arns, M. (2013) *Light pollution*. Working paper. Hanzehogeschool Groningen, Netherlands. DOI: 10.13140/RG.2.2.13587.48163.

Longcore, T. and Rich, C. (2004) Ecological light pollution. *Frontiers in Ecology and the Environment* 2(4), 191–198.

Lovett, G.M., Tear, T.H., Evers, D.C., Findlay, S.E.G., Cosby, B.J. *et al.* (2009) Effects of air pollution on ecosystems and biological diversity in the Eastern United States. *The Year in Ecology and Conservation Biology, Annals of the New York Academy of Sciences* 1162(1), 99–135. DOI: 10.1111/j.1749-6632.2009.04153.x.

Malik, D., Sharma, A., Sharma, A., Thakur, R. and Sharma, M. (2020) A review on impact of water pollution on freshwater fish species and their aquatic environment. In: Kumar, V., Kamboj, N., Payum, T., Singh, J. and Kumar, P. (eds) *Advances in Environmental Pollution Management: Wastewater Impacts and Treatment Technologies*, Vol. 1. Agro Environ Media, Haridwar, India, pp. 10–28.

Mateo-Sagasta, J., Zadeh, S. and Turral, H. (2017) *Water pollution from agriculture: A global review*. Food and agriculture organization of the United Nations, Rome and the International Water Management Institute on Behalf of the Water Land and Ecosystems Research Program, Colombo.

Mukherjee, I. and Singh, U.K. (2018) Groundwater fluoride contamination, probable release, and containment mechanisms: A review on Indian context. *Environmental Geochemistry Health* 40, 2259–2301. DOI: 10.1007/s10653-018-0096-x.

National Oceanic and Atmospheric Administration Marine Debris Program (2017) *Report on Marine Debris as a Potential Pathway for Invasive Species*. National Oceanic and Atmospheric Administration Marine Debris Program, Silver Spring, Maryland.

Norwegian Radiation and Nuclear Safety Authority (2023) Low levels of artificial radioactivity in the sea. Radioactivity in the marine environment. DSA, Østerås, Norway. Available at: https://www.dsa.no/en/radioactivity-in-food-and-environment/radioactivity-in-the-marine-environment (accessed 20 June 2025).

Occhipinti-Ambrogi, A. (2021) Biopollution by invasive marine non-indigenous species: A review of potential adverse ecological effects in a changing climate. *International Journal of Environmental Research and Public Health* 18(8), 4268. DOI: 10.3390/ijerph18084268.

Organisation for Economic Co-operation and Development (2011) *Industrial Biotechnology and Climate Change: Opportunities and Challenges*. OECD, Paris.

Organisation for Economic Co-operation and Development (2013) *OECD Science, Technology and Industry Policy Papers*. OECD Science, Technology and Industry Policy Papers, No. 3, OECD Publishing, Paris. DOI: 10.1787/5k4840hqhp7j-en.

Ortúzar, M., Esterhuizen, M., Olicón-Hernández, D.R., González-López, J. and Aranda, E. (2022) Pharmaceutical pollution in aquatic environments: A concise review of environmental impacts and bioremediation systems. *Frontiers in Microbiology* 13, 869332. DOI: 10.3389/fmicb.2022.869332.

Owa, F. (2014) Water pollution: Sources, effects, control and management. *International Letters of Natural Sciences* 3, 1–6.

Rahman, R., Kozak, M. and Hung, Y. (2014) Radioactive pollution and control. In: Hung, Y., Wang, L.K. and Shammas, N.K. (eds) *Handbook of Environment and Waste Management*, Vol. 2. World Scientific, Singapore, pp. 949–1027.

Rajkhowa, R. (2012) Light pollution and impact of light pollution. *International Journal of Science and Research* 3(10), 861–867.

Rene, E., Bui, X., Ngo, H., Nghiem, L. and Guo, W. (2021) Green technologies for sustainable environment: An introduction. *Environmental Science and Pollution Research* 28, 63437–63439. DOI: 10.1007/s11356-021-16870-3.

Rzymski, P., Drewek, A. and Klimaszyk, P. (2017) Pharmaceutical pollution of aquatic environment: An emerging and enormous challenge. *Limnological Review* 17(2), 97–107. DOI: 10.1515/limre-2017-0010.

Savaşan, Z. (2017) Pollution, land. In: Schintler, L. and McNeely, C. (eds) *Encyclopedia of Big Data*. Springer International, Cham, Switzerland. DOI: 10.1007/978-3-319-32001-4_168-1.

Schlesinger, W.H., Klein, E.M. and Vengosh, A. (2020) Global biogeochemical cycle of fluorine. *Global Biogeochemical Cycles* 34, e2020GB006722. DOI: 10.1029/2020GB006722.

Schoonen, M.A. (2016) Sulfur cycle. In: White, W.M. (ed.) *Encyclopedia of Geochemistry*. Springer International, Cham, Switzerland. DOI: 10.1007/978-3-319-39193-9_73-1.

Ściężor, T. (2019) Light pollution as an environmental hazard. *Technical Transactions* 116(8), 129–142.

Sharma, A. (2020) *Types of Air Pollutants*. Department of Environmental Studies, Shivaji College, University of Delhi, India.

Sharma, A. and Sharma, R. (2020) Water pollution and its effect on aquatic life. *Aayushi International Interdisciplinary Research Journal* VII(XII), 80–93.

Singh, M. and Gupta, A. (2017) *Water Pollution-Sources, Effects and Control*. Nagaland University and Manipur University, India.

Slabbekoorn, H. (2019) Noise pollution. *Current Biology* 29, R942–R995.

Slabbekoorn, H., Bouton, N., Van Opzeeland, I., Coers, A., ten Cate, C. *et al.* (2010) A noisy spring: The impact of globally rising underwater sound levels on fish. *Trends in Ecology & Evolution* 25(7), 419–427.

Sneh (2018) Environmental pollution: Types, causes and consequences. *Journal of Emerging Technologies and Innovative Research* 5(12), 540–552.

Swackhamer, D.L., Paerl, H.W., Eisenreich, S.J., Hurley, J., Hornbuckle, K.C. *et al.* (2004) Impacts of atmospheric pollutants on aquatic ecosystems. *Issues in Ecology* 12, 1–24.

United Nations Environment Programme (2022) *Synthesis Report on the Environmental and Health Impacts of Pesticides and Fertilizers and Ways to Minimize Them*. UNEP, Geneva, Switzerland.

University of California Regents (2023a) *Carbon Cycle*. University of California Museum of Paleontology, Berkeley, California.

University of California Regents (2023b) *Nitrogen*. University of California Museum of Paleontology, Berkeley, California.

Vallero, D. (2008) *Fundamentals of Air Pollution*, 4th edn. Elsevier Academic Press, London.

Vishwakarma, G., Bhattacharjee, G., Gohil, N. and Singh, V. (2020) Current status, challenges and future of bioremediation. In: Pandey, V. and Singh, V. (eds) *Bioremediation of Pollutants*. Elsevier, Amsterdam, pp. 403–415.

von Schneidemesser, E., Monks, P.S., Allan, J.D., Bruhwiler, L., Forster, P. *et al.* (2015) Chemistry and the linkages between air quality and climate change. *Chemical Reviews* 115, 3856–3897.

Weilgart, L. (2018) *The Impact of Ocean Noise Pollution on Fish and Invertebrates*. Oceancare and Dalhousie University, Halifax, Nova Scotia, Canada.

World Health Organization (2022) *A Health Perspective on the Role of the Environment in One Health*. WHO Regional Office for Europe, Copenhagen.

Wu, J. and Strezov, V. (2023) Green technologies and sustainability: A new trend. *Green Technologies and Sustainability* 1, 100008.

Xu, M., Wang, Z., Duan, X. and Pan, B. (2014) Effects of pollution on macroinvertebrates and water quality bio-assessment. *Hydrobiologia* 729, 247–259. DOI: 10.1007/s10750-013-1504-y.

5 Epidemiology, Biosecurity in Aquaculture, Statistics, and Health Economics

Laura D. Urdes[1]* and Chris Walster[2]*

[1]*The World Aquatic Veterinary Medical Association, Romania; Faculty of Veterinary Medicine, University Spiru Haret Bucharest, Romania; Faculty of Management and Rural Development, University of Agricultural Sciences and Veterinary Medicine, Bucharest, Romania;* [2]*The World Aquatic Veterinary Medical Association, UK*

Abstract

This chapter begins by defining epidemiology, statistics, and health economics. This is followed by a section describing the main principles of epidemiology, applied statistics, and the reasoning behind health economics. The chapter also includes a discussion of the impact of biosecurity and aquaculture on sustainable One Health.

5.1 Introduction to Epidemiology

Epidemiology, also known as population medicine, is the discipline that studies the distribution and occurrence of disease in the human population. Epidemiology is a branch of public health.

A relatively new subdivision of epidemiology is *social epidemiology*, which studies the way in which social factors influence health and disease occurrence in human populations. It determines what social factors can cause disease, identifying at-risk communities that are most affected by disease and finding ways to contain the spread of these conditions. Among the risk factors used in social epidemiology are segregation of populations, income inequality, gender, and ethnicity. At the same time, social epidemiology is a tool used to understand ways for reaching out more effectively with medical care services to those communities who need these the most.

In general, the main purpose of epidemiology is to describe how diseases occur in a population in order to determine the causes of their occurrence, and to identify measures to control and prevent them, as well as the economic costs associated with these diseases. Establishing disease causes enables the development and implementation of treatments and preventive measures, the effectiveness of which can then be measured by certain types of epidemiologic studies. The principles and methods of epidemiology are the same regardless of the field in which they are applied.

Epidemiology is mainly a quantitative discipline. However, there are also qualitative epidemiologic investigations, where it is not necessary to measure illness, as in the case of identifying the source of an epidemic.

*Corresponding authors: urdeslaura@gmail.com and chriswalstervet@outlook.com

© CAB International 2025. *One Health Concepts and the Aquatic Ecosystem* (eds L.D. Urdes *et al.*)
DOI: 10.1079/9781800623248.0005

Epidemiologists usually seek to describe how frequently an illness occurs and use this information to understand the causes leading to its occurrence.

Epidemiological investigations seek to answer one or more questions, such as: the number of cases of disease in a population, the factors contributing to their occurrence, the measures needed to prevent the onset or recurrence of disease, the effectiveness of these measures, and so forth. The first step in studying the occurrence of disease is to measure the number of cases and to determine where, when, and in which individuals they occur. These measurements are made using descriptive methods. Sometimes, the most important information is the number of cases of illness in the population, as in the case of determining the risk of illness in the context of social events such as fairs, exhibitions, concerts, etc.

This information allows economic impact assessments to be made and the most appropriate disease control strategies to be devised. In addition, knowing that a disease occurs only in specific individuals, at specific times, and in specific places can help to identify the factors conducive to its occurrence. These factors are known as risk factors for the disease in question. Risk factors may be linked to the environment (e.g. climate), they may be determined by factors related to an individual (genetic predisposition, age-related changes in resistance to disease), or they may be specific infectious or toxic pathogens.

Once assumptions about potential risk factors have been formulated, epidemiologists seek to compare healthy and sick individuals in a population using analytical methods, which involve collecting and analyzing the data to determine whether there are variable factors between these groups that may be associated with the occurrence of disease. This information on risk factors is then used to make recommendations on measures that can be taken to prevent recurrence of the disease in question.

The impact of these measures can be assessed through other types of epidemiologic studies, known as intervention studies. In addition, by studying patterns of disease occurrence, epidemiologists can predict the course of a disease in a population and use mathematical models to test the impact of different control measures that should be taken. The use of theoretical epidemiology can help in deciding on the best control measures.

Epidemiological research is characterized by studies conducted to investigate a disease's frequency, distribution, risk factors, and control measures in a given population.

One Health requires an understanding of the interrelationships and interactions of factors associated with the pathogen, the host, and the environment in the production, spread, treatment, and control of a disease. In the past, the focus was on the relationships between the pathogen and the host organism alone, without taking into account environmental factors, which may in fact influence them both, shaping the course of the disease and often determining whether the disease would occur in a given population. There are three key components that influence the occurrence of a disease; these are the factors depending on: the environment, pathogen, and host. These form the epidemiologic triangle, used to study interactions between direct causes of diseases (Fig. 5.1). The One Health approach expands the concept of the epidemiologic triangle beyond direct causes of diseases to larger interactions and links between human, animal, and environmental factors.

Considered as the medical side of ecology, epidemiology is closely linked to the One Health concept. Statistical models or simulations are used to understand the epidemiology of diseases in populations (i.e. causes, sources of infection, routes of transmission, etc.).

In describing disease patterns, individual characteristics of the individual constituents in the study population are considered, as well as where and when cases occur in the population. Models or patterns are used to represent complex objects or systems. The models can be physical, such as, for example, models/patterns for making machines, building houses, etc.

Epidemiology works with conceptual models, which are a set of ideas designed to explain a specific process. Conceptual models in epidemiology help to investigate the processes by which diseases occur, and to make decisions as to how to plan and interpret studies investigating the causes of disease.

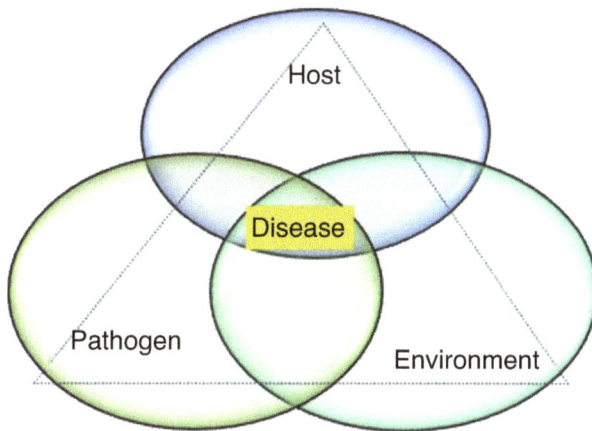

Fig. 5.1. The epidemiologic triangle—key group factors influencing the occurrence of a disease. Figure author's own.

5.1.1 Causal factors

Disease is the opposite of health. When investigating the causes of disease, the aim is to identify and measure each causal factor of the outcome of interest. For practical reasons, not all causal factors evolving in the population at a given point in time can be considered in a single study. Therefore, thresholds delimiting the proportion of the target population in the study population are set, and a list of factors on which to focus the epidemiologic study is drawn up. The selection of factors of interest is based on current scientific data. For this reason, it is necessary to use a concept of causality and a causal model, which is instrumental in determining the frequency of illness and interpreting the associations between exposures and illness.

Irrespective of the stage at which the study takes place, the presence of the disease is established by clinical testing and evaluation. Age, gender, immune status, possible genetic defects (e.g. hereditary diseases), and nutritional factors are among the factors that can determine whether individuals are susceptible to certain diseases, and what form of disease they may develop. If an individual is exposed to a pathogen capable of making them ill, the occurrence of disease will depend on the amount of the pathogen with which the individual has come into contact or to which they have been exposed. If we consider, for example, the consumption of alcohol, a drunk person experiences different stages of drunkenness, depending on the amount of alcohol ingested, from no effects to intoxication to death. In principle, this is the dose-effect relationship. The possibility of an individual becoming ill also depends on the characteristics of the pathogen.

5.1.2 Comparative observations and causality

In field experiments and in observational studies, it is very important that the group with which the results are compared (control), made up of non-exposed individuals, resemble as closely as possible the exposed group in terms of the factors that might influence the outcome of interest (in our case, the outcome of interest is disease). Ideally, these groups should be identical in all respects. This is called a comparison group (counterfactual group). By comparing the frequency of the outcome in two perfectly identical groups, the 'true' causal effect of exposure will be identified. Unfortunately, perfectly identical counterfactual groups do not exist in reality, so some practical alternatives are used. The practice of randomization helps achieve this goal in field experiments. For observational studies, we follow the statistical controls and/or methods used to design the study.

5.1.3 Causal patterns

An outcome can have several causes, just as a specific cause can generate several outcomes. This is why conceptual models of causality have been developed.

5.1.3.1 Component-cause analysis

This model is based on the concepts of 'necessary causes' and 'sufficient causes.' A necessary cause is a cause without which a disease would not occur. It is the factor that will always be present when disease occurs. The sufficient cause always produces the disease. If the factor is present, the disease invariably occurs. In practice it is rare that a factor acting alone is a sufficient cause. More often, factors cluster and combine to become sufficient causes. A component cause is one of a number of factors that combine to form a sufficient cause. Note that not all the factors of sufficient causes are always known. These factors may act concurrently or in a chain of events, sequentially. When there are a number of chains having one or more factors in common, the network of causal chains can be conceptualized (also referred to as a 'causal network').

The factors that make a cause sufficient to cause disease are known as causal complements (complementary causes, risk factors for a disease). For example, lack of immunity (antibodies) to a disease is as important a cause as the presence of the etiologic pathogen of the disease in the population.

5.1.3.1.1 PREVALENCE EFFECT OF THE CAUSAL COMPLEMENT ON THE RISK OF DISEASE. This effect refers to the strength of the associations between exposure and disease, which may be influenced by the prevalence of the other sufficiently causal components. Although the strength of some associations between risk factors and disease is useful in determining the factors that cause disease, the strength of an association detected in a study will depend on the prevalence of other factors, therefore it will be specific to the characteristics of the studied population.

When a disease-causing factor is influenced by the presence of another factor, there is a process called biological synergism. From a biological point of view, we can understand this phenomenon by exemplifying the interaction between two drugs. Drug A and drug B can be administered separately without any risk and with positive effects on the patient (synergism), but if taken together, drugs A and B can have devastating consequences on the body (antagonism). The component-causal model predicts that the ability to detect some statistical interactions will depend on the prevalence of the causal components for the interacting variables, and the prevalence of other sufficient causes of the disease in question.

5.1.3.1.2 IMPORTANCE OF CAUSAL FACTORS. The component-cause model helps us understand how different factors contribute to disease. It shows that a disease usually results from a combination of factors (called component causes). When all the necessary factors are present, they form the so-called sufficient cause and lead to disease. Every factor in that group is linked to the same number of disease cases. This means that if we try to count how many cases are caused by each factor separately, we may end up counting the same cases more than once. As a result, the total number of cases we attribute to these factors may be higher than the actual number of cases in the population.

5.1.3.2 Causal network analysis

This is the second complementary model of causality, which assumes that factors that can influence the occurrence of disease in a population can be connected through a causal network, using temporal relationships between factors. This model is based on a series of directly and indirectly interconnected causal chains (or network structures), which take the factors identified by the sufficient cause approach and link these factors, relating them to a unit of time (Fig. 5.2). Direct causes of disease are closely related to the occurrence of disease over time. These are called proximal causes, which are considered in treatments (e.g. microorganisms, toxic substances). Indirect causes are causes in which the effects of exposure to an outcome are mediated by one or more variables. The causal network model is useful in analyzing and interpreting epidemiological data.

Genetics

Family history
and genetic
predisposition to
diabetes

Lifestyle Factors

Diet, physical
inactivity, obesity,
and smoking

Medical Conditions

Hypertension, high
cholesterol, and
gestational diabetes

DIABETES

Biological Factors

Insulin resistance,
beta-cell dysfunction,
and hormonal
imbalances

Environmental Factors

Socioeconomic status,
access to health care,
and neighborhood
environment

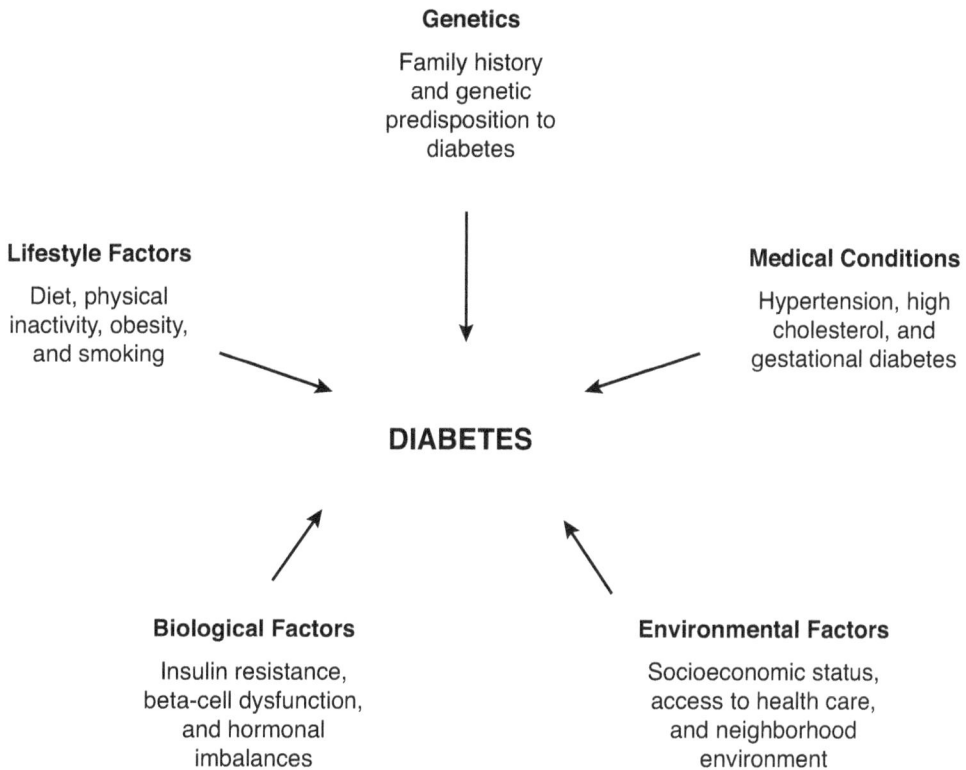

Fig. 5.2. Causal-web analysis—several direct causes of diabetes include genetic factors and pre-existing medical conditions (WHO, 2024). Figure author's own.

5.1.4   Types of epidemiologic studies

Once the pattern of a disease has been established, the next step is to generate hypotheses about the causes, sources of illness, and disease control strategies. This study requires a logical problem-solving approach to disease etiology, predisposing factors, prognosis, and appropriate control strategies. Statistical concepts and theories are used to determine the extent to which observed disease associations or patterns are due to chance. A plan is developed for implementing and analyzing observational and experimental studies, retrospectively, prospectively, or in a cross-sectional setting.

5.1.4.1   Observational (descriptive) study

The observational study describes how a disease occurs and spreads in a population, based on

characteristics of the individuals in the study population and where and when the disease occurs. For example, the description of cases of COVID-19 in a country will require knowledge of the degree of population involvement, the geographical distribution of cases in the country where the study is being conducted, which serotype strains are present in the territory, etc.

5.1.4.2   Analytical study

During the description of a disease, a number of questions arise, which are then formulated as working hypotheses about the characteristics of the disease process in the study population. For example, one seeks to identify whether there are differences in the degree of resistance to disease between different sub-populations of the study population. Other working hypotheses may relate to the identification of a particular systematic pattern in the geographical distribution of the disease, or the existence of national geographical barriers (mountains, rivers, oceans) that halt the spread of the disease.

The analytic study tests the extent to which observed differences between cases of disease may be due to chance or whether there are other factors involved in disease transmission in the study population. In this case, simple statistical tests are performed to determine the probability that observed differences between two groups are due to chance. Some more complex statistical models may also test whether the data fit a system of hypotheses about different causes and effects that may interact with each other.

5.1.5 Intervention study

By testing various hypotheses about the origin, transmission routes, and risk factors, conclusions are subsequently reached on what would be effective measures to control the disease under study. In most cases, several intervention options are identified, each of which has advantages and disadvantages. Intervention trials are used to identify the best control measures. These studies test different interventions (e.g. vaccination, serologic or molecular testing, etc.) in small groups of individuals representative of the population in which the study is being conducted. These tests reveal the effectiveness of the interventions being studied and any practical problems that may arise in implementing them in the study population. Typically, intervention studies also include a control population to which the intervention being tested is not applied. In this case, the use of a control eliminates the possibility that factors other than those identified by the study may alter the frequency of illness in the study population.

5.1.6 Variables utilized in one health studies

Two types of variables are used to measure the number of illnesses in a population: categorical and numerical.

Categorical data describe attributes that can be expressed by a limited number of values. Categorical data are also called qualitative data and do not involve physical measurements. The simplest categorical data are binary variables, which are expressed as any one of two values. Examples of binary variables (also known as 'dichotomous data,' 'yes/no,' or '0–1') are: male/female, dehydrated/not dehydrated, anaemic/non-anaemic, etc. Binary variables help to divide the population into two distinct groups according to a common characteristic. There are many observations that can be expressed as categorical data, e.g. country of origin, age, ethnic group, etc. These are also called nominal data. These data do not indicate whether one value is better than another, but only group individuals on the basis of group characteristics. Scales of 1–3, 1–5, or 1–10 are used when the intention is to classify a disease in relation to its severity. These data are ordinal data, also called ordered categorical data.

Numerical data are variables that can be measured or counted (they have numerical values). Numerical variables are also called quantitative variables (quantitative data). Numerical data can be discrete or continuous. Discrete data are integers. For example, a pregnancy can be with 1 or 2 fetuses, never 1.5 or 2.5 fetuses. If weight is measured instead, values appear as whole numbers and fractions of a kilogram, depending on the accuracy of the weighing equipment. These are continuous data.

5.1.7 Factors influencing safety and health

The factors on which the occurrence and transmission of diseases depend (determinants) are related to the characteristics of the pathogen, the host, and the environmental conditions. In this context, the host is the recipient of the disease, the pathogen is the component responsible for inducing the infection or disease into the host, and the environment is all the surrounding elements that facilitate the interaction between host and pathogen.

These factors are sometimes viewed as discrete and independent entities, but they often combine to contribute as a group causing disease or infection. In other words, most diseases are multifactorial in nature. In order to control and prevent the spread of a disease, it is necessary to describe that disease in detail. To describe a disease in detail, it is necessary to know the determinants of the disease, i.e. which host-dependent, pathogen-dependent, and environmental factors interact to produce the disease (Fig. 5.1). Once these are identified, the impact of the disease is determined by establishing the characteristics of these factors and their interactions. Often the pathogen alone cannot cause disease in all individuals and under all environmental conditions. An example would be those diseases that become evident only under certain conditions, such as poor nutrition or poor living conditions, which increase the microbial load per unit area and lead to decreased immunity of the population concerned.

5.1.7.1 Pathogen-dependent factors

The factors dependent on the pathogen are:

- Host specificity—the broader the host specificity, the greater the chance of the pathogen surviving.
- Infectivity—the ability of the pathogen to enter, grow, and/or multiply and to cause changes in the host.
- Infectious/infective dose—the amount of pathogen required for an infection to occur.
- Contamination—the presence of infectious pathogens on the external surface of the body, on bandages, equipment, food, clothing, footwear, etc.
- Pollution—the presence of offensive matter, including non-infectious matter, in the environment.
- Pathogenicity—the ability of the pathogen to cause clinical disease. To produce clinical disease, the pathogen must first infect the host.
- Virulence—a measure of the degree of aggressiveness of the pathogen. In most cases, this term is used when the pathogen is capable of producing lethal infections.
- Immunogenicity—the ability of pathogens to stimulate immune response in infected hosts. The immune response is usually associated with infection or exposure to the pathogen.

- Antigenic stability—the likelihood that the genome that governs the antigenic structure of the pathogen will undergo antigenic changes.
- Viability—the ability of a pathogen to withstand stress from environmental conditions.

The most important elements are infectivity, pathogenicity, and virulence. The level of infectivity and pathogenicity will indicate the control and prevention measures most appropriate to the particular disease in question. Pathogens with high infectivity and pathogenicity usually produce considerable epidemics in susceptible populations. If the virulence of the pathogen is also high, high mortality will also occur in susceptible populations. In aquatic animals there are numerous examples, where preventive measures are often aimed at precluding the introduction of the pathogen into a population and detecting outbreaks as early as possible. Infectious pancreatic necrosis virus (IPNV) causes a widespread acute systemic disease in young trout and Atlantic salmon in fish farms.

In humans, severe acute respiratory syndrome coronavirus 2 (SARS–CoV–2) is a relevant example. Detection of the disease may consist of the detection of clinical symptoms, with laboratory confirmation of suspect cases. Control measures are aimed at eliminating the pathogen from the population and reducing the spread of the disease through measures such as demarcation of areas where infected individuals and their possible contacts are found, and containment of those infected or suspect of having been infected. If vaccination is practiced in these areas, it should be continued for a long period of time, as the spread of the pathogen may continue for a long time in vaccinated individuals.

A pathogen with high infectivity and low pathogenicity spreads rapidly in a susceptible population, but these diseases are not always obvious because the likelihood of individuals developing clinical forms is very low. Although the outbreak of such a disease may appear favorable, rapid detection and elimination of the pathogen from the population is much more difficult to achieve than in the case of highly pathogenic agents. Therefore, detection of the disease cannot be based solely on the clinical symptoms of the disease, which are usually lacking in this case. Tests such as serologic and molecular investigations are needed to detect asymptomatic carriers. An example of a highly infectious low-pathogenic virus of fish is *Lymphocystivirus* (*Iridoviridae*, DNA virus), causing a benign, chronic infection of lymphocytic cells leading to gross hypertrophy of cutaneous, muscle, peritoneum, or visceral organs but having little impact on the host.

In these cases, eradication of infection requires a sustained effort to detect asymptomatic carriers. Vaccination or other measures that neutralize the effect of the pathogen after it infects the host can be considered.

A pathogen with low infectivity and high pathogenicity may require control measures that reduce host exposure to the pathogen, instead of immunizing the host by vaccination. *Mycobacterium fortuitum* and *M. marinum*, the causative pathogen of tuberculosis in fish and other aquatic species, are examples of pathogens with low infectivity and high pathogenicity. To detect infected individuals, specific tests are required.

Infectious pathogens of low infectivity and low pathogenicity are less important than those presented above. However, they may also cause isolated cases of disease if the population or geographical area has particular susceptibility to infection and/or disease. In the case of terrestrial species, tick-borne pathogens, such as those causing babesiosis and Lyme disease (also known as borreliosis), commonly give rise to sporadic cases, usually of mild illness. In these cases, management emphasizes the reduction of risk factors predisposing to infection and clinical disease rather than direct control of the pathogen.

5.1.7.2   Host-dependent factors

These factors influence the extent of exposure, susceptibility, and the response to pathogens. The host-dependent factors can potentiate or, conversely, they can limit the occurrence of disease. Host factors can be intrinsic and extrinsic. Intrinsic factors are the permanent, unchangeable characteristics of the host. These factors are: age, sex, immune status, and genetic makeup. Identifying these factors enables the understanding of risk within groups of individuals. Extrinsic factors are factors that change over the lifetime of individuals.

5.1.7.3   Environment-dependent factors

These factors include the physical environment (geography and seasons) and the biological climate (diet, lifestyle, type of living environment, nutrition, etc.). These factors are classified into macro- and micro-environment. The macro-environment is the physical environment, while the micro-environment is the elements of the surrounding space in the immediate vicinity of an individual or a population. Macro-environmental factors include geography, air quality, seasons, and the presence of vectors. Micro-environmental factors include diet, housing conditions, and other quality-of-life and stress-inducing factors. Stress increases susceptibility to infection and clinical disease. Thus, both macro- and micro-environments can influence how a disease may develop in a population.

5.1.8   The epidemiologic chain and disease transmission patterns

In order to control the transmission of pathogens in populations, it is important to identify critical points in the epidemiologic chain. An epidemiological chain consists of the following elements: one or more reservoirs, an entry gate, an exit gate, one or more modes of transmission, and one or more susceptible hosts. The reservoir, exit gate, mode of transmission, and entry gate determine the ease and speed with which the infectious agent is transmitted in a susceptible population. This characteristic is known as transmissibility or infectivity.

Reservoir can be a living organism or a fomite (an inanimate object such as, for example, clothing, equipment, tools, etc.) in which an infectious agent can multiply or survive, and from which it can be transmitted further. When the reservoir is localized and persists for a long time (i.e. there is a spatially defined focus of transmission), it is called a nidus. If the reservoir is either an animal or a human, the reservoir is called the maintenance/preservation host of the pathogen. The degree of spread of the pathogen in a reservoir depends on the extent to which the susceptible population is at risk of exposure to infection with that pathogen.

The fact that a pathogen can be isolated from a particular host does not necessarily mean that the host is the reservoir of the pathogen. There are rare infections that occur sporadically in some individuals that do not belong to species known to be susceptible to the pathogen. These hosts are called accidental hosts.

Source is where the pathogen passes directly to the susceptible host. The source may also be the reservoir of the pathogen.

Carriers are infected individuals who transmit the pathogen in the absence of overt clinical signs of disease, and who serve as a potential source of infection for other individuals with whom they come into contact. Throughout the course of infection, the carrier status may evolve in apparently healthy, asymptomatic individuals, or during incubation or convalescence. It follows that carriers can be of many types:

- Incubation phase—they spread the pathogen before clinical signs appear.
- Convalescent phase—they spread the pathogen for a short period of time after a variable length of time during which clinical signs of disease occur. This also includes intermittent disseminators, who spread the pathogen from time to time for a variable time after recovery from disease.
- Chronic carriers—these spread the pathogen for a long time after recovery from disease.
- Asymptomatic (clinically healthy) carriers—these are carriers who spread the pathogen without having experienced the disease, and without ever having shown symptoms of the disease.

Three categories of *modes of disease transmission* are recognized:

- Primary sources or vehicles—these can be secretions, excretions, or other fluids or tissues from infected individuals (e.g. skin mucus, saliva, urine, feces, blood, etc.).
- Secondary sources or vehicles (fomites)—these are inanimate objects and materials that become contaminated through contact with primary sources (e.g. fishing nets, water, air, food, etc.).

- Vectors—these are generally invertebrate animals responsible for the transmission of an infectious agent (e.g. hematophagous insects, molluscs, etc.).

Diseases can be transmitted directly or indirectly. Direct transmission occurs through physical contact of a susceptible individual with an infected individual. Many directly communicable diseases pass to other individuals via primary sources, through skin mucus, nasal secretions, saliva, urine, and feces. This category also includes directly transmitted diseases that require a specific type of contact. This includes sexually transmitted diseases and vertically transmitted diseases, the latter assuming transmission from mother to fetus. A general characteristic of directly transmitted pathogens is that these pathogens have low resistance in the environment.

Indirect transmission involves an intermediary between the host and the susceptible individual. These intermediaries are secondary sources and vectors, such as individuals resistant to the infection concerned, animals, vehicles, contaminated food, water and air contaminated with secretions of sick individuals, etc. Airborne transmission and ingestion of pathogens are the most important modes of indirect transmission.

5.1.8.1 Airborne transmission

Air can become contaminated with droplets of secretion from exhalation or during breathing or speaking. Dust particles in the air can also become contaminated with pathogens from infected hosts, thus becoming sources of infection for susceptible individuals.

Dust is formed by breaking down larger materials into solid particles, and always has a microbiological content, consisting mainly of bacteria and fungi. Dust particles vary in size from $10\ \mu m$ to $100\ \mu m$. Contaminated airborne dust gets into the mucus in the upper respiratory tract and can cause localized or generalized infections, depending on its microbiological load. More resistant microbes are generally transmitted via airborne dust.

Droplets of saliva from deep exhalation, sneezing, coughing, and released during speech produce aerosols by the atomization of these secretions, which are forcefully sprayed outwards from nose and mouth.

Aerosols can be about $100\,\mu m$ in diameter and are usually emitted over a distance of up to 1 m. In air, depending on the environmental conditions (damp or dry), these droplets increase or decrease in volume (in the latter case, by evaporation). It is important to note that aerosols from infected individuals are the most important means of spreading pathogens with an affinity for the respiratory tract. Aerosols contaminated with pathogens that are more resistant to environmental conditions end up contaminating the air and soil dust, and are further transmitted via dust. Aerosol-transmissible pathogens require a high density of susceptible individuals in order to be transmitted, producing in this case a transmission pattern similar to that found in directly transmissible diseases.

Spray nuclei are particles of smaller diameter $(2–10\,\mu m)$, which are formed by the rapid evaporation of aerosols from saliva and droppings when the environment in which they are projected is dry. If these nuclei reach the respiratory tract, where the environment is moist, they become rehydrated. Once in the alveoli of the lungs, the spray nuclei stick to the surface of the alveoli. If the aerosols have a pathogenic microbial load, they can facilitate the transmission of the pathogen to a new host. In contrast to aerosols, spray nuclei have the potential to be emitted over much greater distances. Good air ventilation and the use of ultraviolet light sources destroy aerosol nuclei.

Spray nuclei (vapors and gases) cause diseases that resemble the transmission pattern of directly transmitted diseases. Again, a population's density plays an important role in the risk of microbial transmission.

5.1.8.2   Digestive transmission

The fecal-oral route is a closed transfer cycle. In humans it occurs through ingestion of food and water accidentally contaminated with feces or via ingestion of food of animal origin derived from diseased animals (e.g. as in the case of trichinosis, tuberculosis, salmonellosis, fungal mycotoxicosis, etc.). Waterborne disease transmission is generally the result of accidental contamination of municipal waters with wastewater from human settlements or animal farms.

As in the case of illnesses caused by eating contaminated food, waterborne pathogens produce group (or community) diseases with a single, common source of infection.

In susceptible populations, contaminated food can cause extremely severe, sometimes lethal, outbreaks of disease. Pathogens that are transmitted by the digestive tract remain viable outside the host, and the food in which they are found provides a medium in which they can be stored and possibly multiply. There are many foodborne pathogens that can be transmitted to humans, either from the raw material (e.g. carcasses from anthrax-positive animals) or introduced during improper food processing, packaging, or storage (e.g. *Salmonella* spp.). Milk, in particular, is a good growth medium for bacteria.

5.1.8.3   Vector transmission

Vectors are living invertebrate organisms that can transmit a pathogen to susceptible individuals. Vectors can be mechanical or biological.

Mechanical vectors are organisms in which the pathogen remains unchanged (they do not multiply or develop further). The range of transmission from one host to another by mechanical vectors is usually short and depends on the ability of the pathogen to survive in, or on, the body of the mechanical vector.

Biological vectors are organisms in which changes in the pathogen occur, such as multiplication, maturation, reproduction, or a combination of these. In contrast to mechanical vectors, biological vectors cannot transmit the pathogen immediately upon contact, in which case a prepatent (or extrinsic incubation) period is required.

Some pathogens are highly adaptable to their specific vectors. Examples of such relationships are the transovarial and trans-stadial transmission of pathogens in their vectors. In transovarial transmission, the pathogen is transmitted from the female vector to her egg, whereas in trans-stadial transmission the pathogen survives through the developmental stages from larva to adult.

According to their degree of mobility, vectors are divided into flying vectors and non-flying vectors. These characteristics are of great epidemiological importance, as they reflect the ability of these pathogens to spread in relation to their ability to move (or geographical range), thereby influencing the speed at which an infection can spread. Flying vectors, such as mosquitoes and other hematophagous insects, actively seek out (vertebrate) hosts and can transmit infections over considerably greater distances than non-flying vectors. The latter depend on passive contact with their hosts. However, to compensate for this disadvantage, many species of non-flying vectors have evolved the ability to transmit infectious pathogens transovarially and/or trans-stadially. In the case of aquatic vectors, the infectious pathogen will be released by them into the water, from where they passively disseminate to susceptible animals. In this case, vertebrate infection always occurs at a distance from the vector.

5.1.8.4 Disease description data

To achieve effective disease management, it is necessary to describe as accurately as possible the way in which diseases occur in a population. Such a description should include details about the number of individuals and their characteristics (e.g. elderly, pre-existing health conditions, predominantly female, etc.), the variation in the number of outbreaks over time, and whether there is a temporal or spatial relationship between cases.

5.1.8.4.1 TEMPORAL DATA (TIMING OF DISEASE EVO-LUTION). In an outbreak, newly registered cases rarely occur in a random sequence. Therefore, depending on the time at which illnesses are recorded, it is possible to determine the nature of the disease evolving in a population at that point in time. Depending on the timing of new cases in the outbreak, three patterns of disease evolution can be identified:

* Sporadic—cases appear unrelated to each other.
* Endemic—cases occur and follow each other regularly in the population, and these events can be included in a pattern because they manifest in a similar way.
* Epidemic—the number of cases exceeds the normal number of cases in the population, a given area, or season. When the epidemic

takes on international proportions, the term 'pandemic' is used.

Thus, communicable diseases can be sporadic (isolated), endemic (stable, predictable), or epidemic (unpredictable). An effective means of representing the temporal distribution of illness in a population is to construct an epidemic curve, which can illustrate both the magnitude of the problem, given by the number of new cases, and the speed at which the epidemic is progressing in the population.

5.1.8.4.1.1 Epidemic curve. The epidemic curve is a graphical representation of new cases per unit of time (Fig. 5.3). The slope of the epidemic and the speed at which the epidemic is progressing can give an indication of the type of exposure and the way in which the pathogen is transmitted in the outbreak. If transmission is rapid, the upslope is steep. In epidemics involving pathogens with a low transmission rate, or where the incubation period is long, the slope of the ascending curve is not as steep. Short-term exposure of a large number of individuals to a pathogen at the same time, as in food poisoning, will generate a single-source epidemic curve (Fig. 5.3). As shown, the upward slope is almost vertical before reaching the epidemic peak. In this graph, there is a rapid increase in the number of cases, and most cases occur within a very short period of time (within 6 hours).

When the pathogen is transmitted by contact or by vectors, the epidemic curve reflects a propagated epidemic curve. In this case, the steepness of the slope also depends on some characteristics of the pathogen, such as its ability to survive outside the host, as well as some factors related to contact rate and population density. The duration of the plateau phase and the slope of the downward sloping branch of the curve generally depend on the structure of the population at risk in terms of immune status (i.e. the existence of so-called herd immunity) or on some interventions, such as vaccination or treatment.

A second peak in the epidemic curve may occur if:

* susceptible individuals enter the population that has become infected; or

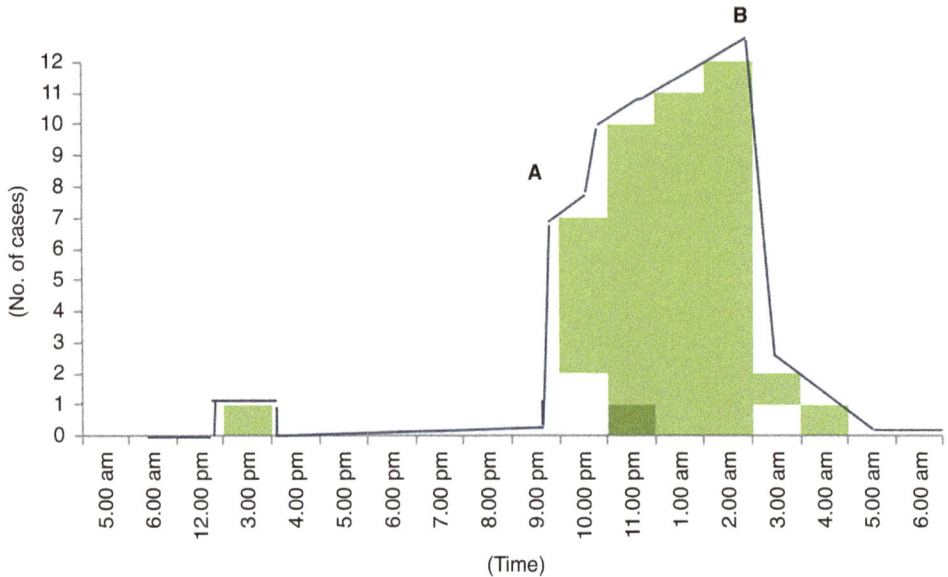

Fig. 5.3. Single-source epidemic curve. Figure author's own. (A) rapid rise; (B) peak of the epidemic.

- individuals in the population that has been infected come into contact with individuals from another population susceptible to that infection.

The main peak of the epidemic may be preceded by a smaller peak, which could be the index case—the first case at the beginning of the epidemic. The interval between case 0 and beginning of the upward slope toward the epidemic peak could indicate the incubation period, when clinical manifestations of the disease are absent.

When the epidemic curve extends over a long period of time and is based on observations at short time intervals, it may indicate patterns of seasonal variations, cyclical fluctuations, or secular trends.

Seasonal variations have an alternating (sometimes increasing, at other times decreasing) periodic character, often corresponding to the change of the seasons. A seasonal variation may last from 1 week to 1 year, depending on the biological phenomenon.

Fluctuations are considered cyclical when these variations occur at regular intervals, and they are usually longer than seasonal fluctuations.

Secular trends are long-term changes involving decades or hundreds of years. The graphical representation of these trends shows a secondary curve that is sometimes rising and sometimes falling, and a main curve (trend) that is rising or falling. In Fig. 5.4, the epidemic trend (i.e. every 100 years) is ascending.

5.1.8.4.2 SPATIAL DATA (WHERE DO CASES OCCUR?). To obtain indications of the way in which a disease is spread in a population, the geographical location ('spatial location') of the population is analyzed. Occurrence of cases near tourist areas, markets, and where there are generally crowds of people will indicate what sources of infection may be involved. By comparing the geographical distribution of the disease with the environmental and climatic characteristics of the geographical area, clues can be obtained to some of the risk factors involved in disease transmission.

The spatial distribution of the disease is obtained by marking on a map, with a writing instrument or with pins, the places where the cases have been reported. This procedure helps us to get an overview of the affected areas and some of the risk factors related to the geography and climate of the area. This method is, however,

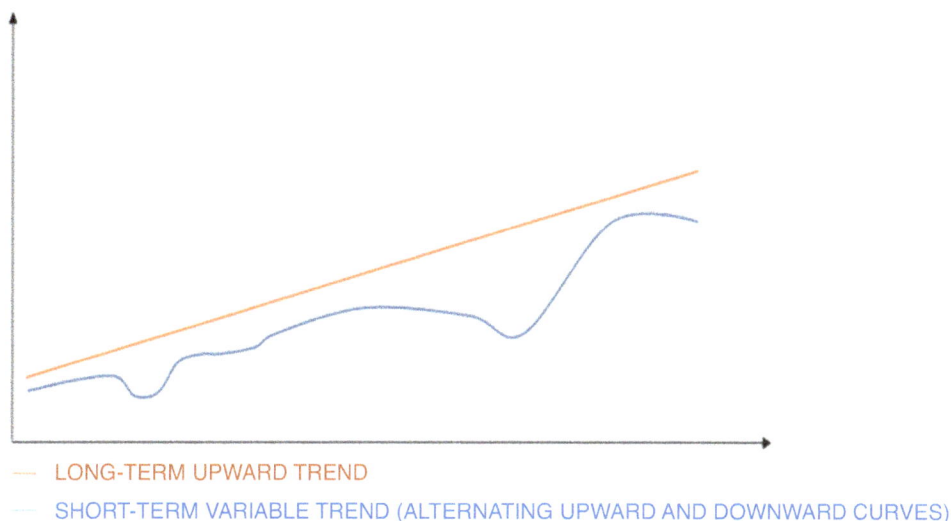

— LONG-TERM UPWARD TREND

— SHORT-TERM VARIABLE TREND (ALTERNATING UPWARD AND DOWNWARD CURVES)

Fig. 5.4. Cyclic trend, every 100 years. Figure author's own.

not effective because it cannot be recorded and passed on to other specialists involved in disease control. Other options to effect spatial mapping are: point maps, distribution maps, proportional circle maps, choropleth maps, and isopleth maps.

Point maps are easy to generate, do not require geographic information systems (GIS) knowledge, are intuitive, and allow a graphical visualization of the area where the cases occur (not the number of cases, however). These maps are useful when the number of cases is relatively small compared to the geographical area affected. On the other hand, maps that use dots to represent the locations of cases can lead to erroneous conclusions because they do not give information about the number of sick individuals or the susceptible population. For example, case locations may represent where the majority of the population is found.

Distribution maps are suitable for observing the disease status of a geographical area, but are not useful for epidemics because they do not include the exact location of the sick individuals.

Proportional circle maps are similar to point maps and are very useful for visualizing disease outbreaks where cases occur in the same place or area. These maps can include indications about the incidence of cases in the locations of interest by enlarging the circles proportional to the incidence of disease (Fig. 5.5).

Choropleth maps are suitable for indicating prevalence and incidence patterns of endemic diseases. They are not as intuitive as point maps, nor do they help to analyze the risk factors present in a small geographic area, because the exact location of cases cannot be identified on the map (Fig. 5.6).

Isopleth maps may contain a great deal of information, but they are less intuitive. These maps are the ideal way to visualize the number of cases and susceptible categories of individuals in the population (Fig. 5.7).

A more efficient alternative for spatial description is the use of electronic maps. Electronic tools that are used in epidemiologic analysis of the spatial distribution of diseases are referred to as GIS.

When a large number of cases occur in a defined geographical area, the term 'cluster' is commonly used. Applications are now available to statistically assess to what extent geographical clusters occur by chance or whether there are factors influencing their occurrence. In diseases that affect large territories over a long period of time, in addition to the spread of the epidemic in time and space, the interdependence of the temporal and spatial spread of the disease is analyzed. There are also seasonal patterns in

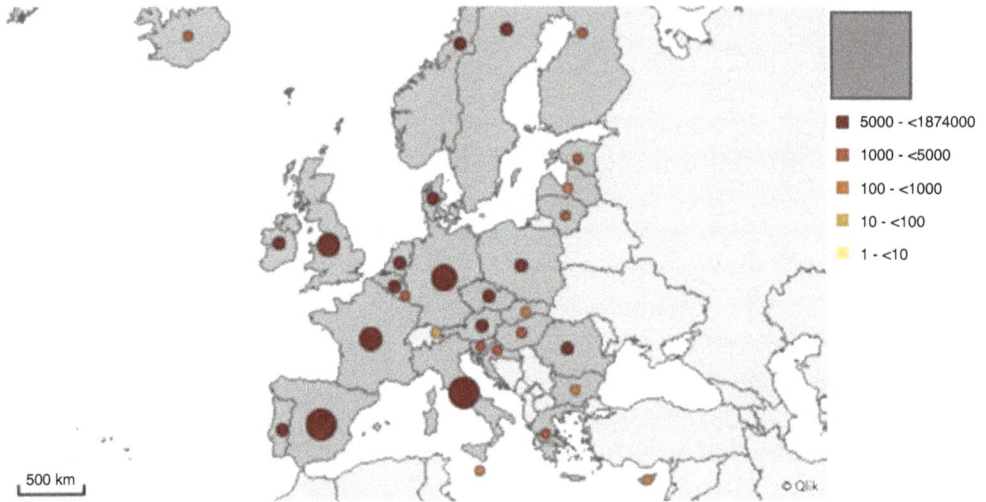

Fig. 5.5. Distribution of COVID-19 cases in Europe, April 14, 2020. From ECDC – EU (https://data.europa.eu/en/publications/datastories/covid-19/global-and-european-dashboards-mapping-spread-covid-19).

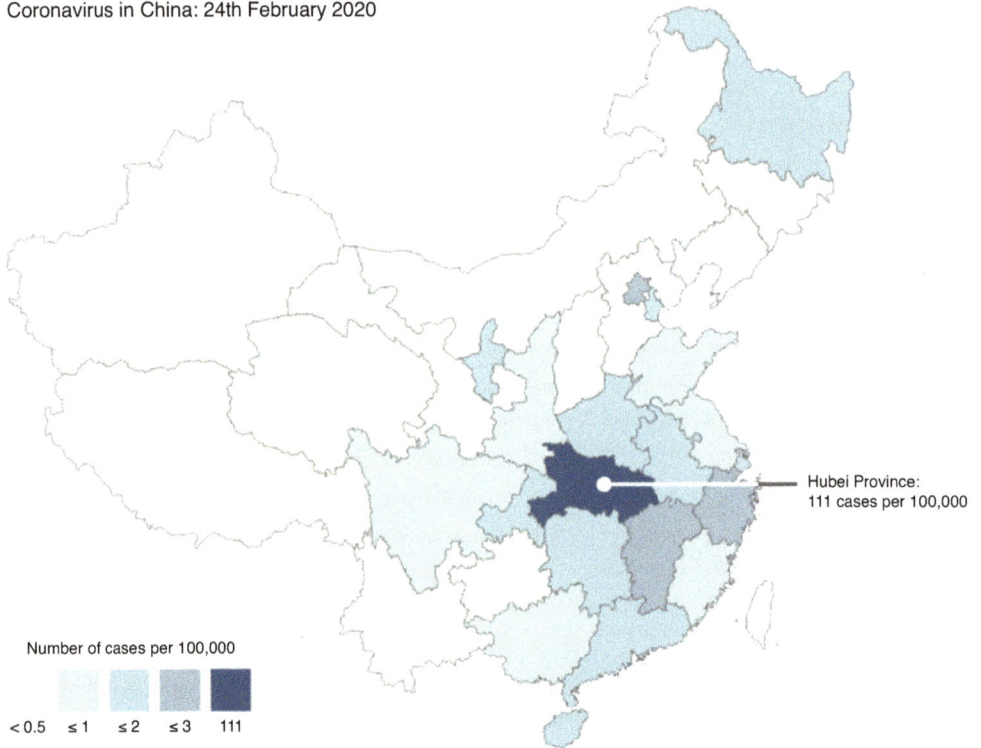

Fig. 5.6. Prevalence of COVID-19 cases in China, February 24, 2020. From ArcGIS – ESRI (https://www.esri.com/arcgis-blog/products/product/mapping/mapping-coronavirus-responsibly/).

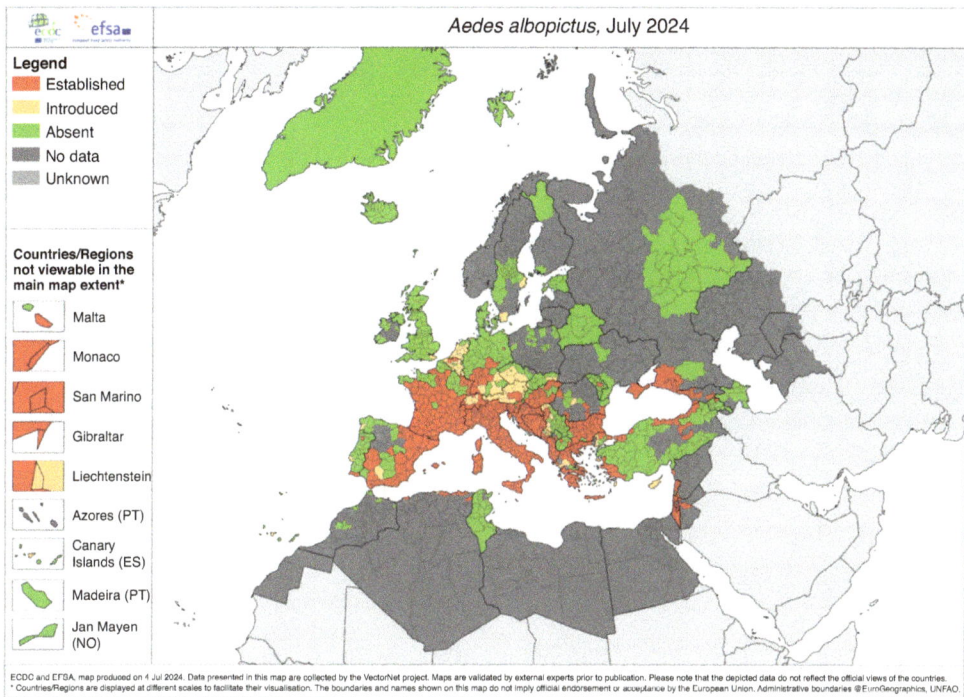

Fig. 5.7. Isopleth map of *Aedes albopictus*, considered to transmit chikungunya, dengue, and Zika. From ECDC – EFSA (https://www.ecdc.europa.eu/en/publications-data/aedes-albopictus-current-known-distribution-july-2024).

the occurrence of cases in different geographical areas, which may be linked to climatic conditions that are favorable for the vectors involved in disease transmission.

5.2 Epidemiology and One Health

These are two interconnected approaches of which the common point is that they address the health of populations, but they focus on different scopes. To understand and manage health risks that affect humans, animals, and the environment simultaneously, it is critical to be able to combine these two.

One Health recognizes the interconnection between the health of humans, animals, and ecosystems and is especially important for understanding diseases that can be transmitted between animals and humans, as well as the environmental factors that influence health. Key areas of One Health include zoonotic diseases,

healthy environments, antimicrobial resistance (AMR), and food safety.

In recent years, epidemiology has expanded to consider the interconnectedness of human and animal health, and a healthy environment. Operationalizing an ecosystem through the One Health approach should be based on a social-ecological system and resilience framework that recognizes key components in human and natural systems in order to recognize emerging health and environmental risks and to promote global health.

Epidemiology plays a critical role in implementing the One Health approach. Currently available epidemiological tools and study methods can be applied to identify the way in which diseases spread between animals, humans, and the environment, identify risk factors, and develop strategies to prevent or control outbreaks. As explained in the previous section, the key components of epidemiology include examining the distribution of diseases by time, place, and person (descriptive studies), investigating the causes or determinants of health outcomes, often through case-control or cohort studies (analytical studies), and developing strategies to control or prevent disease spread, often through public health initiatives or clinical interventions (interventional studies).

Determining the presence of a disease in a population, estimating the frequency of disease, studying the determinants and contributing causes by comparing the frequency of illness among different groups of individuals in the population, and determining the likelihood that a treatment or measure can be effective are the subject of epidemiological investigations. These are objectives that epidemiological studies are designed to achieve on the basis of valid assumptions (hypotheses).

5.2.1 Epidemiologic studies (surveys)

Epidemiologic studies have two main functions:

- to describe the occurrence of disease by measuring the frequency of illness in a population (this is achieved using cross-sectional investigations or surveys and can be used to generate hypotheses about causes of diseases); and
- to enable understanding of the underlying causes of diseases by testing hypotheses about causes of diseases (for example, vaccination prevents severe disease in vaccinated individuals, etc.).

Studies used in epidemiologic surveys are divided into observational and intervention studies.

Observational studies do not interfere with the situation in the field, but simply record what is actually changing within population. Observational studies are subdivided into ecological, cross-sectional, case-control, and cohort studies.

Intervention studies aim to test the effect of an action on the frequency of illness by changing the level of exposure of individuals in the population included in the study. Intervention studies are of two types: therapeutic and prophylactic. Therapeutic studies aim to test the effectiveness of a treatment on cases, and prophylactic studies test the effectiveness of interventions, such as

vaccination, in protecting healthy individuals in the population where the disease is occurring.

To carry out these studies, five steps are necessary:

1. Establishing the objectives so that it is clear which individuals need to be examined and which measurements need to be performed.
2. Sample selection (selection, from the population of interest, of the individuals to be included in the study). This stage involves knowing the number of individuals in the population and the identity of the study population and specifying the methods for selecting the individuals that constitute the study sample. When specifying the selection methods, reference is also necessarily made to the criteria for exclusion from the study and inclusion in the study of the individuals in the population. In intervention studies, the selected individuals are then allocated to other treatment groups.
3. Collection of the necessary information about exposure to risk factors and disease occurrence in the population, and in the case of comparison groups, ensuring that this process is identical across all groups entering the study.
4. Use of appropriate analytical methods to estimate the frequency of illness or associations between disease and the risk factors of interest.
5. Interpretation of the study results.

5.2.1.1 *Step 1: Setting study objectives*

This is the most important step in epidemiologic investigations. Good data collection depends on how clearly the aims and objectives of the investigation have been formulated. A specific objective should have sufficient information to enable people involved in the implementation of the study to decide what information is needed and from which individuals it can be obtained. For example, setting an objective such as investigating the occurrence of dermatitis in humans is not a well-defined objective because, as formulated, it does not give sufficient information to allow those conducting the study to determine what data is needed for the study, or from which individuals this data should be obtained.

Some additional data, such as the category or categories of individuals affected, what type of dermatitis is being targeted, and what factors of interest are involved, would be needed for such a study.

The main purpose of a study must allow the following data to be collected:

- what the study is intended to achieve— what the intended outcome of the study is;
- what the risk factors of interest in the study are; and
- what unit of analysis is being used— whether the results of the study will be subsequently reported at individual, group, or population level.

5.2.1.2 Step 2: Sample selection

Regardless of the type of study intended to be implemented, only part of the total population may be included in the study in order to meet the proposed objectives. This approach is based on reasons of cost-effectiveness, which often make it unfeasible to collect data from the entire population. Therefore, data will be collected from a sample of the population of interest and then extrapolated to the entire population of interest. Epidemiologic studies identify three categories of populations:

1. The target population—this is the population to which the results obtained from the study will be reported/extrapolated.
2. Study population—this is the group of individuals in the target population accessible to the investigators. This is the population on which the study will be conducted.
3. The sample—this represents the individuals who are selected from the study population to be entered into the epidemiologic study.

Before selecting the individuals to be included in the study, it is important to establish the unit of analysis that will be used and to ensure that this is included in the objective-setting phase of the study.

5.2.1.2.1 RANDOMIZED SAMPLE SELECTION. In order to extrapolate the results obtained in the selected sample from the study population, it is necessary that the sample be representative of that study population. This assumes that there is no systematic variation between the characteristics of the individuals selected in the sample and those of the individuals not selected in the sample but who are part of the study population. The results obtained from the investigation of a sample will differ from the results that would have been obtained if the sample had included the entire study population, based on randomized variation (which is based on chance phenomena) alone. The impact of this randomized variation on the results obtained in the study is generally determined by statistical tests. The best measure that will ensure that the sample selected is representative of the study population is randomized sampling, in which every individual in the study population has an equal chance of being selected for inclusion in the study.

There are several selection strategies that ensure randomized selection of individuals, which depend on the nature of the study population. These selection strategies include:

- Simple randomized selection—individuals are selected based on a list using numbers in a randomized sequence. This is the easiest method to obtain a randomized sample, but it assumes the existence of a list of all individuals in the population (e.g. census-based).
- Systematic selection—in this type of selection, the total number of individuals in the study population is divided by the number of individuals needed in the sample to obtain the selection fraction (n). The first individual is selected at random, from the first n available individuals, and then every n-th individual is selected. This type of selection usually produces results similar to those obtained in simple randomized selection and can be used when a list of all individuals in the study population is not available at the start of the study but all members of the population are eligible for the study.
- Cluster selection—this type of selection can be used when, for logistical reasons, a randomized selection of all individuals cannot be practiced. In this case, groups

of individuals are selected which either all enter the study or subsequently go through another selection process of individuals from within the group formed by the previous selection.

- Stratified selection—this is used when the study population can be grouped into several clusters on the basis of characteristics common to the individuals in the cluster that may influence the likelihood of these individuals developing the disease. This type of selection seeks to ensure that a sufficient number of individuals are included in the sample so that all the characteristics of the groups in the study population are represented (found) in the sample.

5.2.1.2.2 RANDOMIZATION AND SAMPLE SIZE. Because generally only a sample of the study population is included in the study, the results obtained from determining the frequency of illness and measuring associations between potentially implicated factors will not be exactly the same as the values that would have been obtained if the entire population had been examined. This is due to chance (randomized) variation between individuals. Data obtained on correctly obtained samples from the study population can generate, by statistical analysis, a wide range of disease frequency values and measures of associations between pathogenic factors, taking into account actual data on the study population. The most commonly used scale of values is the 95% confidence interval, which is interpreted as the 95% probability that the true population values lie within the range indicated by the confidence interval. An alternative to the confidence interval is to perform a statistical analysis to determine the probability (P-value) that the difference between the groups identified in the study sample would be greater than, or equal to the difference observed if the groups in the study population were identical.

These two methods of presenting statistical results are linked. If the 95% confidence interval of an association measure does not include the value 1, the P-value will be less than 0.05. Confidence intervals are most commonly used because they provide more data than the P-value.

Statistical analysis is also used to determine the number of individuals to be included in the study, to ensure that the frequency of illness will be measured with the necessary precision and that associations present in the study population will be detected, should these exist. It depends, however, on the objectives of the study and how accurately the prevalence of disease and the exposure of individuals in the study population to risk factors are estimated. These two elements are determined when calculating the sample size, which should allow a sufficiently large number of individuals to be included in the study to achieve the study objectives without wasting resources, especially in large or long-duration studies.

5.2.1.3 Step 3: Information gathering

Whatever the epidemiological study, it is necessary to collect data from all participants in a standardized way. This is generally accomplished using standardized protocols for examination and sample collection through diagnostic tests and questionnaires planned for this purpose. All data collection methods should be pre-tested to ensure that participants can provide all the necessary data. Section 5.2.2 'Types of epidemiologic studies used in epidemiologic investigations,' shows how to evaluate the results obtained from the diagnostic tests and how to select tests adapted to the different circumstances.

5.2.1.4 Step 4: Methods of analysis

Once the data have been collected and verified, the appropriate method of analysis is selected. These analyses shall be used to estimate the frequency of diseases and to determine the associations between exposure to the risk factor(s) and disease occurrence in the study population.

5.2.1.5 Step 5: Interpretation of results

The last step in conducting epidemiologic investigations is the interpretation of the results obtained to ensure that all statistical associations detected are valid.

5.2.2   Types of epidemiologic studies used in epidemiologic investigations

5.2.2.1   Ecological studies

This type of study uses group-level data to formulate hypotheses about the occurrence of disease in individuals in the study population. These studies use routine data collected by different sources, which do not provide additional data on the characteristics of the individuals or on the exposure to risk factors of the individuals in the population. These studies, however, are a first step in the investigation of diseases, suggesting the presence of causal factors that can be further investigated in studies where data is obtained from individuals. Ecological studies are useful for investigating associations with factors that do not vary considerably between groups or that cannot be easily determined at the individual level.

ADVANTAGES.   These studies can be carried out at minimal cost and in a short time.

DISADVANTAGES.   Because the associations observed at the group level may not be the same as those detected at the individual level (often the differences observed are due to differences between populations, and not to the factors of interest to the study), the results should be interpreted with caution. Comparative analysis of groups of individuals should take into account spatial (if comparing disease occurrence in different regions) and temporal (if comparing the lag time from exposure to risk factors to disease onset) correlations, as well as the size of the groups under study. In addition, studies investigating changes in disease over time should also take into account changes in diagnostic methods.

5.2.2.2   Cross-sectional studies

These studies provide a snapshot of the diseases that evolve in a population. Cross-sectional studies are also called prevalence studies. Cross-sectional studies determine at the same time the current status of individuals in the population, i.e. whether an individual is ill or at risk of illness (healthy at the time of the study but at risk of disease). The particularity of this study is that each individual will be classified as ill or at risk of illness at the time of assessment.

Results from cross-sectional studies are generally used to determine the prevalence of a disease in populations where individuals are exposed (or not) to the risk factors of interest. These results can also be used to determine associations between disease occurrence and the risk factors of interest (odds ratios).

ADVANTAGES.   Since the selection of individuals is done without knowing in advance whether individuals are ill or at risk (exposed), these studies are more precise than cohort and case-control studies. Cross-sectional studies provide objective data on the prevalence of disease and frequency of exposure. Data from cross-sectional studies can be used with confidence to investigate causality when studies look at unchanging factors such as genetic makeup. In addition, several diseases and risk factors can be investigated at the same time, and the costs of these studies are relatively low compared to other studies.

DISADVANTAGES.   Because disease prevalence is very low in rare and short-lived diseases, it is necessary to sample large numbers of individuals. Also, because cross-sectional studies determine the current status of individuals in the study population, these studies are of limited value in determining the cause of the disease, because the current status determined by the study may be a consequence of the disease. For example, one may suspect a link between poor physical condition and the hip pain a patient experiences while walking. However, it cannot be determined whether the pain is the result of the poor physical condition or whether the poor physical condition is the result of the pain.

Another disadvantage is that the results of these studies are often influenced by factors which determine survival rates. For example, the number of cases of malaria in the population of town X may be higher than in the population of town Y because more individuals with malaria die in town Y than in town X.

5.2.2.3 Case-control studies

Case-control studies are observational epide-miologic studies in which a sample of individu-als with the disease of interest (called cases) is selected. The history of exposure to one or more risk factors of interest for the study is then com-pared with a group of individuals who do not have the disease (called controls). In this study, the control individuals are used to estimate the prevalence of exposure in the population from which the cases were selected. After obtaining the samples of case and control individuals, the history of each individual is investigated to reveal the past exposure of each individual. Case-control studies are therefore also called retrospective studies.

ADVANTAGES. Case-control studies are useful for diseases with long latency or incubation. These studies are useful in rare disease investigations, and allow the study of the effect of a large number of potential risk factors on a disease. These studies are suitable in the early stages of investigating new diseases, are relatively inexpensive, and are quick compared to cohort studies.

DISADVANTAGES. Both the disease and the risk factors have already manifested themselves by the time individuals enter the study, which means that we cannot be sure whether or not exposure to the risk factors preceded the onset of the disease. Case-control studies cannot be used to estimate disease prevalence and inci-dence. In addition, case-control studies are often subject to selection bias, where the individuals who are selected for inclusion in the study are not the same as those in the study population. The correct selection, on objective criteria, of a group of individuals who constitute the case group is not always randomized. If, for example, cases are selected from individuals who come to see the doctor, it will be difficult to select a population of controls that accurately reflects the exposure status in the population of individuals from which the cases were selected. This is because it is very likely that the factors to which the individuals who go to the doctor are exposed are different from the factors to which those who do not go to the doctor are exposed. In

addition, if only individuals who present to the clinic are retained for the study, they may not reflect the exposure conditions in the popula-tion in which the cases occurred; it is likely that these individuals were more courageous or had other personal motivations that led them to come to the clinic. For these reasons, the groups (cohorts) of individuals who constitute cases may be different from the control individuals. When selecting control individuals, it should be kept in mind that, for any individuals that did come to the clinic, the reason why they did so may be associated with exposure to the risk factors of interest for the study.

5.2.2.4 Cohort studies

As with case-control studies, cohort studies are observational studies. Individuals who are selected for this type of study must be free of disease at the start of the study. Individuals in the sample are grouped into two cohorts, based on their exposure status, and are then observed over a period of time. The frequency of disease in the exposed and unexposed cohorts is then compared.

ADVANTAGES. Because all individuals are disease-free from the start of the study, the time sequence of exposure to the risk factors of inter-est and the outcomes of these exposures can be determined. Consequently, cohort studies are more suitable for investigating disease etiology than case-control or cross-sectional studies. When studying rare risk factors, cohort studies are particularly useful as only a small number of animals are exposed. Cohort studies can be used to investigate associations between exposure to a risk factor and several diseases at the same time. In addition, cohort studies allow us to calculate the incidence rate and cumulative incidence of disease in cohorts exposed and unexposed to risk factors.

DISADVANTAGES. These types of studies can be time-consuming, especially for diseases with long incubation and latency periods. These studies are not practical for rare diseases because they require large samples, which in such cases are not available in the population. The propor-tion of exposed and unexposed individuals in the study population cannot be estimated in studies in which exposed and unexposed cohorts are selected, although this would be possible if a population were selected and divided into cohorts based on exposure status established after the selection stage. In addition, it may be difficult to follow certain studies over time, especially long ones. Cohort studies are slow and costly compared to case-control studies. Additionally, ensuring unbiased outcome assessment is challenging if the investigators are informed of the patient's exposure status.

5.2.2.5 Intervention studies

These are the only studies in which the investigators interfere to influence exposure to putative risk factors through deliberate attempts to modify disease occurrence. These studies are similar to cohort studies, in which individuals with different levels of exposure to the factors of interest are observed over time. Since the exposure status of the individuals in the study is deliberately changed, and those conducting and participating in these studies are usually aware of the working hypothesis of the study, it is very important to ensure that the outcome of interest is obtained under exactly the same conditions in both groups.

ADVANTAGES. These are similar to those presented in cohort studies, where individuals with different levels of exposure to the factors of interest are tracked over time. These studies provide the strongest evidence that the associations between risk factors and disease occurrence are causal (that there is a link between the two). These studies use inert treatments (placebo) in the control group to keep individuals' exposure status secret for the duration of the study.

DISADVANTAGES. Ethical issues need to be considered before proceeding to manipulate the status of all individuals in the study.

5.2.3 Validity of diagnostic tests

In this section, we will discuss how to assess the validity of data obtained from diagnostic tests

and the choice of diagnostic tests according to the field situation.

5.2.3.1 Detecting infections

Various methods or tests can be used to determine whether an individual is sick or infected. Any process that detects or quantifies the presence of a sign, substance, or tissue change can be a test. For example, the intradermal tuberculin test is a classic test involving the injection of antigen to elicit a cell-mediated response. The local reaction (where the allergen has been inoculated) is then assessed by palpation or measurement with an instrument. Although it is a widely used test for *Mycobacterium* spp. infection, it is widely accepted that its accuracy is not ideal, because not all individuals who test positive for tuberculin are actually infected, nor do all those who are infected test positive.

5.2.3.2 How to assess test validity

Before choosing a diagnostic test, it is necessary to know the accuracy of the test. The accuracy of a test is the ability of the test to recognize asymptomatic and symptomatic individuals from uninfected (healthy) individuals. The ability of a test to accurately determine an individual's status is called test validity. The validity of a test can be determined with a binary result using two measures: test sensitivity and test specificity.

5.2.3.2.1 TEST SENSITIVITY AND SPECIFICITY. Assuming there is a gold standard test (e.g. computerized tomographic examination) that allows us to determine the true status of each individual in a group and we use another available test, we can assess the validity of the latter by comparing the results with the gold standard test. If we present the data obtained from the test in tabular format (Table 5.1), it can be seen that, according to the gold standard, only individuals a and c are sick (true positives, +).

Of these individuals, group a also tested true positive on the alternate test, individuals in group c tested false negative on this test.

Sensitivity (SENS) is calculated using the formula:

$$SENS = a/ (a + c)$$

Table 5.1. Test sensibility and specificity

		ACTUAL DISEASE STATUS (GOLD STANDARD)	
		POSITIVE	NEGATIVE
DISEASE STATUS	POSITIVE	*a*	*b*
	NEGATIVE	*c*	*d*

and the obtained result can be expressed proportionally (with values from 0 to 1) or as a percentage (with values from 0% to 100%).

The sensitivity and specificity of a test are the essential elements to take into account when choosing the test to use in practice. There is, however, the problem of determining the real (true) status of the individuals tested, in addition to the need to meet the requirement of representativeness of the study population when selecting the sample of individuals from the population to be tested.

Specificity is determined especially for low-frequency diseases. The specificity of a test implies that all positive test reactions are false positives.

A test is designed to have either high specificity (detects false positives) or high sensitivity (detects false negatives).

5.2.3.2.2 TESTING STRATEGIES. The test to be used is chosen according to the type of investigation envisaged. Since the sensitivity and specificity of a test are inversely proportional (a test with high specificity will have low sensitivity, and vice versa), it must be decided beforehand which of the two elements, sensitivity or specificity, is more important in the given situation.

A high *sensitivity* test is used to exclude disease in the following situations:

- at the beginning of the epidemiologic investigation of a disease, when clinical signs could be due to several different pathogens;
- when it is important to identify all infected animals, e.g. emerging diseases and zoonoses;
- when disease prevalence is low (in screening programs); and

- in the early stages of control programs, when disease prevalence is high, and when it is important to remove infected individuals from the population as quickly as possible.

A high *specificity* test is used to screen for disease in the following situations:

- to confirm diagnoses, when only individuals positive to a previously applied test of higher sensitivity are retested with a test of higher specificity;
- when false positive reactions have severe consequences, such as total quarantine; and
- in more advanced stages of control programs, when the prevalence of a disease is low, and when it is desirable not to record too many false positive results.

5.2.3.3 Factors utilized to describe disease

As already mentioned, environmental, pathogen, and host factors often influence occurrence and manifestation of diseases. In this section, the methods used in describing illnesses in a population will be presented in order to generate hypotheses about the causes of the occurrence of these illnesses.

Describing the distribution of disease cases in a population helps to understand the factors that predispose to or cause disease.

5.2.3.3.1 INDIVIDUAL VARIABLES. Differences between individuals affected by the same pathogen may give information about host characteristics and environmental factors. If there is a different incidence of disease from one gender to another, for instance, it may be suspected that there is some influence from the hormonal structure of the individuals concerned. When individuals from different categories are affected, apparently unrelated to the individual, environmental risk factors may be suspected. In conclusion, it is necessary to consider all factors that may be responsible for the variations in an outbreak from case to case or from one group of people to another. Once these variations in the population have been detected, hypotheses can be formulated and subsequently tested.

The individual factors to be taken into account during an epidemiologic investigation are: age, gender, physiologic status, immune status, body conformation, weight, and height.

5.2.3.3.2 TEMPORAL VARIABLES. When cases of illness vary over time, exposure of affected individuals to certain environmental factors or to the etiologic pathogen may be suspected. Seasonal and cyclical patterns may be related to climate or to the dynamic balance between host and infectious pathogen. As shown in Section 5.1.8.4, 'Disease description data,' in order to understand the time-linked patterns of disease occurrence, graphical representations or mapping of the cases would be the most appropriate.

Temporal models classify diseases into:

- sporadic diseases: disease occurs irregularly, by chance;
- endemic diseases: disease has a stable, predictable course; endemics may be high or low, and may follow a seasonal (repeating every 12 months) or cyclical pattern (recurring in the population after a longer period of time);
- epidemic diseases: disease is produced by an infectious organism (or non-infectious pathogen) and has a frequency significantly higher than usual; and
- pandemic diseases: disease affects entire populations in different countries or continents (also known as cross-border epidemics).

5.2.3.3.3 SPATIAL VARIABLES. The locations in which cases of illness occur can give clues to the sources or causes of these illnesses. In general, abiotic factors (soil structure, air, water, climate) or biotic factors (vegetation or the distribution of vectors or pathogens) are involved and may suggest that infectious pathogens or environmental factors influence the occurrence of disease. As indicated, maps are an effective means of understanding the spatial and temporal distribution of a disease.

5.2.4 Factors influencing the validity of epidemiologic studies

The validity of a study is the extent to which the conclusion/hypothesis being drawn has a sound, realistic basis. The internal validity of a study determines the extent to which the frequency of the disease or associations detected in the study sample can be equated with the frequency of the same disease present in the study population from which the sample was drawn. In other words, validity determines whether the selection of participants and the comparisons made in the epidemiologic study show that the differences found are due to the hypotheses made and not to chance. The external validity of a study determines whether the frequency of disease or associations detected in the study population can be the same as in the target population, where the study results can be extrapolated. Of the two types of validity, the most important is to ensure internal validity, whereby the results of the study can be applied to the study population.

There are three reasons why the results obtained in one sample may differ from the results that would be obtained if the entire study population were tested, thus raising the question of the internal validity of the study. The three causes are:

- chance;
- subjectivity (bias); and
- confusion.

To check whether the results obtained in the study are due to *chance or coincidence*, we use statistical tests to determine the probability that the difference (e.g. differences in means and proportions) is as large in the sample as the difference in the study population. The likelihood of a statistically significant association increases with the number of risk factors being investigated. Therefore, associations detected in studies investigating a limited number of a priori hypotheses will produce stronger evidence of causal factors than associations detected in studies investigating a large number of causal factors.

There are two types of *subjectivity (bias)*: (i) selection and (ii) information.

Selection subjectivity occurs when the individuals selected or chosen to take part in a study are not representative of the rest of the individuals who could have taken part in that study. Selection subjectivity is especially important in studies where it is necessary to determine the frequency of illness, because the selection or response of study participants should not be related to the outcome of the study. Subjectivity of selection is also important in case-control studies, where the selection of controls should not be related to the risk factor under study. In cohort studies, selection bias occurs rarely, provided that selection of the participants in the study is not made on the basis of the probability of the participants becoming infected.

Comparing the characteristics of those included in the study with the individuals in the study population can help us determine to what extent selection bias may be introduced. Any differences found between these two populations may indicate that our sample is not representative of the study population.

Information bias occurs when the information collected from study participants varies (differs) from one study group to another. This type of bias is especially important in case-control studies, where the outcome of interest has already been identified so that the information obtained from the case groups may differ from that obtained from the control group. When information bias arises in all groups included in the study, non-differential bias is present, which reduces the likelihood of detecting an association between disease and exposure to the factors of interest. If, however, the bias does not differ between groups, the study may fail to detect any association or may record associations that do not actually exist in the study population.

If information collection or selection of participants is done on criteria that introduce information bias (subjectivity), these data cannot be used in the analysis of the data collected. Therefore, bias should be avoided.

Confounding occurs when one variable, other than the risk factor of interest for the study, is associated with both the study outcome and the risk factor of interest.

If data on potentially confounding variables have been collected, the effect of these variables on the associations between disease occurrence and the risk factors of interest can be accounted for during data analysis by stratifying the data or using multivariate models. Another method that can be used to control the effect of confounding variables, in addition to stratified analysis, is standardization. Standardization involves the use of the illness rates or population distributions reported at the levels of confounding variables in a standard population to calculate those rates in the populations of interest, in which the differences caused by the confounding variables have been removed. The effects of confounding can be controlled at the study design stage by selecting all participants to have the same level of confounding variables, by selecting controls with confounding variable values that match the case variable, or by randomizing individuals into treatment groups in the case of intervention studies.

5.2.5 Selection of the type of epidemiological study

5.2.5.1 *Ecologic studies*

These are used in the initial stages of investigations of diseases or other epidemiologic problems in order to generate hypotheses. The main problem with this type of study is that associations at the individual level may differ from associations detected at the group level.

5.2.5.2 *Cross-sectional studies*

Like ecological studies, these studies are used in the initial stages of investigating diseases or other epidemiologic problems, generating some evidence of causal associations. Cross-sectional studies also investigate the mode of occurrence of several diseases and/or risk factors, but cannot be used for rare diseases or diseases that have a short duration in the population. Cross-sectional studies can investigate the effect of risk factors that do not vary over time. To correctly estimate disease frequency, the study samples must be representative of the study population. Cross-sectional studies can detect associations with factors influencing duration and with factors influencing incidence. For time-varying risk factors, it may be difficult to establish the order between exposure and disease (i.e. whether exposure to the risk factors of interest caused the disease or whether disease caused exposure to the risk factors of interest).

5.2.5.3 *Case-control studies*

These identify causal associations and investigate rare diseases with multiple risk factors and the effect of risk factors that do not vary over time. For these studies to be valid, it is necessary to ensure that the markers are selected from the population in which the cases under study occurred, and that the accuracy of the information collected from cases and controls is the same. If some time-varying risk factors are included in the study, one cannot be certain that exposure to the risk factors caused the disease and not vice versa.

5.2.5.4 *Cohort studies*

These are considered the best type of observational study because they produce evidence of causal association. Cohort studies help to investigate risk factors that are less common in populations and investigate multiple effects at the same time. However, these studies are expensive and require follow-up and determination of the effects of associations in both exposed and unexposed groups of individuals.

5.2.5.5 *Intervention studies*

These produce the best evidence to confirm causal associations but are costly, and ethical issues need to be considered. Intervention studies require follow-up over time and, as with cohort studies, need to determine the effects of associations in both exposed and unexposed groups of individuals.

5.2.6 Characteristics of infectious and parasitic epidemics

Infectious epidemics are caused by infectious agents (viruses, bacteria, prions) and parasitic epidemics are caused by parasites.

In endemic outbreaks, the number of host individuals and the prevalence of the pathogen are in balance, so that a predictable pattern of disease occurrence can be described. Most parasitic diseases are endemic. In epidemics, cases of disease occurring in a population emerge over a given period of time, with a higher than expected incidence. For an epidemic to occur, the pathogen and the susceptible host must be present in the same place, at the same time, and in sufficient numbers. In addition, the pathogen must be transmitted to the susceptible host via a source. Endemic outbreaks are outbreaks in which infectious pathogens evolve.

In general, epidemics occur when changes in the pathogen, the host, or contact between them happen under the following circumstances:

* when a pathogen is introduced for the first time into the population, or when an already existing but undetected pathogen in the population suddenly acquires high virulence, thereby producing clinical forms of disease (e.g. in the case of pathogens that may undergo genetic mutation);
* when the susceptibility of the host population to the pathogen changes;
* when a new route of transmission of the pathogen is established or when host-to-host transmission increases (e.g. when population density increases); and
* when changes in climate or vegetation take place, resulting in changes in population behavior or in the distribution and persistence of vectors, reservoirs, and the pathogen.

As shown previously, the temporal pattern of disease is described by epidemic curves, which record the number of cases occurring in the population at a given time period. From the resulting pattern of illness, inferences can be made about the nature of the infectious process occurring in the body of infected individuals and the type of exposure to the pathogen. The main characteristics of the infectious process occurring at the level of the individual that influence the pattern of disease are:

- the latent period (latency), that is, the period between exposure to the infection and excretion of the pathogen by the host;
- the incubation period, that is, the time elapsed between exposure to the infection and clinical onset of disease (onset of symptoms); and
- the infective (infestation, in the case of parasitic diseases) period, that is, the time from when excretion of the pathogen begins until the end of the disease, when, after a period of convalescence, the patient recovers (recovery may be with or without sequelae), dies, or progresses to the chronic (clinically inapparent) form of the disease.

In the case of short-term host exposure to a source of infection, the epidemic produced by this source is a point-source epidemic (point sources), and the epidemic curve will reflect the incubation period of the disease, with a steep upward slope and a gradual decline. Exposure to an infectious agent, particularly through contaminated food or water, can produce this pattern of illness in a population (Fig. 5.8).

When individuals in a population are exposed for a long time to a common source of infection, the epidemic is common-source (continuous or widespread source) and will generally evolve as a gradually increasing slope with a plateaued peak, reflecting the time margin between exposures and incubation periods (Fig. 5.9).

The shape of the epidemic curves that occur in propagated (widespread) infections, in which the infectious agent is transmitted between individuals, depends on the nature of the infectious process that is taking place in the body of the infected individuals and the transmission process. There may be several peaks between successive temporal clusters of cases, reflecting the length of the incubation period.

Combined epidemics occur when a common source infection is followed by a widespread epidemic. Note that common-source or spreading epidemics do not occur for vector-borne pathogens.

Epidemic curves are influenced by control measures taken (vaccination, isolation of susceptible individuals, etc.) in the epidemic.

In this case, outbreaks decline more rapidly than if these measures are not applied, due to the decrease in the number of susceptible individuals in the population.

The distribution of host and pathogen populations and the relationships between host, pathogen, and environment determine the pattern of disease in the host population. These data are used to predict and control future disease occurrence.

5.2.7 Disease transmission and factors influencing transmission

Transmission of an infectious disease requires the presence of:

- the pathogen;
- a reservoir of infection in which the pathogen lives and multiplies; this may be the host species, other species, or the environment;
- a means of transmission from the reservoir to a susceptible host; this depends on the mode of transmission of the pathogen between individuals, which may be vertical or horizontal:
  - o Vertical transmission involves the spread of the pathogen from one generation to another, while horizontal transmission may result from direct or indirect contact between susceptible individuals. Direct transmission occurs when individuals are in close proximity to each other, in which case the pathogen is transmitted through skin or mucous membrane contact. Indirect transmission occurs via aerosols, contaminated food, water, and inanimate objects, or via vectors, which can be mechanical or biological.
- a susceptible host, in which the pathogen replicates and from which it then excretes, regardless of whether clinical manifestations of disease are present.

The likelihood of an infection being transmitted between individuals depends on the characteristics of the pathogen, the host, and the contact between two or more susceptible hosts. Further, the distribution of the pathogen

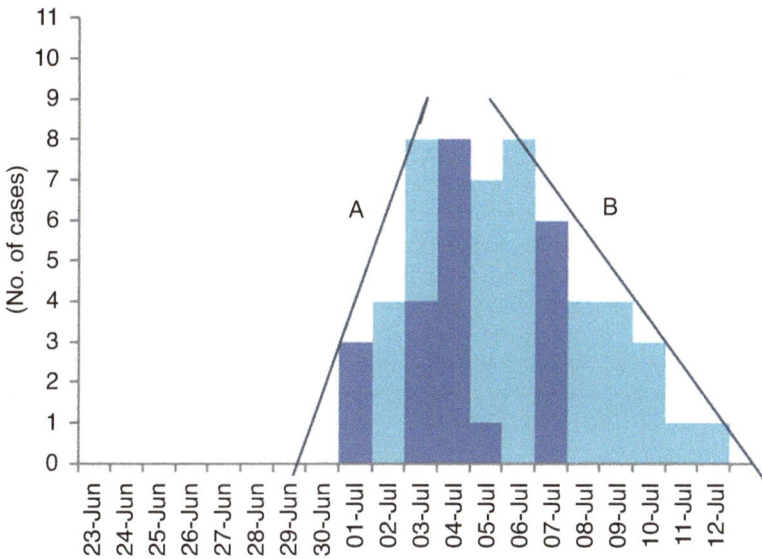

Fig. 5.8. Point-source outbreak: (A) steep ascending slope; (B) gradual decrease. Figure author's own.

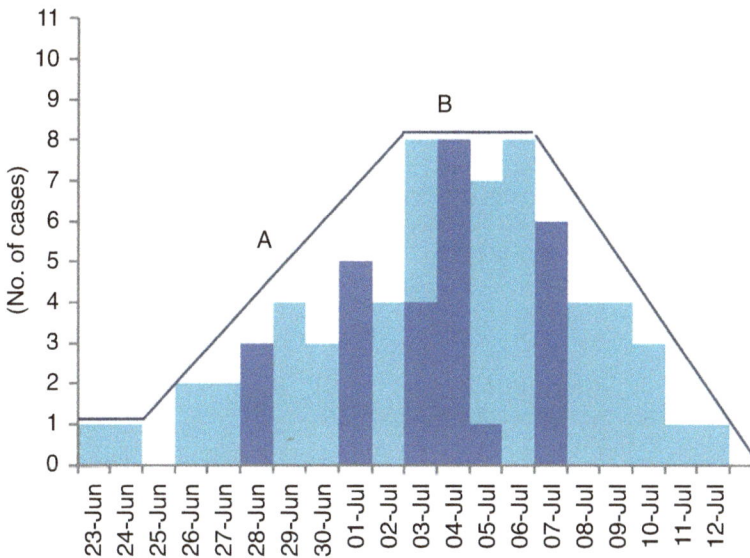

Fig. 5.9. Common-source outbreak: (A) slow rising slope; (B) peak 'in plateau.' Figure author's own.

and the type of contact between individuals may be influenced by environmental factors.

Pathogens exhibit numerous characteristics that allow them to persist for a variable time in a population. These characteristics include:

• the possibility that they are directly transmissible, which excludes the influence of environmental factors from the chain of transmission, as for example in influenza, measles, chickenpox, and hepatitis;

- the ability to develop resistant forms in the environment (anthrax spores, helminth eggs);
- rapid infection/infestation and excretion of the pathogen, before host immunity produces specific immune response (antibodies); crowding (high population density per unit area) and the ability of the pathogen to undergo rapid genetic mutation, as in SARS-CoV-2 and influenza viruses, prevent the development of long-term immunity in the host population;
- persistence in the host body by suppressing the immune response, by long incubation periods, or by producing intermittent shedding carrier status, as in tuberculosis and transmissible encephalopathies; and
- broad host species specificity, which provides the pathogen with the means to survive, and sometimes multiply, in several population types, including animal populations (zoonoses).

5.2.8 Determination of the transmission potential and dissemination rate of the pathogen

The transmission potential of a pathogen in a recipient population is determined using the basal reproduction number (R0) of the pathogen, which measures how many new infections or infestations occur as a result of infecting or infesting an individual in a susceptible population. In other words, R0 indicates how contagious a disease is. R0 depends on:

- the probability of contact between individuals, which is a function of the mode of transmission and the density of the susceptible population; the type of contact between individuals required for infection or infestation to occur depends on some characteristic of the pathogen;
- infectivity of the pathogen, which determines the probability of infection or infestation; and
- duration of the infectious period, which depends on the latent and infectious periods.

If R0 is less than 1, the infection or infestation cannot be maintained in the population,

because in this case an infected individual cannot transmit the pathogen further to another individual. The value of R0 is used to determine the contact rate (susceptible population density) and duration of infection that are necessary to reduce or prevent transmission of the pathogen in the population. To prevent the transmission of an infection or infestation in a population, R0 must be reduced to less than 1. For example, the R0 for influenza is around 2, and for COVID-19 the rate is above 3.

5.2.9 Mathematical epidemiology

The representation of events in quantitative mathematical terms, using mathematical models, makes it possible to predict the effects of actions within the systems being analyzed. Theoretical studies are used in epidemiology when field experiments and observations cannot be applied. Theoretical investigations are useful in the study of disease control measures in human and animal populations.

Mathematical models are used for:

- formulation of hypotheses—with the help of mathematical models one can organize existing data and describe how certain factors influence the occurrence of disease;
- better understanding of existing information and determining what new data are needed;
- hypothesis testing—mathematical models can be used to test the level of knowledge about how a system works;
- predicting disease effects at the population level—this requires knowledge of the biology of the disease and valid data; and
- supporting decisions on future disease control measures by predicting the effects applicable in different situations.

5.2.9.1 Classification of mathematical models

According to their degree of specificity, mathematical models can be strategic or tactical.

- Strategic (generalist) models are used to study general problems related to the spread of disease.

- Tactical (detailed) models are used to solve specific problems, such as controlling an epidemic in a population.

In addition to these, the following types of mathematical models are also utilized:

- density and prevalence models;
- deterministic and stochastic models;
- models including linear and non-linear relationships between variables;
- static and dynamic models;
- models using different equations and dynamic (time-varying) models;
- simple and complex models; and
- homogeneous and heterogeneous spatial models.

5.2.9.2 *Setting up mathematical models*

The steps involved in building a mathematical model are as follows:

- setting the objectives of the investigation;
- gathering information about factors influencing the occurrence of illness and data about these factors;
- developing a model;
- checking the concordance and conformity of the results with the objectives of the investigation;
- conducting sensitivity analysis to determine the degree to which the results of the study vary when there are changes in the variables used or when more variables are introduced into the system;
- validation, which involves a process of evaluating the biological and mathematical sensitivity of the model; and
- experimentation, in which stage the model is used to determine the effect of different disease control options.

The validity of these models can be determined by the following characteristics:

- generality—whether the model can be used to investigate different diseases or systems;
- simplicity—which depends on the number of variables that are considered;
- accuracy of the estimated model effects;
- appropriateness—which refers to whether there are any indications that the

model's estimates or structure might be incorrect;
- the extent to which the model has been tested against independent datasets; and
- usefulness—whether better decisions can be made with or without these models.

5.2.10   Disease surveillance and monitoring (MOSS)

Surveillance and monitoring systems (also known as MOSS) are epidemiologic tools used to obtain representative data about the health status of a population in a structured way. A well planned MOSS system can increase the value of the data that are collected. MOSS systems describe the tasks involved in surveillance and monitoring activities at the population level.

Monitoring describes ongoing efforts to assess the health or disease status of a population at the national, regional, or local level. In monitoring, the collection of study samples (i.e. individuals in the population) is done repeatedly or continuously. In other words, monitoring is a continuous and adaptable process of collecting data about diseases and their causes in a given population.

Surveillance describes a more active system than monitoring and involves the existence of targeted actions that are implemented when field data indicate a certain level of prevalence or incidence of a disease. Surveillance involves knowledge of the causative pathogen and is directed toward investigating its prevalence or incidence in the population (i.e. it is directed at a specific disease or pathogen). As in the case of monitoring, the selection of individuals from the population to determine the health or disease status of the study population is done repeatedly or continuously, and the population may be represented in the sample at national, regional or local level. Surveillance is a specific type of monitoring in which control and eradication measures are implemented whenever there is a change in disease status or when the predefined level of disease frequency in the population is exceeded. In this context, three main components are described for surveillance systems:

- a defined disease surveillance system;
- a predefined critical level or clear limit for disease frequency above which action is required; and
- a predefined targeted action (also known as intervention), achieved through a disease control program.

There are two types of surveillance programs: disease control and disease eradication programs. Disease control programs are part of the surveillance system and use long-term intervention strategies aimed at reducing the frequency of a disease. Disease eradication programs aim to eliminate a disease (or its cause) from the population.

Disease surveillance is a measure that provides early warning of changes in the health status of populations. Of particular importance in this respect is the provision of data on the absence of disease (when the population is free from the pathogen for which MOSS has been established) or, when the pathogen is present in the population, on the frequency of disease.

Epidemiological surveys, investigations, or studies differ from MOSS. Epidemiologic studies are based on a set of systematically collected data for a purpose, based on a conceptual hypothesis, and the duration of these studies is limited and specific. MOSS monitoring and surveillance systems involve the systematic and continuous collection of data and information over long periods of time. While epidemiological surveys or studies are generally used to answer specific research questions for exploratory scientific purposes, MOSS is of a rather practical nature, being used to demonstrate that a population is below specific prevalence or incidence levels of the monitored disease, for early detection (through early warning) of increases in these frequencies, and for effective implementation of disease prevention and control measures. As stated above, surveillance is an integral part of any control program for diseases at population level.

5.2.10.1 Data collection methods

Data collection for MOSS can be done passively or actively.

Active data collection for MOSS involves the systematic or regular recording of cases strictly for monitoring or surveillance purposes. Active data collection assumes that the population is defined in space and time. The identification of the population and the types of samples to be used depend on: the objectives set for the MOSS, the event of interest, the prevalence and incidence of the disease (these are estimated in advance), and the diagnostic tests available at the time. Depending on the proposed MOSS objectives, data may be collected from every individual in the target population or from a sample of individuals in the study population (Fig. 5.9). Verification of patients' individual records or of pathogens existing in biobanks are examples of active data collection for MOSS. Basic rules for MOSS include the following:

- Each individual in the population must have an equal chance of being included in the program (randomized sample collection).
- The most susceptible individuals, who are at risk of becoming infected (the elderly, the poor, people close to infected areas, those in contact with certain sources or reservoirs of the pathogen, etc.), should have a greater chance of being included in the sample than those at low specific risk (targeted sampling).
- There must be repeated sampling of a specific group of sentinel individuals at regular time intervals (sentinel networks). Sentinel individuals are randomly selected from the high-risk sub-population or from volunteers.

Passive data collection for MOSS is based on informal notifications of cases and suspicions of disease reported by people in the social environment of a sick person (family doctors, testing laboratories, employers, etc.). Information about cases or suspicions of illness can also be provided by interview, questionnaires, or email. Biological samples can be collected during the visit to the family doctor, at testing laboratories, or by a specialist.

5.2.10.2 Objectives of MOSS systems

The objectives of MOSS systems are as follows:

- determine the prevalence and incidence of disease and describe changes occurring over time;
- detect potential sources of infection;
- develop an early warning system for cases where new pathogens enter the population, or where cases re-emerge after eradication of a disease; and
- demonstrate that a population is 'free from disease,' meaning the disease prevalence or incidence is below a predefined level. The freedom from disease status is achieved through rigorous control programs.

A specific way to achieve 'disease-free' or 'zero-risk' status is *eradication*, which involves the total elimination of the pathogen from all host species, all potential vectors, and the environment. Specific disease eradication programs are used to completely eliminate the pathogen.

5.2.10.3 Quality assurance in MOSS systems

The validity and accuracy of the information obtained can be improved by using a combination of different methods of data collection, sampling, and diagnostic testing. This approach increases MOSS system costs. The more diseases to be monitored in the population, the more difficult it becomes to maximize the quality of surveillance data for each individual disease. In any case, prioritization is always based on the required accuracy of data for different diseases. An important element in the quality assurance of MOSS systems is to identify strengths and limitations as well as possible elements that may affect the validity of the data to be obtained by MOSS.

Quality assurance establishes a framework that allows a systematic evaluation of the design and implementation of a surveillance program. Quality assurance requires that the MOSS fulfills certain conditions. They should:

- be useful (contribute to early detection of disease);
- take into account the degree of representativeness (often difficult to achieve);
- be current (cases should be detected as close as possible to the time of infection);
- be simple (easy to implement);
- be flexible (additional samples or additional information can be added along the way);
- ensure good quality data;
- be stable (implies that data collection can be planned and data quality can be controlled);
- give predictable values for positive test results (these depend on the capacity of the test used);
- be acceptable (depends on how aware the population is of the disease and the consequences of detecting/reporting cases); and
- be sensitive (depending on pathogenicity and virulence of the pathogen).

5.3 Introduction to Biosecurity

Effective biosecurity requires knowledge of the causal relationships between exposure and disease, since the occurrence and consequences of infectious diseases involve a complex interplay between multiple disease determinants. The epidemiology and relative risk for each disease are assessed to determine how best to allocate resources for effective disease control. In addition, statistical epidemiology can be used to determine how best to contain the financial risk posed by a pathogen. Thus, among the important epidemiological factors that are considered for biosecurity measures are: sources of infection, modes of transmission and spread of the pathogen, and favorable factors.

5.3.1 What is biosecurity?

There are several definitions of biosecurity. Some of the most helpful definitions are provided below, whilst biosecurity and its applications to One Health are described in Chapter 6, 'Public Health.'

Biosecurity is a strategic and integrated approach that encompasses the policy and regulatory frameworks (including instruments and activities) for analyzing and managing relevant risks to human, animal and plant life and health, and associated risks to the environment. Biosecurity covers food safety, zoonoses, the introduction of animal and plant diseases and pests, the introduction and release of living modified organisms (LMOs) and their products (e.g. genetically modified organisms or GMOs), and the introduction and management of invasive alien species. Thus, biosecurity is a holistic concept of direct relevance to the sustainability of agriculture, and wide-ranging aspects of public health and protection of the environment, including biological diversity.

The overarching goal of biosecurity is to prevent, control and/or manage risks to life and health as appropriate to the particular biosecurity sector. In doing so, biosecurity is an essential element of sustainable agricultural development. (FAO Biosecurity Toolkit, 2007, p. 3)

The biosecurity concept covers implementation of a set of programs and procedures preventing the entry, establishment or spread of unwanted pests and infectious disease pathogens in people, animals, plants or the environment. Biosecurity is the state of having applied appropriate measures to prevent or limit the possibility of pathogens entering populations from an extraneous source. It may be applied at various levels: international, national, region, farm and even between particular holding facilities. (EFSA, 2008)

Biosecurity is a set of management practices that collectively reduce the potential for the introduction or spread of animal disease-causing organisms onto and between farms. (The Scottish Government, 2022)

The doing of bio-security is a nuts-and-bolts part of livestock management, largely involving careful hygiene practices. (Donaldson, 2008)

Biosecurity is often seen as a relatively recent concept, and the term probably came into general use in the early 1990s, although other agriculture sectors such as intensive poultry practiced it for several decades prior. However, it is a very old concept, going back to ideas of 'clean and unclean' that are at least 2000 years old.

Biosecurity impacts on human health (e.g. zoonoses, foodborne illness), animal health, animal welfare (e.g. stress through increased handling, improved health), the environment (e.g. waste discharges), biodiversity (e.g. introduction of alien species or aquatic nuisance species), and production economics (e.g. decreased disease = increased production). **Biosecurity is one tool for implementation of the One Health concept**.

There is no widely accepted definition of biosecurity, partly because any given definition tends to reflect the purpose or field of the author. Despite the range of impacts that biosecurity can have, the focus for veterinarians is on the prevention, control, or eradication of infectious diseases. Therefore, a simple definition might be keeping the pathogens out, and if that fails, stopping them from spreading, or more scientifically, the concept of biological risk management (BRM).

The key to understanding biosecurity is the concept of an EpiUnit and that biosecurity consists of multiple layers. An EpiUnit can be anything that can be spatially defined or is a contiguous area in which an infectious or contagious disease can be transmitted between individuals of a population and in which meaningful systems can be implemented to prevent introduction or to control or eradicate disease. Thus, an EpiUnit can range from the whole world down to a single tank. It could be defined geographically, such as a country, watershed, or offshore island. Alternatively, it could consist of a supply chain or the various sites of an integrated business. This concept is reflected in the seven layers of biosecurity that are found.

1. Globally, there is the Agreement of the Application of Sanitary and Phytosanitary Measures (SPS Agreement) of the World Trade Organization (WTO, 1994). The WTO's SPS Agreement requires 'to harmonize sanitary and phytosanitary measures on as wide a basis as possible, Members shall base their sanitary or phytosanitary measures on international standards, guidelines or recommendations' (WTO, n.d., Article 3) The World Organisation for Animal Health (WOAH) is named as the standards-setting body for animal health. The standards set by the WOAH pertaining to aquatic animals are contained in the WOAH *Aquatic Animal Health Code* (WOAH, 2024) and the WOAH *Manual of Diagnostic Tests for Aquatic Animals* (WOAH, 2022). From a veterinary standpoint, the SPS Agreement and the WOAH *Code* and *Manual* are the most relevant of global agreements, but there are several other global agreements that impinge on biosecurity, such as the Convention on Biological Diversity (CBD), and discussion

on this point can be found in Chapter 2 of the Food and Agricultural Organization of the United Nations (FAO) publication *Development of an Analytical Tool to Assess Biosecurity Legislation* (Manzella and Vapnek, 2007, pp. 17–35).

2. Then there are groupings of countries for mainly trade purposes, such as the European Union (EU) and the Regulation (EU) 2016/429 of the European Parliament and of the Council of March 9, 2016, on transmissible animal diseases and amending and repealing certain acts in the area of animal health ('Animal Health Law'). Again there is further legislation which impinges on biosecurity, such as the Regulation (EC) No 852/2004 of the European Parliament and of the Council of April 29, 2004, on the hygiene of foodstuffs, Regulation (EC) No 853/2004 of the European Parliament and of the Council of April 29, 2004, laying down specific hygiene rules for food of animal origin, and Regulation (EU) 2017/625 of the European Parliament and of the Council of March 15, 2017, on official controls and other official activities performed to ensure the application of food and feed law, rules on animal health and welfare, plant health, and plant protection products. An EU Directive requires Member States to achieve a particular result without dictating the means, while an EU Regulation dictates what must be done and does not require any implementing measures by the Member State.

3. At the national level, compliance with international legislation is carried out by the government through the competent authority, which may be the government veterinary services or a separate aquatic animal health service. There will be national legislation controlling the movement of aquatic animals, and diseases additional to those of international concern may be controlled for import purposes when a country has been shown to be free of a particular disease, for example the additional guarantees required by the UK for *Gyrodactylus salaris*, which is not listed by either the WOAH or the EU. Government activities will be focused on listed, notifiable, or reportable diseases and will include surveillance and monitoring

both internally for disease and also for potential introduction at ports of entry, Import Risk Analysis (IRA) are carried out where concerns have been raised over potential routes of introduction of a pathogen and in order to determine an Acceptable Level of Protection (ALOP), and for disseminating information to stakeholders.

4. Within a country, there may or may not be a regional layer of biosecurity which may be at the level of state, province, county, or geographical region. This will be classed as a zone, which is clearly geographically defined (i.e. an offshore island or a watershed which acts as an effective barrier to disease incursion), or a compartment, which will possibly be geographically defined, but is aimed at a business or group of businesses with a higher health status than the surrounding area that can show separation or isolation from the adjacent area.

5. At the level of the individual business or unit, although the above layers will all impact on the health management of the unit.

6. A single unit can and should be broken down into single systems within the unit. This is particularly important should the unit include a broodstock facility, hatchery, or different year classes (stages of production). It is also of importance in the ornamental trade, where wholesalers and retailers will probably purchase from several sources but those sources should be kept in separation.

7. The individual pond or tank, which is the smallest practical level at which biosecurity can be practiced.

Each layer is not a solid barrier to disease incursion and should be considered porous. It is extremely unlikely that, with appropriate biosecurity in place at each layer, the incursion will flow up or down through seven layers. At most, contingency planning or containment will prevent the incursion breaching two layers, although these may not be contiguous layers. Incursions occur through illegal trading, ignorance of the disease risks (both public and industry), or either failure to act in enforcing biosecurity or reacting to an event too slowly.

Each EpiUnit and layer interacts with each other. An EpiUnit can act individually, but to ensure there is no duplication of effort and to put the necessary controls in place, there is a need to be aware of what is happening in the unit's surroundings. A simple example would be an individual unit implementing a biosecurity plan while being ignorant of the diseases found within the region or even the next door unit. There is a critical need for good communication and intelligence gathering.

Although biosecurity principles apply across all types of EpiUnits, there is not one single biosecurity plan that is applicable to all. Each plan must be tailored to the specific EpiUnit and, indeed, to the specific diseases of concern to the unit, and at the individual unit level or below, to the acceptable level of risk (ALOR) that the unit is prepared to take.

Biosecurity at the international or national level mainly consists of movement controls, monitoring of endemic diseases, and surveillance for emerging diseases. Below the national level, the EpiUnit can take a more holistic approach. Biosecurity is not just the washing of hands or the placement of disinfectant mats, but is an integral part of any unit's animal health plan and includes all areas of the business activities. Where the unit is a new build, biosecurity should be considered in the initial planning stages. Biosecurity starts with considering building materials, water sources, contact with wild animals, feed sources and storage, vaccination programs, monitoring and surveillance programs, and livestock purchasing and movements, among many other considerations. Above all, biosecurity is about identifying the risk factors and putting in place controls that prevent, contain, or eradicate disease incursions.

5.3.2 Biosecurity—what does it accomplish?

As biosecurity contributes greatly to disease prevention and control, it should be part of routine farming practice. (HMSO, 2002, p. 148)

There is no argument that biosecurity is essential to controlling endemic disease in aquatic animals and preventing the emergence of new diseases or re-emergence of old diseases. Where the effectiveness of biosecurity has been questioned, it is likely to have been due to a failure in understanding the concepts involved,

inadequate enactment, or not including all stakeholders in the decision processes and reaching a consensus. Properly planned biosecurity can accomplish all of the above benefits—improved animal health through disease prevention and control, increased welfare through better management of stock, increased production through less energy expended on eliminating pathogens, and better viability and profitability of the enterprise through access to larger markets and increased ability to compete globally.

Aquatic animals are permanently surrounded by a medium that readily supports the growth of microorganisms, many of which are potential pathogens. For some known pathogens, the routes of transmission have yet to be clearly worked out and there is considerable debate as to what species are susceptible to these pathogens. Thus, known and unknown risks are an ever-present issue. Biosecurity itself is a risk, as one can never be certain that the costs are outweighed by the benefits. However, biosecurity provides risk mitigation at the very least and can remove the risk totally. It is easier for an EpiUnit to operate with a known cost (the cost of biosecurity) on an annual basis than suffer a devastating loss of stock or production on an unknown magnitude. It is the assurance that biosecurity provides to the EpiUnit and its trading partners that is the most important aspect that biosecurity can accomplish.

5.4 Economic Assessment

Economics is not concerned principally with money but with making rational choices/decisions in the allocation of scarce resources for the achievement of competing goals. Monetary units are simply used as a yardstick to compare the different resources and goals involved in the decision. (Rushton, 2009)

5.4.1 Introduction

There are four reasons why an EpiUnit might adopt and develop a biosecurity plan. These are: public good, improvement in animal welfare, increased productivity, and increased profitability.

A *public good* could be where a farm or EpiUnit decides to be the first in an area to undertake a disease control strategy, such as vaccination. Whilst the EpiUnit gains a benefit, there is also a benefit to surrounding farms due to the reduction in the prevalence of disease in the area, leading to lower infection rates. This external benefit to surrounding farms is an example of an *externality* and can lead to the problem of abuse. Because of this issue, public goods are often funded through public means, such as taxes or producer levies.

Animal welfare and health are linked with improvements where one improves the other. Animal welfare criteria are often part of quality assurance schemes, and consumers are said to be willing to pay a premium for a product they perceive as being produced in a 'welfare friendly' fashion. Additionally, the EpiUnit may wish to have a reputation for taking 'the best care' or want to exhibit a certain level of pride in their work. While these *social benefits* may be hard to quantify, they are part of any economic benefit. However, in economic terms, farmers sell a *homogeneous product* (similar appearance and quality), making it difficult for them to differentiate their product, so any increased value due to better welfare may simply be due to the farmers' increased pride in the job.

Productivity is a measure of efficiency based on the variation in output as additional equal units of variable input are used. There are three types of input:

- Fixed inputs do not vary over the period of analysis (i.e. buildings or labor costs).
- Variable inputs vary with the amount of product produced and can be controlled by the EpiUnit (i.e. health care).
- Random inputs are usually treated as constant and can be ignored. An example of a random input would be the weather.

Decreasing exposure to endemic diseases improves the feed conversion ratio (FCR), weight gain, and organoleptic qualities and enhances the ability to assess genetic merit correctly. All of these will increase productivity, which in turn increases profitability. But consider what happens if all farmers take measures to increase productivity, which in turn leads to sufficient increase in production: there is a decrease in the price paid for the product. This could cause a fall in profits, although in economic terms, the benefit is transferred to the consumer, who now pays a lower price for the product.

Increased profitability would appear to be the natural outcome of any disease control strategy. Disease control strategies will produce a net economic benefit, with the only issues being who pays the cost and who gains the benefit. There are, though, additional issues that affect the uptake of biosecurity plans, and these are discussed in the next section.

5.4.2 Opportunity costs, rational choices, and marginal costs and benefits

5.4.2.1 Opportunity costs

An almost ubiquitous story that most veterinarians will recognize is that of farmers who increase the chance of oestrus detection by observing their animals during the evening. However, many farmers prefer to spend this time with their family. This is an example of an *opportunity cost*, where the decision to take one action removes the option to take another action, with consequences that can be termed as costs. Opportunity costs are not necessarily financial, they can be emotional or of convenience.

To evaluate the opportunity costs properly, individual producers need to know how much effort they must invest in disease control, where information is lacking and how much it will cost to collect that information, and, ultimately, how to design disease control strategies. Optimal economic strategies may not be identical for all businesses, as the interpretation of what provides the greatest value in terms of 'costs' will vary. In developing a biosecurity plan, consideration must be given to the needs, wishes, and aims of the EpiUnit.

5.4.2.2 Rational choices

An economist would say a rational choice has been made when the decision creates the largest benefit relative to the cost. Decisions should not be made on cost alone and should also compare the likely benefits to be derived. People making rational choices know the cost and understand the value of the decision. People making poor decisions could be said to know the cost of everything but not necessarily the value of those decisions. In encouraging an EpiUnit to make the rational choice, we are required to provide not only as accurate an estimate of cost as possible but also the value of the likely returns. Rational choice focuses on the comparison of total costs vs total benefits.

5.4.2.3 Marginal costs and benefits

A fish farm produces X_1 amount of product (output or 'benefit') for Y_1 costs (inputs) but can, with additional costs (such as additional labor, time, feed, vaccinating stock), produce X_2 for Y_2 costs. Is it worth adopting these additional procedures? This involves assessing the marginal costs (Y_2-Y_1) of the extra inputs vs the marginal benefits (X_2-X_1) they create. Even though there may be an additional benefit, it may not warrant the additional costs.

5.4.3 Economic analysis of the EpiUnit biosecurity plan

Any analysis requires *a baseline of disease prevalence* and estimates for the level of production of the EpiUnit with or without the disease. Two questions that arise regarding data are, how accurate does it need to be, and is it available? Carrying out a literature search of published work on a disease will often reveal a surprising amount of data on the economic impact, but where data is missing, it is perfectly acceptable to use expert opinion. Analysis, however, is only as good as the data used. Where there is doubt, analyzing and comparing the information generated within a couple of production cycles between productivity of animal groups with and without disease can provide a good indication of the production loss due to disease or the expected gains due to control measures. A final general consideration is whether the control measure is technically feasible and socially acceptable.

There are three methods that can be used individually or in combination that will assist the economic decision at the farm or EpiUnit level. These are:

- partial budgeting;
- decision tree analysis; and
- cost–benefit analysis.

All three methods are relatively easy to carry out using computer spreadsheets. There are some further terms to introduce, as follows.

5.4.3.1 Discounting

Is $1000 received today the same as receiving $1000 in 2 years' time? The answer is no and can be calculated by the following formula:

$$PV = \frac{FS}{(1+r)^n}$$

Where:
  PV = Present Value
  FS = Future Sum
  r = periodic interest (discount) rate
  n = number of periods

i.e. $1000 received 2 years hence discounted at 7% $PV = \frac{1000}{(1+0.07)^2} = \frac{1000}{1.145} = \$873.36$

5.4.3.2 Net Present Value (NPV)

For a cost–benefit analysis the NPV is used, which is the sum for each time period (usually 1 year) of the project, being the total benefits received during the period minus the total costs, discounted by the appropriate discount factor to convert each total to the present value:

$$PV = \sum_{(t=0)}^{t} \frac{(B^t - C^t)}{(1-r)^2} t$$

where
  B^t = monetary benefits received in any year t
  C^t = costs incurred in any year t
  r = discount rate

From this, the higher the discount rate used, the lower the present value of future cash flows. The discount rate to use is often debated, but the real interest rate (the cost of borrowing money minus the inflation rate) is a common starting point.

5.4.3.3   Internal rate of return (IRR)

The IRR is calculated by finding the discount rate that would provide a net present value (NPV) of zero and can be used to compare alternative uses of the funds. To manually calculate the IRR requires a trial-and-error approach. The simplest way to calculate both the IRR and the NPV is by using a computer spreadsheet (see Fig. 5.10).

5.4.3.4   Economic assessment methods

5.4.3.4.1 METHOD 1: PARTIAL BUDGETING.   Partial budgeting is only concerned with the variable cost (e.g. disease control) and the revenue affected by the proposed intervention. Partial budgeting uses four headings:

- Additional income (e.g. increased number of fish produced);
- Reduced expenses (e.g. savings on treatment costs);
- Reduced income (e.g. possible decreased market value if production increased sufficiently); and
- Additional expenses (e.g. staff training, equipment).

The advantage of partial budgeting is that it is quicker to carry out than a complete budget and focuses on the issues of interest. There are several disadvantages, which really mean it is only useful for initial appraisal. Should a project pass this first hurdle, then one of the following techniques should be used.

5.4.3.4.2 METHOD 2: DECISION TREE ANALYSIS.   Decision tree analysis allows you to incorporate the uncertainties inherent in any animal health decision and the economic consequences. The decision tree is a pictorial representation of the logical flow of events and can be a good way of communicating veterinary opinion to the farmer.

The initial step in building a decision tree is to identify all the possible courses of action that might be used to address the problem. The first node of a decision tree is always a decision node, which is represented by a rectangular box. There is one separate branch for each possible decision, with each branch leading to either a chance node (circle) or a terminal node (triangle) (Fig. 5.11). The chance nodes describe the probability of each possible outcome, and the probabilities at each chance node must sum to one (or 100%). The terminal nodes provide information on the value at the end of the branch.

The preferred course of action is determined by a process called 'folding back.' Here, the monetary value of each terminal node is multiplied by the probability given at the preceding chance node. The products of the branches originating from the same node are summed up to provide the expected value (EV) of the node. The folding back process continues until a single EV is obtained for each branch originating from the decision node at the source of the decision tree.

Decision tree analysis makes you look at all the available options, including timing choices (e.g. when to cull), and also incorporates the uncertainties involved in making the decision. Probabilities are not only used to state the likelihood that an animal/group of animals will live or die but can also be used to explore whether an EpiUnit will experience a clinical outbreak of disease.

5.4.3.4.3 METHOD 3: COST–BENEFIT ANALYSIS.   Cost–benefit analysis (CBA) is used to compare different disease control strategies over a period of years, say over 5–20 years. It could simply compare no control with a proposed control program at the EpiUnit over a period of 5 years. As you extend the analysis further into the future, the less accurate it will become. CBAs do not mean you know all the figures, and they frequently involve some form of subjective judgments. If the control program requires finance, then it is adequate to project over 5 years.

A CBA can be summarized as four steps specifying:

- the flow of costs—what is the cost of the project for each time period? You may need to quantify non-monetary costs into a monetary value;
- the flow of returns—what is the monetary value of the benefits?;
- deciding on appropriate discount rate—commonly the real interest rate; and
- the decision criterion—usually the NPV, benefit–cost ratio (BCR), IRR, and payback period (PBP).

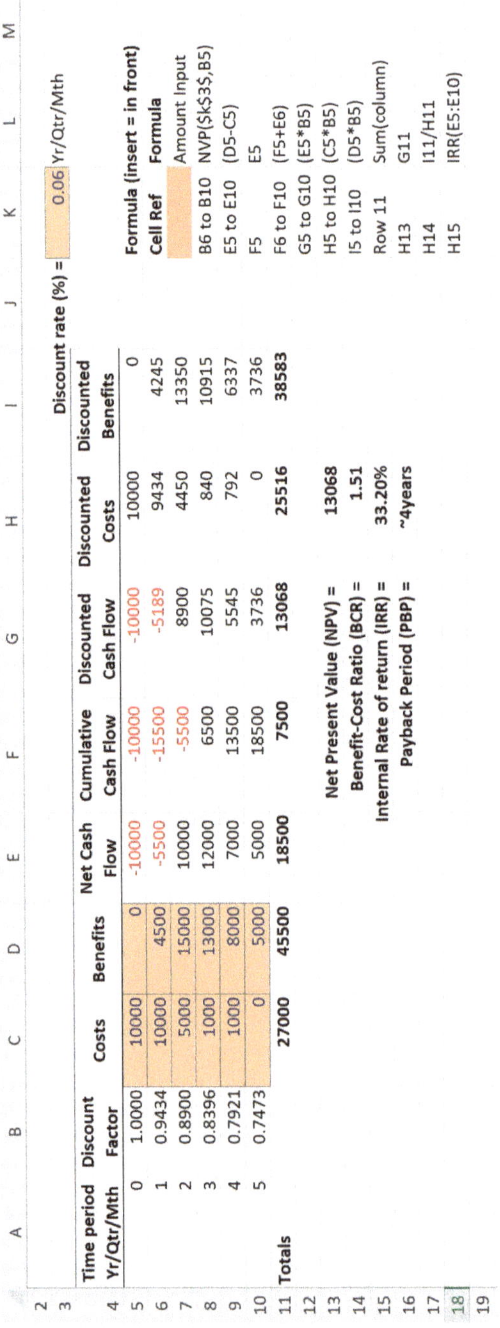

Discount rate (%) =   0.06 Yr/Qtr/Mth

Time period Yr/Qtr/Mth	Discount Factor	Costs	Benefits	Net Cash Flow	Cumulative Cash Flow	Discounted Cash Flow	Discounted Costs	Discounted Benefits
0	1.0000	10000	0	-10000	-10000	-10000	10000	0
1	0.9434	10000	4500	-5500	-15500	-5189	9434	4245
2	0.8900	5000	15000	10000	-5500	8900	4450	13350
3	0.8396	1000	13000	12000	6500	10075	840	10915
4	0.7921	1000	8000	7000	13500	5545	792	6337
5	0.7473	0	5000	5000	18500	3736	0	3736
Totals		27000	45500	18500	7500	13068	25516	38583

Net Present Value (NPV) =   13068
Benefit-Cost Ratio (BCR) =   1.51
Internal Rate of return (IRR) =   33.20%
Payback Period (PBP) =   ~4years

Formula (insert = in front)

Cell Ref	Formula
	Amount Input
B6 to B10	NVP(K3$,B5)
E5 to E10	(D5-C5)
F5	E5
F6 to F10	(F5+E6)
G5 to G10	(E5*B5)
H5 to H10	(C5*B5)
I5 to I10	Sum(column)
Row 11	Sum(column)
H13	G11
H14	I11/H11
H15	IRR(E5:E10)

Fig. 5.10. How to calculate the net present value (NPV), benefit-cost ratio (BCR), and internal rate of return (IRR) of a project. Figure author's own.

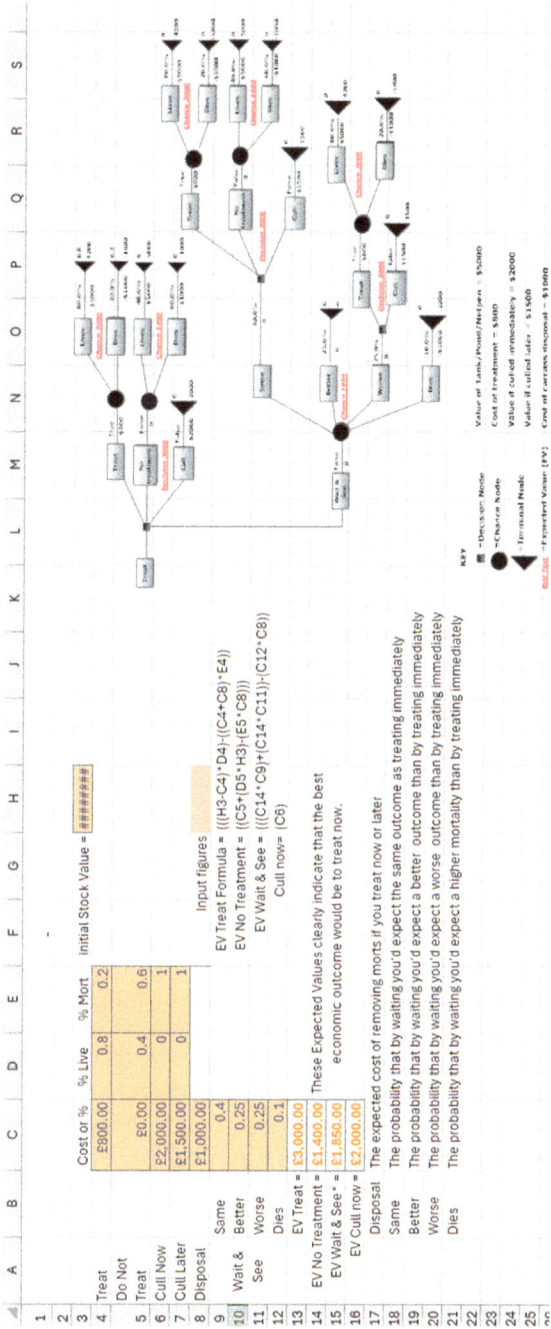

Fig. 5.11. Example of a decision tree with expected outcome values. Figure author's own.

Decisions are based on the following:

- If NPV > zero, then the return is greater than the opportunity cost (discount rate).
- If BCR > 1, then NPV must be > 0.
- IRR (NPV = 0): this allows comparison with alternative uses of the funds. It may vary over the lifetime of the project and may not exist if net cash flow is positive for each year of the project as NPV will never be zero.
- PBP = (costs – benefits = 0): this is calculated ignoring the time value of money and ignores any future returns.
- Whether sufficient capital is available for the project.

References

Donaldson, A. (2008) Biosecurity after the event: Risk politics and animal disease. *Environment and Planning A* 40, 1552–1567.

EFSA (2008) Scientific opinion of the Panel on Animal Health and Welfare on a request from the European Commission on Animal welfare aspects of husbandry systems for farmed Atlantic salmon. *The EFSA Journal* 736. Available at: https://efsa.onlinelibrary.wiley.com/doi/epdf/10.2903/j.efsa.2008.736 (accessed 14 June 2025).

FAO Biosecurity Toolkit (2007). FAO, Rome. Available at: www.fao.org/4/a1140e/a1140e00.htm (accessed 14 June 2025).

HMSO (2002) *Foot and Mouth Disease 2001: Lessons to be Learned Inquiry Report*. Doc No HC 888. London: HMSO. Available at: http://webarchive.nationalarchives.gov.uk/20100807034701/http://archive.cabinetoffice.gov.uk/fmd/fmd_report/report/index.htm (accessed 14 June 2025).

Rushton, J. (2009) *The Economics of Animal Health and Production*. CAB International, Wallingford, UK.

Manzella, D. and Vapnek, J. (2007) Development of an analytical tool to assess *Biosecurity* legislation. FAO Legislative Study 96. Available at: www.fao.org/docrep/010/a1453e/a1453e00.htm (accessed 14 June 2025).

Scottish Government (2022) Biosecurity practices for animal health: Guidance. Available at: https://www.gov.scot/publications/biosecurity-practices-for-animal-health-guidance/ (accessed 14 June 2025).

WHO (2024) Causal web analysis: A model approach to joint programme planning. Available at: https://apps.who.int/iris/bitstream/handle/10665/204848/B0242.pdf?sequence=1&isAllowed=y (accessed 14 June 2025).

World Organisation for Animal Health (WOAH) (2024). Aquatic Animal Health Code. Available at: https://rr-europe.woah.org/en/news/new-aquatic-code-available-for-download-in-pdf/ (accessed 29 July 2025).

World Organisation for Animal Health (WOAH) (2022). Aquatic Animal Health Manual. Available at: https://rr-europe.woah.org/app/uploads/2024/08/en_csaa-2022.pdf (accessed 29 July 2025)

Agreement on the Application of Sanitary and Phytosanitary Measures, 15th April 1994 (1867 UNTS 493, WTO Doc LT/UR/A-1A/12), OXIO 269 https://www.wto.org/english/res_e/publications_e/ai17_e/sps_art3_jur.pdf (accessed 29.07.2025)

Further Reading

FAO/NACA/UNEP/WB/WWF (2006) *International Principles for Responsible Shrimp Farming*. Network of Aquaculture Centres in Asia-Pacific (NACA), Bangkok. Available at: https://library.enaca.org/Shrimp/Publications/International_Principles_for_responsible_shrimp_farming_Draft_25Jan2006.pdf (accessed 14 June 2025).

Rushton, J. (2009) *The Economics of Animal Health and Production*. CAB International, Wallingford, UK.

Scarfe, A.D., Lee, C.S. and O'Bryen, P.J. (eds) (2006) *Aquaculture Biosecurity: Prevention, Control, and Eradication of Aquatic Animal Diseases*. Blackwell, Oxford, UK.

6 Public Health

Abstract

The chapter defines Public Health and describes the concept in an interdisciplinary manner, with a focus on the disease threats and covering, among other topics, antimicrobial resistance, aquaculture biosecurity, EU legislation on food safety and antimicrobial use, and the SPS Agreement of the World Trade Organization.

6.1 Introduction

Public health is concerned with the protection and improvement of the health of populations by early detection, prevention, and response to infectious diseases, and by promoting healthy lifestyles.

Public health has two main types of functions: core and support functions.[1]

Core functions are based on three pillars: *prevention* of illnesses, injuries, or death, *protection* of vulnerable groups, and *promotion* of laws, guidelines, and regulations aimed at improving health and safety. Core functions involve detection, confirmation, analysis, and response actions. Support functions refer to training, supervision, communications, and resource management.[2]

PART I: ANTIMICROBIAL RESISTANCE (AMR): CHALLENGES AND IMPLICATIONS FOR FOOD-PRODUCING SYSTEMS

Pablo Jarrin-V.[1], C. Alfonso Molina[2] and Gabriela N. Tenea[3]*

[1]*Dirección de Innovación, Instituto Nacional de Biodiversidad, Ecuador;* [2]*Instituto de Investigación en Zoonosis (CIZ), Universidad Central del Ecuador, Ecuador;* [3]*Biofood and Nutraceutics Research and Development Group; Faculty of Engineering in Agricultural and Environmental Sciences, Universidad Técnica del Norte, Ecuador*

6.2 Concepts and Transfer Mechanisms

Antimicrobial resistance (AMR) has been described as one of the most critical worldwide public health threats of this era (Prestinaci *et al.*, 2015). This resistance jeopardizes human and animal health, and impacts the environment, demanding a comprehensive One Health strategy to address this complex problem.

The discovery of antibiotics, while a triumph of biotechnology, has been accompanied by the alarming emergence of antibiotic-resistant bacteria. Resistant strains and resistance determinants are thought to be well-distributed in aquatic environments, with profiles being largely shaped by anthropogenically derived contamination of the environment (Grilo *et al.*, 2020). Notably, insufficiently treated wastewater draining into natural waterways was suggested

Corresponding author (part 1): gntenea@utn.edu.ec; Corresponding authors (part 2): urdeslaura@gmail.com and chriswalstervet@outlook.com

© CAB International 2025. *One Health Concepts and the Aquatic Ecosystem* (eds L.D. Urdes *et al.*)
DOI: 10.1079/9781800623248.0006

as one of several of the major anthropogenically derived sources of AMR bacteria in domestic and wildlife ecosystems (Guenther *et al.*, 2011; Dolejska and Papagiannitsis, 2018; Dolejska and Literak, 2019).

The rise of AMR, fueled by factors such as the misuse and overuse of antimicrobials, presents a formidable challenge to the efficacy of treatments, making infections more difficult to treat and increasing the risk of severe illness and even death (Verraes *et al.*, 2013). From a public health perspective, AMR is understood as the change in microorganisms that is evident in a lack of response to medicines, making infections difficult to treat, with the consequential increased risks of contagion, illness, and death (Tiseo *et al.*, 2020). Human, animal, and environmental health are interdependent, and the well-being of each is essential for addressing complex global health challenges. AMR is a phenomenon detected across humans, animals, and ecosystems, the three pillars of the One Health approach. Under the One Health framework, AMR is recognized as a critical global challenge with profound impacts on human health, animal welfare, and environmental sustainability. One Health emphasizes the interconnectedness of these three components and the need for interdisciplinary approaches to address AMR. Antimicrobial use in food animals and crops poses risks to human health due to the potential spread of resistance from food production to humans, highlighting the interconnectedness of AMR within the One Health paradigm. Antimicrobial-resistant zoonotic pathogens present on food constitute a direct risk to public health (Founou *et al.*, 2021). AMR occurs when microorganisms, such as bacteria, develop the ability to survive exposure to antimicrobials, like antibiotics, that would normally kill them or inhibit their growth. This resistance can arise through various mechanisms, including genetic mutations and the acquisition of resistance genes from other microorganisms. Understanding the transfer mechanisms of AMR is crucial for comprehending how AMR emerges and disseminates within food-producing environments. The implications of AMR in food-producing systems are far-reaching, including reduced efficacy of antimicrobial treatments in livestock, increased risk of foodborne illnesses

in humans, and substantial economic losses (Founou *et al.*, 2021). These implications extend beyond food production to endanger public health and global economies.

The World Health Organization (WHO) identifies AMR as a global issue driven by poor hygiene and improper use and overuse of antibiotics. The primary challenges posed by AMR include rising morbidity and mortality rates, increased healthcare costs, and prolonged infectious diseases. Since AMR stems from genetic mutations, resistant bacterial strains can be found worldwide, originating from diverse sources such as humans, animals, plants, water, and soil (Aghamohammad and Rohani, 2023). Recent studies suggest that antibiotic resistance remains an ongoing issue that has extended for decades. With the onset of the antibiotic industry for treating infectious diseases came an improvement in human quality of life and life expectancy, the betterment in health and welfare of animals, and challenges to the responsible use of antibiotic technologies (Nwobodo *et al.*, 2022; Ding *et al.*, 2023). As part of an unfortunate feedback loop, improvement in quality of life and steep population growth led to an increasing demand for animal protein and crops, which incentivized competitive and intensive livestock production through antibiotic overuse (Van Boeckel *et al.*, 2015, 2019). This chapter aims to explore the challenges posed by AMR in food production and discuss the implications for food safety, public health, and the sustainability of food-producing systems.

6.2.1 Food as an antibiotic pathway for AMR transmission and resistance genes

As highlighted by organizations such as the FAO (FAO, 2021) and recent publications (Adamie *et al.*, 2024), the complexity of AMR transcends national frontiers and requires integrated action. Food-producing systems serve as critical points of convergence. Multiple studies show that antibiotic-resistant microbes and AMR genes in humans are also present in animals that have had no direct contact with humans. This indicates that AMR may be transmitted to humans through the consumption of contaminated food and inadequate food handling practices (Samtiya *et al.*, 2022).

The divide between the meat food industry and the plant food industry dissolves when antibiotics are considered, as both lines of food production are plagued by the overuse or misuse of antibiotics and become significant sources for the reinforcement of antibiotic resistance in microorganisms. Both animal-derived and plant-based foods, as well as processed products, could act as reservoirs and vectors for antimicrobial-resistant bacteria (ARB) and genes (ARGs) (Sweileh, 2021). AMR is a crucial issue in human health, which is why it is also a significant research topic; however, there remains a worrying knowledge gap regarding its prevalence and mechanisms in food-producing animals and plants (Jacobsen et al., 2023). This gap is concerning, given the vital role of agriculture in food security and public health. Understanding the contributions of these sectors to the global AMR burden is imperative for developing targeted mitigation strategies (Agga and Amenu, 2024).

Antibiotics seem to be everywhere. They permeate the industrialized food and livestock chains, through industrial waste they contaminate soil and water, and through food they pollute the human microbiome, contributing further to bacterial environments rich in selective pressures for developing resistance. Humans have contributed to force microorganisms into mechanisms of survival that have become an economic burden, a global health risk, and resulted in the rise of the superbugs.

When measuring the characteristics of the AMR problem and its related costs, we should consider antibiotic production and consumption, resistance rates, resistance-associated deaths, and resistance-associated healthcare costs, among other parameters whose values are often so shockingly large that they are difficult to envision. The economic consequences of AMR are also substantial, with projected significant losses in livestock production and global gross domestic product (GDP). These gargantuan numbers reflect the magnitude of the global food industry and the severity of the problem we have created with our dependency on antibiotic technology. As a global health issue, antibiotic consumption and antimicrobial-resistance-associated deaths echo global inequalities in access to proper health services, economic disparities among countries and regions, and levels of political and societal organization. Although yet unclear, as more research is necessary, a regrettable pattern may be visible, where rich countries consume larger amounts of antibiotics, and the poor ones suffer more antimicrobial-resistance-related deaths and disease.

AMR can enter the human system through the consumption of contaminated food, contact with farm animals and produce-resistant bacteria, and environmental exposure at farms and production facilities (Choy et al., 2024). Animal waste and feces, a by-product of the livestock industry, can contain resistant bacteria, including 'superbugs' resistant to multiple antibiotics (multi-drug resistance, MDR). These bacteria can contaminate the environment, including soil and water used to grow crops. Fruits and vegetables can become contaminated through contact with this contaminated soil or water. Resistant bacteria can spread from one type of food to another or from the environment to food during processing and handling. This can occur through the use of contaminated equipment or contact with contaminated surfaces (Fig. 6.1).

Food is often processed through beneficial microorganisms that can be hosts to resistance genes; starter cultures, probiotics, and biopreserving microorganisms can potentially transfer resistance after the ingestion of food. Once in the human digestive system, genetic mechanisms via conjugation are frequent and can transfer AMR to pathogens, although other mechanisms of gene transfer, such as transformation or transduction, are also possible (Rolain, 2013; Verraes et al., 2013; Samtiya et al., 2022).

6.2.2   The livestock Issue

Food serves as a significant pathway for transmitting resistant bacteria and resistance genes, impacting human, animal, and environmental health, thus highlighting the need for a One Health approach to address this challenge. Food can become contaminated with AMR bacteria and resistance genes anywhere along the food chain, from primary production to consumption. AMR genes may also contaminate food; for example, a previous study showed that

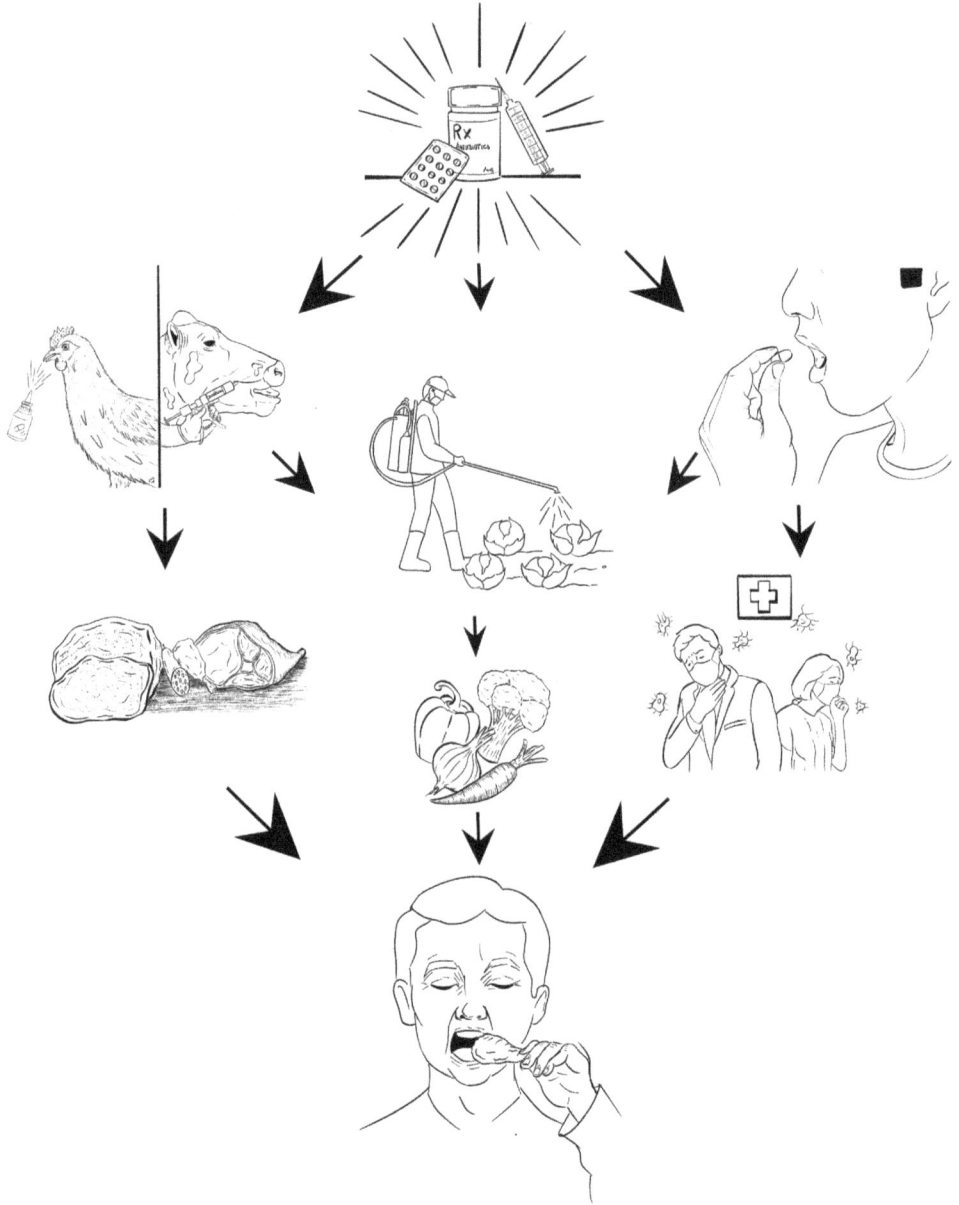

Fig. 6.1. Basic pathways of AMR transmission. Illustration by Pablo Mayorga, 2025. Figure used with permission from Pablo Mayorga.

plasmid-borne ampicillin-resistant genes transfer into *Escherichia coli* K12 from *Salmonella typhimurium* DT104 in ground meat and inoculated milk (Samtiya *et al.*, 2022).

Contamination in livestock production can occur through various pathways, including:

- Contamination during animal slaughter and processing: When animals are slaughtered and processed for food, AMR bacteria can contaminate meat or other animal products. This can occur through contact with contaminated processing equipment, storage bins, or the animal's feces.

- Contamination by infected food handlers: Food can be contaminated with AMR bacteria by infected food handlers who do not observe good hygienic practices.
- Cross-contamination during food processing: Improper food processing or unhygienic food preparation environments can also lead to contamination. Food processes that kill bacteria in food products decrease the risk of transmission of AMR.

Food-producing systems, particularly animal farming, often employ subtherapeutic doses of antimicrobials for growth promotion and disease prevention (Almansour et al., 2023). This practice, while economically beneficial, exerts selection pressure on microbial populations, fostering the emergence and dissemination of resistant strains. As reported by Agga and Amenu (2024), such environments exhibit low but persistent antibiotic selection pressure, which facilitates the horizontal transfer of ARGs among bacteria.

AMR is prevalent in various types of livestock, including cattle, pigs, and poultry (AbuOun et al., 2021). For cattle, the prevalence of AMR is generally higher in those raised in intensive production systems, where animals are often kept in crowded conditions and antibiotics are used routinely for disease prevention and growth promotion (Ager et al., 2023). Pigs have been identified as a significant reservoir of AMR genes, with a high proportion of isolates harboring resistance genes (AbuOun et al., 2021). The prevalence of AMR in poultry is also of concern, with studies showing high rates of resistance to antibiotics such as tetracycline and ciprofloxacin (Vezeau and Kahn, 2024). Wildlife populations can also act as reservoirs and spreaders of AMR, with potential for transmission to domestic animals and humans (Vezeau and Kahn, 2024).

Animal manure, a by-product of farming, is a significant medium for ARG dissemination. Often stored for use as fertilizer, manure can contaminate water sources and promote the spread of AMR in agricultural ecosystems (Jacobsen et al., 2023). Evidence also suggests that vectors, such as flies (Diptera), act as mechanical carriers of ARGs, effectively bridging the gap between different hosts and environments (Caderhoussin et al., 2024).

Specific studies underscore the breadth of the problem. For example, investigations into cow-calf operations reveal a concerning prevalence of resistant bacteria. In a recent study assessing baseline AMR in fecal bacteria from pre-weaned calves, samples of E. coli were found to show resistance to tetracycline, third-generation cephalosporin, and extended-spectrum beta-lactamase (ESBL). Factors such as grazing type and age were found to influence resistance profiles. Wheat grazing was associated with a higher prevalence of resistance genes, such as tet(A)xz (Agga et al., 2022).

In South America, studies have documented the prevalence of AMR in pathogens isolated from animal-derived products. In Brazil, it has been reported that 38% of Salmonella spp. isolates from swine and poultry products exhibited resistance to at least one of eight tested antibiotics, with 55% of these strains demonstrating multi-drug resistance (Maciel et al., 2019). These findings emphasize the regional variation and complexity of AMR dynamics.

6.2.3   The plant issue

Crops, essential for human and animal sustenance, can inadvertently serve as conduits for AMR transmission. The crop environment, encompassing soil, water, and the plants themselves, provides a complex ecosystem where resistant microbes can thrive and spread. Plant-based foods and fresh produce, though often overlooked, are also significant reservoirs of ARGs and can harbor resistant bacteria due to contamination during irrigation or handling. For example, streptomycin and oxytetracycline are commonly used to control fire blight in apple and pear orchards, potentially leaving residues that contaminate crops and leach into the surrounding soil and water (Brunn et al., 2022).

A recent study identified 91 ARB isolates from fresh produce, predominantly cephalosporin-resistant Enterobacterales and carbapenem-resistant Pseudomonas aeruginosa. Notably, 95% of the samples contained ARGs such as *sul*1, *bla*TEM, and *erm*B, highlighting a strong correlation between these genes and AMR indicators (Sweileh, 2021). Antimicrobial-resistant bacteria can be transferred from compost to crops; thus, swine compost increased ARG abundance in soil and rice roots, while rice grains showed no significant ARG differences across treatments, highlighting the role of irrigation water in the transmission of AMR (Xu *et al.*, 2024b). Moreover, processed foods are not immune to AMR contamination. Cross-contamination during production and inadequate cooking practices can intensify the problem, making the entire food supply chain a critical area for intervention.

After the success of antibiotic use in bacterial infections in humans and other domestic animals, crop growers have used antibiotics extensively on plants and their products (Batuman *et al.*, 2024). Streptomycin was the first antibiotic used in plant agriculture, to control *Erwinia amylovora*, a Gammaproteobacteria that causes disease in pome fruit and resulted in serious economic loss in the USA. The pome fruit industry flourished again due to the efficacy of streptomycin against *E. amylovora*. From the middle of the 20th century, crop growers began using streptomycin to control fire blight and other bacterial diseases in commercial orchards of high-value crops (Batuman *et al.*, 2024). A few years later, in 1962, the first plant-pathogenic bacteria resistant to streptomycin was reported (Stall and Thayer, 1962; Thayer and Stall, 1962). To control and treat streptomycin-resistant *E. amylovora* strains, growers used oxytetracycline (OTC) on pear and peach, registered as a pesticide for the first time in 1974 (Batuman *et al.*, 2024).

Globally, antibiotics are utilized in diverse crop systems, including apples, pears, and citrus, revealing their role in agricultural disease management. Despite representing less than 0.5% of total antibiotic use in the USA, agrarian applications warrant continued vigilance to mitigate potential environmental residues and resistance risks, ensuring sustainable disease management strategies (Batuman *et al.*, 2024). The use of streptomycin in crop production, particularly as a treatment against bacterial diseases, highlights both its potential and the concerns associated with

its application. While methods such as foliar spraying and injection have been effective, particularly in managing Huanglongbing (HLB, also known as citrus greening) caused by *Candidatus Liberibacter asiaticus*, they raise significant issues regarding environmental impact and resistance development. Only three antibiotics—streptomycin, oxytetracycline, and kasugamycin—are registered for agricultural use in the USA (Batuman *et al.*, 2024). This limited antimicrobial arsenal underscores the importance of optimizing application methods and rigorously monitoring resistance in bacterial populations.

6.3   Darwin's Nightmare: The Evolution of AMR

One fundamental property of life is the capacity to evolve adaptations when facing challenging and changing environments, through biological mechanisms such as natural selection, genetic drift, mutation, gene flow, and non-random mating. AMR arises when microorganisms—such as bacteria, viruses, fungi, or multicellular parasites—develop the ability to avoid the expected effects of antimicrobial agents intended to kill or inhibit them, including antibiotics, antivirals, antifungals, and antiparasitic drugs (Aghamohammad and Rohani, 2023). As a result, these treatments become less effective, making infections more difficult to treat and increasing the risk of severe illness, extended recovery periods, and even death. AMR is a product of evolutionary processes that include genetic changes in microorganisms and several biological mechanisms, some of which are relevant to the appearance of AMR:

- Mutation: Random changes in microbial deoxyribonucleic acid (DNA). Although mutations are often deleterious, some could eventually enhance survival in the presence of antimicrobials.
- Horizontal gene transfer: The transfer of resistance genes between microbial species or strains via plasmids, transposons, or other mobile genetic elements.

- Natural selection: The use and misuse of antimicrobial drugs in human health, animal husbandry, and agriculture create selective pressure, encouraging the survival and proliferation of resistant strains.
- Genetic drift: Random fluctuations in the frequency of alleles within a population and through reproductive generations could eventually contribute to an increasing frequency of genes conducive to antibiotic resistance, and the potential evolution of resistance after selection pressures appear.

When thinking of evolutionary processes, ages are often considered. However, microorganisms, particularly bacteria, can acquire new characteristics surprisingly fast. The use of mobile genetic elements, which can dynamically occupy regions of the genome according to the specific needs of gene expression and jump from one cell into another, together with integrons, specialized genetic platforms to acquire and mobilize genes, make bacteria especially suited for rapid evolution and almost instant adaptation to antibiotic fluctuating environments (Johnsen et al., 2021; Souque et al., 2021). Experiments, such as the evolution of adaptive niche construction, have shown that bacteria can evolve environment-shaping strategies after only 30 days of selective pressure, becoming a ubiquitous characteristic after 120 days; this corresponds to between 100 and 400 generations (Callahan et al., 2014). Similar rates of evolution have been proposed for mutator-strain dynamics (Denamur and Matic, 2006) and gene transfer swap experiments (Shi et al., 2021). Radical experiments of ultra-fast evolution, through centrifugal microfluidics and a highly condensed bacterial matrix, have shown that E. coli can develop AMR in the first 10 hours of exposure to antibiotics and, after 48 hours of experimental treatment, evolve a 64–128-fold reduction in sensitivity to ciprofloxacin, amikacin, and disinfectants such as triclosan (Xu et al., 2024a). Thus, the biological evolution and origin of AMR is a real and urgent problem, fueled by the dynamism and adaptability of microbial genomes, and the widespread use of antibiotics in the health and food industries that extends for almost a century since the release of the first antibiotic in 1935 (Wainwright and Kristiansen, 2011).

6.4 Quantifying the Threat: AMR in Numbers

Antibiotics play a crucial role in combating bacterial infections by either eliminating bacteria or hindering their growth. This, in turn, contributes to improved animal health, growth promotion, and enhanced feed efficiency (Jia et al., 2023). Global antibiotic consumption, however, has seen a marked increase, rising by 46% between 2000 and 2018. This translates to a shift from 9.8 defined daily doses (DDD) per 1000 people per day in 2000 to 14.3 DDD in 2018 (Browne et al., 2021). Consumption patterns vary geographically, with the lowest levels observed in sub-Saharan Africa and the highest in Eastern Europe and Central Asia (Browne et al., 2021), hinting at economic and cultural factors as determinants of antibiotic use. This disparity highlights the need for further investigation into the economic, cultural, and political factors driving antibiotic use in different regions.

According to the estimates for AMR rates in world regions (Adamie et al., 2024), South Asia consistently leads with the highest rate overall, with Latin America and the Caribbean having the lowest rate (Fig. 6.2); regarding livestock production, cattle is the production sector with the lowest AMR when compared to chicken and swine (Fig. 6.2). A thorough assessment of the impact of AMR by Murray et al. (2022), across 204 countries and 471 million individual records, from a diverse set of sources, including hospital and laboratory data, scientific datasets, and surveillance networks, estimated an effect for the year 2019 of 4.95 million (3.62–6.57) deaths related to bacterial AMR. The rate of AMR-related death is about four times higher in poor regions of the world, such as sub-Saharan Africa, when compared to developed regions such as Australasia (Murray et al., 2022). The European Centre for Disease Prevention and Control (ECDC) has reported an annual rate of 670,000 AMR-related infection cases in the EU (Tang et al., 2023).

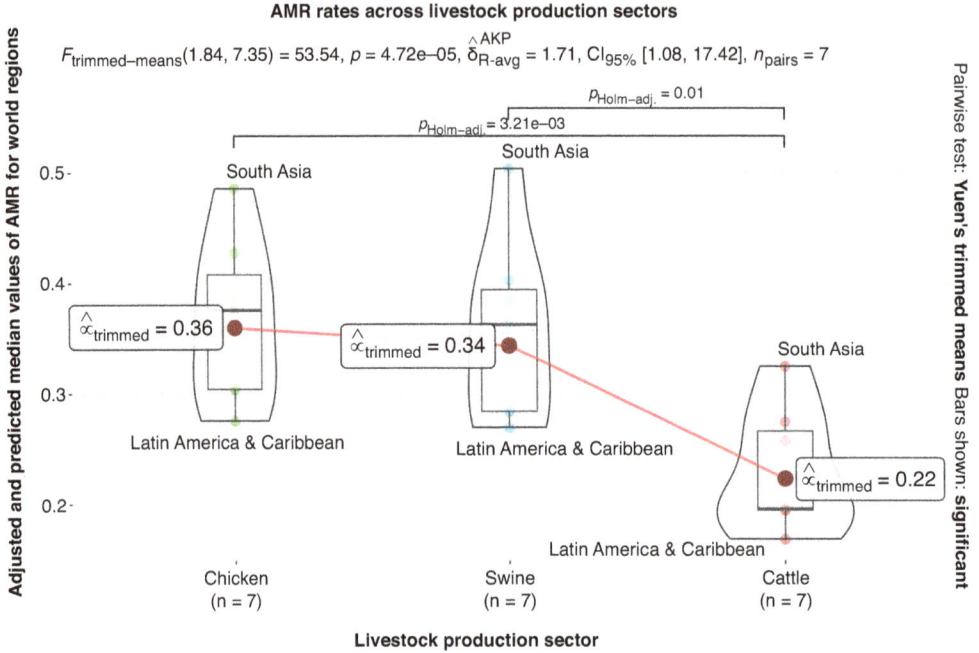

Fig. 6.2. Graphical representation of the estimates made by Adamie *et al.* (2024) on average AMR rates across livestock production sectors. Each dot represents a world region: South Asia, Middle East & North Africa, East Asia & Pacific, Sub-Saharan Africa, North America, Europe & Central Asia, and Latin America & Caribbean. No statistical differences are found in the contrast between chicken and swine on AMR rates, with cattle being significantly different from the other two species and having the lowest AMR rates. Horizontal bars indicate significant contrasts between groups (Yuen's trimmed means and Holm–Bonferroni correction). Robust ANOVA revealed a strong livestock production effect on the adjusted and predicted median values of AMR for world regions (significant p-value with effect size = 1.71). Highest (South Asia) and lowest (Latin America & Caribbean) data points are mentioned in the figure. Figure author's own.

If we think of antibiotics as environmental pollutants and toxins that interfere with the biological cycles of the microbial biosphere (beneficial bacteria and algae), their economic and social impact acquires further significance, as these substances, often extraneous to ecological networks and feeding chains, and often in unprecedented concentrations in water and soil, could disrupt ecosystem services and impact biodiversity (Cycoń *et al.*, 2019; Han *et al.*, 2022). Animal intestines have limited absorption capacity for antibiotics; between 30% and 90% of ingested antibiotics are excreted as contaminants into the water and soil of the environment (Jia *et al.*, 2023).

At least 73% of the world's antimicrobial production is destined for livestock production, with increasing proportions of antimicrobial compounds reaching resistance higher than 50% (P50) (Van Boeckel *et al.*, 2019). Countries that have not regulated the use of antimicrobials to boost livestock growth have an estimated 45% higher antimicrobial use per kilogram of animal biomass than countries that have established effective regulations and controls (Adamie *et al.*, 2024). An intensity of antibiotic use for humans of 118 mg/kg and a larger intensity for livestock of 133 mg/kg has been proposed (Ritchie and Spooner, 2024). When these estimates are multiplied by the estimated biomass of humans and animals, and translated to global antibiotic consumption, it results in 35,000 t/year for humans and 85,000 t/year in livestock (Ritchie and Spooner, 2024), a difference of nearly 243%. For the year 2013, Van Boeckel *et al.* (2017) proposed an estimated average of antimicrobial consumption for livestock at 131,109 t (95% CI: 100.812–190,492). Wide variation was detected by country, with extremes represented by Norway (8 mg/Population Correction Unit) in the lower bound and China (318 mg/PCU) in the upper bound (Van Boeckel *et al.*, 2017). In an investigation of the relationship between AMR and antimicrobial consumption across countries, Ajulo and Awosile (2024) found a statistically significant correlation between beta-lactam/cephalosporin and fluoroquinolones consumption and AMR in specific bloodstream bacterial samples. For every unit increase in DDD of these antibiotics, the recoveries of resistant *E. coli* and *Klebsiella* increased by between 11% and 40%.

With the presentation of the first global map of antibiotic consumption and resistance in livestock for 228 countries, Van Boeckel *et al.* (2015) projected a rise in consumption of 67% by the year 2030, with an expected growth of 99% in BRICS countries (Brazil, Russia, India, China, and South Africa). In a more recent study, Tiseo *et al.* (2020) estimated that the global consumption of antibiotics in livestock averaged 93,309 t (95% CI: 64,443–149,886) in 2017, with a projected increase of 11.5% by 2030. This reported overuse of antibiotic dosage for the meat and dairy industry is motivated by production practices that seek to maximize return and minimize cost, such as antibiotic overdosing as compensation for lack of adequate livestock living environments and hygienic conditions, with the consequent higher risk of AMR evolution and AMR-associated disease (Van Boeckel *et al.*, 2017). The practice of antibiotic overuse in the meat industry to maximize returns could also result in its ultimate demise (Van Boeckel *et al.*, 2019).

Livestock is often associated with cattle, pig farming, and poultry production, with very little attention on aquaculture (Van Boeckel *et al.*, 2015); given the limitations of language to encompass all forms of animal production, we hereby refer to all forms of reported production on land vertebrates as livestock. Obstacles to providing information on antimicrobial consumption and sales, such as the lack of publicly funded surveillance systems and limited collaboration by the stakeholders of the antibiotic industry, such as animal feed producers and veterinary pharmaceutical companies, make information on antibiotic consumption for humans and livestock scarce and difficult to obtain (Van Boeckel *et al.*, 2015).

There are, however, significant efforts to systematize global data on antimicrobial consumption and resistance. One such effort is resistancebank.org (Criscuolo *et al.*, 2021), a geodatabase platform based on 1285 literature surveys from the years 2000–2019, and specialized in low- and middle-income countries, which often lack robust surveillance systems; it includes 22,403 records on resistance rates for pathogens isolated from livestock. Another source of data for global antimicrobial use (AMU) and AMR in humans is the repository of the Global Antimicrobial Resistance and Use Surveillance System (GLASS) of the WHO; it provides valuable insights into resistance trends, antimicrobial consumption, and the status of national surveillance systems (Ajulo and Awosile, 2024).

The data in these repositories was used to establish general patterns and trends in global AMU and AMR. The high-income group of countries, as defined by the World Bank, consumes almost twice the amount of antibiotics for human health than the rest of the income groups individually, with differences in magnitude that are statistically sound (Fig. 6.3). However, when considering world regions, the human consumption of antibiotics is overall the same, as there are no detectable statistical differences (Fig. 6.4). Yet, with due caution for small sample size, regions dominated by poor or developing countries, such as the African Region, the Americas, the Eastern Mediterranean Region, and the Western Pacific Region, are all noticeably distant in terms of their central tendency to the European and South-east Asia regions (Fig. 6.4). The average consumption for all countries is also the same throughout time (2016–2022), and no discernible trend in either higher or lower use is detectable (Fig. 6.5). The species that are the most common part of livestock and food production do not have differences in terms of the percentage of bacterial isolates with AMR. This trend is based on observations across measured countries (Fig. 6.6).

Thanks to the monumental contribution of the team behind resistancebank.org (Van Boeckel

Total DID across World Bank income groups

$F_{\text{trimmed-means}}(1.86, 7.46) = 15.69$, $p = 2.31\text{e--}03$, $\hat{\delta}_{R\text{-avg}}^{AKP} = -1.19$, $\text{CI}_{95\%}$ [--25.26, 4.67], $n_{\text{pairs}} = 7$

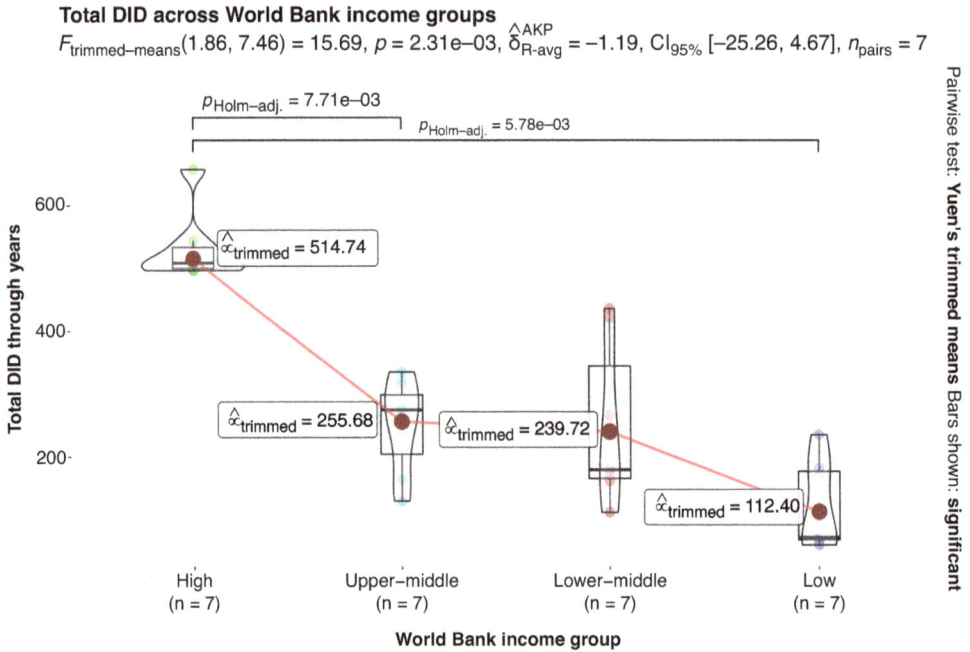

Fig. 6.3. High-income countries consume nearly twice as many antibiotics per capita as other income groups, according to WHO GLASS data (2016–2022). The defined daily dose per 1000 inhabitants per day (DID) reflects human antibiotic use across World Bank income groups. Dots represent yearly totals, while horizontal bars indicate significant contrasts between groups (Yuen's trimmed means and Holm–Bonferroni correction). Robust ANOVA revealed a strong income group effect on DID (significant p-value with effect size = --1.19). Figure author's own.

et al., 2019; Schar *et al.*, 2021; Mulchandani *et al.*, 2024), we can propose an overview of how four common pathogens, *Salmonella* spp., *Staphylococcus aureus*, *E. coli*, and *Campylobacter* spp., form part of the world's livestock production with their associated resistance (Fig. 6.7). The exclusive presence of *E. coli* in horse and buffalo, with the absence of the other three recorded pathogens, is noteworthy, particularly considering that these species do not form part of mainstream food supply chains (see the low sample sizes in Fig. 6.8). However, *Salmonella* has noticeable proportions in poultry, while *S. aureus* is present in cattle and sheep. For these four common pathogens, the quinolones antibiotic group is the most frequently associated with elevated resistance, except for Aminoglycosides in *S. aureus* (Fig. 6.8). Together, quinolones, aminoglycosides, and aminopenicillins constitute an average of 72.25% (SD = 7.97%) of total registered antibiotic use in worldwide livestock production. These are relevant patterns, based

on 10,717 studies and surveys recorded at resistancebank.org, and registering 50% or more of bacterial isolates from livestock with AMR.

6.4.1 Resistance in the genomes of food-associated bacteria

Studies on bacterial genomes associated with crops, fruits, and vegetables have commonly reported the presence of genes associated with resistance to various antimicrobial drug classes (Tenea and Ortega, 2021; Molina *et al.*, 2024). A large-scale analysis of closed bacterial genomes revealed that over 95% of the genomes harbor genes associated with resistance to various antimicrobial drug classes, including disinfectants, glycopeptides, macrolides, and tetracyclines; each of the analyzed genomes was found to encode an average of nine classes of antimicrobial drugs (Domingues *et al.*, 2023). More

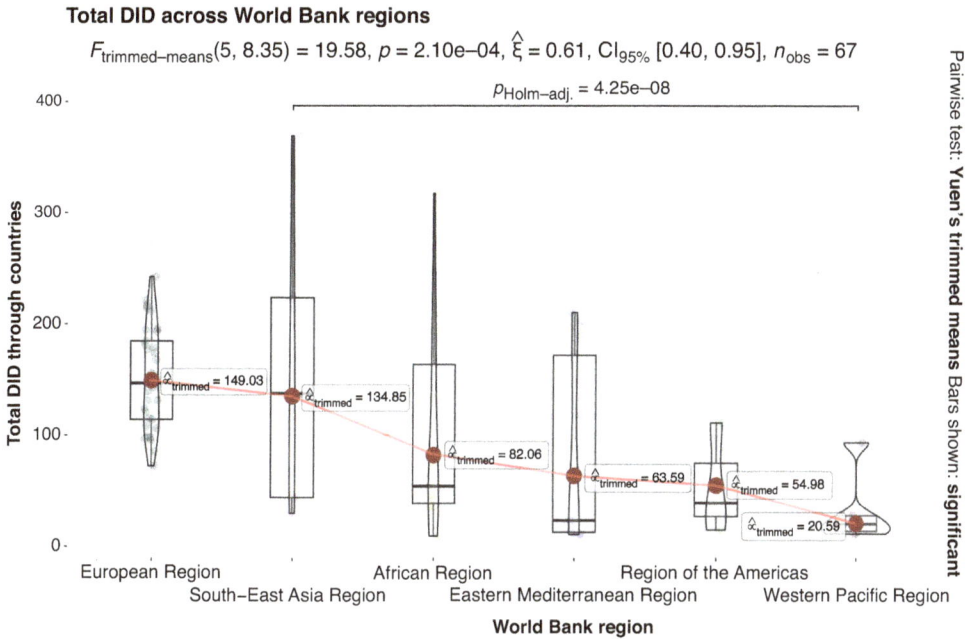

Total DID across World Bank regions

$F_{\text{trimmed-means}}(5, 8.35) = 19.58$, $p = 2.10\text{e-}04$, $\hat{\xi} = 0.61$, $\text{CI}_{95\%}$ [0.40, 0.95], $n_{\text{obs}} = 67$

$p_{\text{Holm-adj.}} = 4.25\text{e-}08$

$\hat{\mu}_{\text{trimmed}} = 149.03$
$\hat{\mu}_{\text{trimmed}} = 134.85$
$\hat{\mu}_{\text{trimmed}} = 82.06$
$\hat{\mu}_{\text{trimmed}} = 63.59$
$\hat{\mu}_{\text{trimmed}} = 54.98$
$\hat{\mu}_{\text{trimmed}} = 20.59$

Total DID through countries

European Region   African Region   Region of the Americas
      South–East Asia Region   Eastern Mediterranean Region   Western Pacific Region

World Bank region

Pairwise test: **Yuen's trimmed means** Bars shown: **significant**

Fig. 6.4. Defined daily dose per 1000 inhabitants per day (DID) applied to humans across World Bank regions. Measured countries through 2016–2022 are represented as dots. There are no statistical differences across world regions in terms of antibiotic usage (Robust ANOVA was not significant). Horizontal bars indicate significant contrasts between groups (Yuen's trimmed means and Holm–Bonferroni correction). Data from WHO GLASS (2016–2022). Figure author's own.

interestingly, Domingues *et al.* (2023) found that the higher-than-expected co-occurrences of resistance genes in priority and critically important pathogens appeared in plasmids, but not in chromosomes, pointing toward associated phenomena of potential resistance dissemination, and probably recent epidemic events. This widespread prevalence of resistance genes highlights the magnitude of the antibiotic resistance problem.

Given that the genome contains most, if not all, of the codified instructions that determine bacterial metabolic responses to the environment, including AMR, understanding the information in this molecule is vital for strategies around the problem of resistance. The Whole-Genome Sequencing for Antibiotic Susceptibility Testing (WGS-AST) is emerging as a powerful tool for predicting antibiotic resistance. WGS-AST involves analyzing the entire genome of a bacterium to identify resistance genes and predict its susceptibility to different antibiotics. This technology has the

potential to revolutionize antibiotic susceptibility testing and guide treatment decisions (Su *et al.*, 2018). WGS-AST provides a complete picture of the resistance genes present in a bacterial isolate, including known and novel genes, and, therefore, accurately predicts resistance phenotypes, enabling timely treatment decisions. Additionally, WGS-AST can be used for surveillance purposes, tracking the emergence and spread of resistance genes in bacterial populations (Argimón *et al.*, 2020; Ren *et al.*, 2021). It is important, however, to acknowledge the challenges in predicting resistance based only on genome sequence; factors such as gene expression, epigenetic modifications, and interactions between different genes can influence resistance phenotypes and may not be fully captured by WGS alone (Su *et al.*, 2018).

Two recent and contrasting studies on the presence of genes for AMR may serve as examples of the importance of WGS technologies for elucidating characteristics in bacterial

Total DID across years

$$F_{\text{trimmed-means}}(6, 86.12) = 0.66, p = 0.68, \hat{\xi} = 0.20, CI_{95\%}\ [0.11, 0.29], n_{\text{obs}} = 340$$

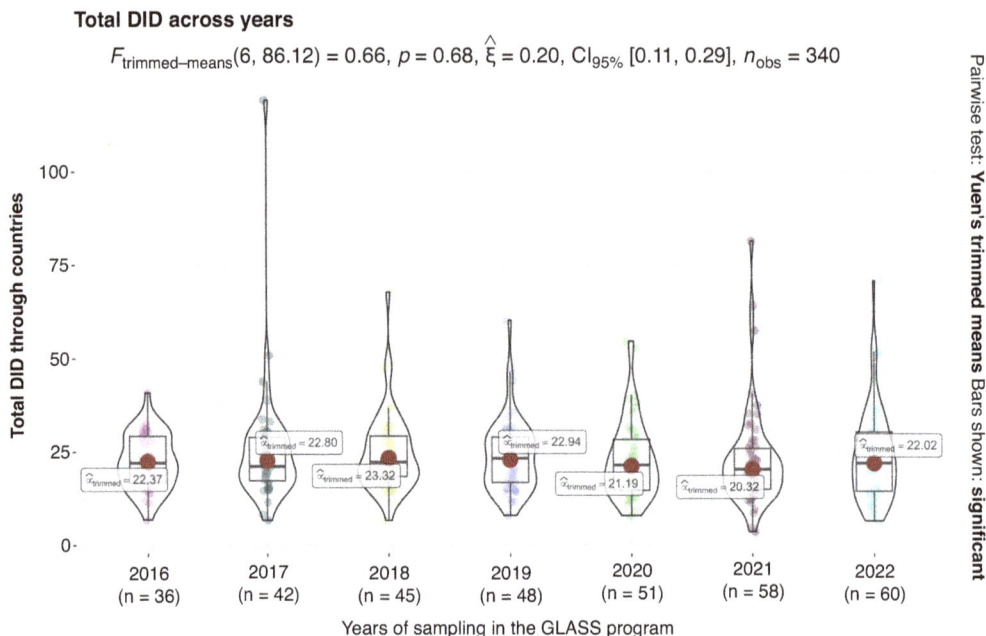

Fig. 6.5. Defined daily dose per 1000 inhabitants per day (DID) applied to humans through 2016–2022, across all recorded countries in the GLASS. Each dot is a recorded country. Given the robust ANOVA test, no statistical differences are found across years. Data from WHO GLASS (2016–2022). Figure author's own.

genomes of relevance to human health. Tenea and Ortega (2021) explored the genomic characterization of the *Lactiplantibacillus plantarum* strain UTNGt2, derived from wild white cacao in the Ecuadorian Amazon; they found that the genome contains 3115 genes, including 3052 protein-coding genes, and elements involved in carbohydrate transport and metabolism, transcription, defense mechanisms, and secondary metabolites biosynthesis. Notably, the genome indicates the presence of genes responsible for riboflavin and folate production and the absence of acquired antibiotic resistance genes, virulence factors, and pathogenic factors. These latter characteristics provide value to *L. plantarum* strain UTNGt2, suggesting the strain's economic potential, safety, and value as a probiotic. On the opposite side, Molina *et al.* (2024) investigated a multi-drug-resistant strain of *Escherichia coli* (strain L1PEag1) that was isolated from a patient in Ecuador. This strain carries the blaKPC-2 gene, which makes it resistant to carbapenems, a class of antibiotics used to treat serious bacterial infections. The genome analysis revealed that L1PEag1 had acquired several genes responsible for resistance to various antibiotics, including beta-lactams, aminoglycosides, tetracyclines, and sulfonamides. These genes were located on mobile genetic elements like plasmids and transposons, which can easily transfer between bacteria. The study aimed to understand the genetic makeup of this strain and its relation to other *E. coli* strains, especially those from the ST258 clade, which are high-risk clones often associated with the *bla*KPC-2 gene and are known to cause healthcare-associated infections (Molina *et al.*, 2024). The findings contribute to a better understanding of the genomic characteristics of multi-drug-resistant *E. coli* and can help with developing strategies to control its dissemination. Thus, these contrasting examples (Tenea and Ortega, 2021 vs Molina *et al.*, 2024) illustrate how genome analysis can differentiate between beneficial bacteria with probiotic potential

% of isolates resistant to antimicrobial compounds by species

$F_{\text{trimmed-means}}(6, 119.8) = 143.12$, $p = 0.00$, $\hat{\xi} = 0.31$, $CI_{95\%}$ [0.20, 0.46], $n_{\text{obs}} = 33{,}187$

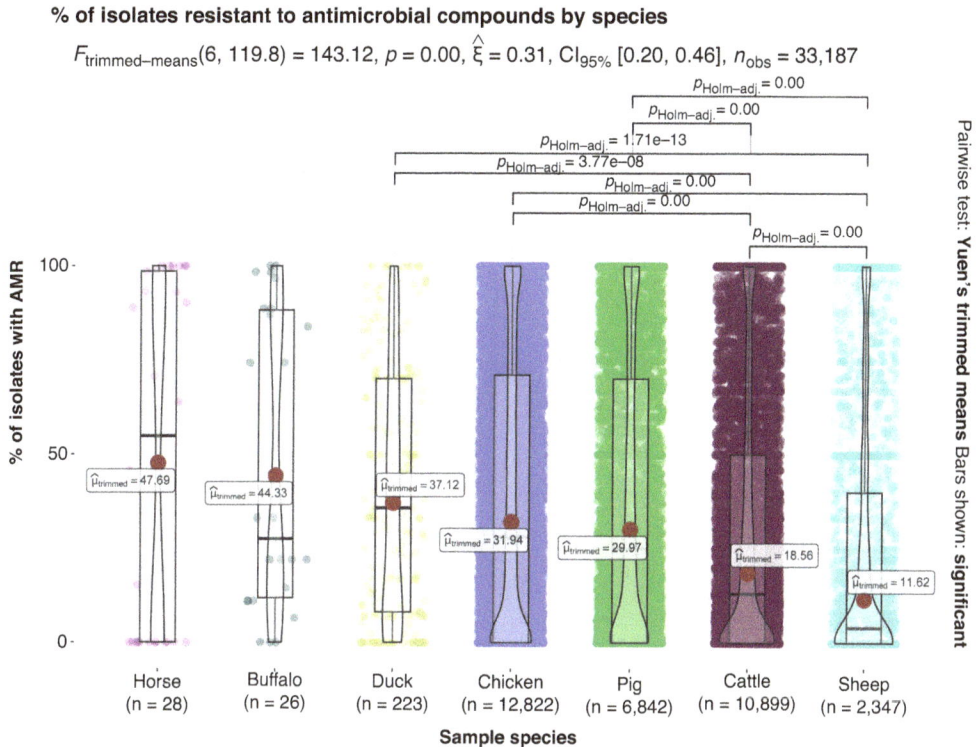

Fig. 6.6. A worldwide summary of AMR across common livestock species. The variation in the percentage of bacterial isolates with AMR resistance is statistically significant across species; however, the observed differences and the moderate effect size reported should be accounted for because of the large sample size ($\hat{\xi} = 0.31$). The horizontal bars show significant contrasts between species (Yuen's trimmed means and Holm–Bonferroni correction). A circumstantial diminishing trend in the central tendency is observed across species, with horses having higher values than sheep at the extremes of the trend. Variation is wide-ranging, with reports across the scale of percentages of isolates with AMR. Each dot represents a point survey or study of AMR in the world. For further details of the source data, refer to resistancebank.org. Figure author's own.

and those posing significant risks due to AMR, highlighting the importance of this technology in safeguarding human health.

The power of sequencing methods also includes the possibility of assessing whole collections of DNA sequences, such as metagenomics. After combining conventional bacteriological culture methods (*in vitro*) with metagenomics, a non-culture method (*in silico*), Angamarca *et al.* (2023) found that commercial ready-to-eat avocado fruits contained detectable levels of ARG bacteria, such as methicillin-resistant *Staphylococcus*. Evidence based on multiple sources and experiments contributes to the body of evidence for understanding foodborne

pathogens as vectors for the spread of antibiotic resistance.

6.4.2 Human guts are a propitious ecosystem for AMR transmission and resistance genes

The challenge posed by AMR has spurred on reinvigorated processes of scientific inquiry and technological innovation in our fight for health. As technology evolves and finds solutions, the economic and human costs will eventually be contained; yet, solutions should

Species of livestock and associated pathogens with AMR
Selected samples with % of resitant isolates >= 50%

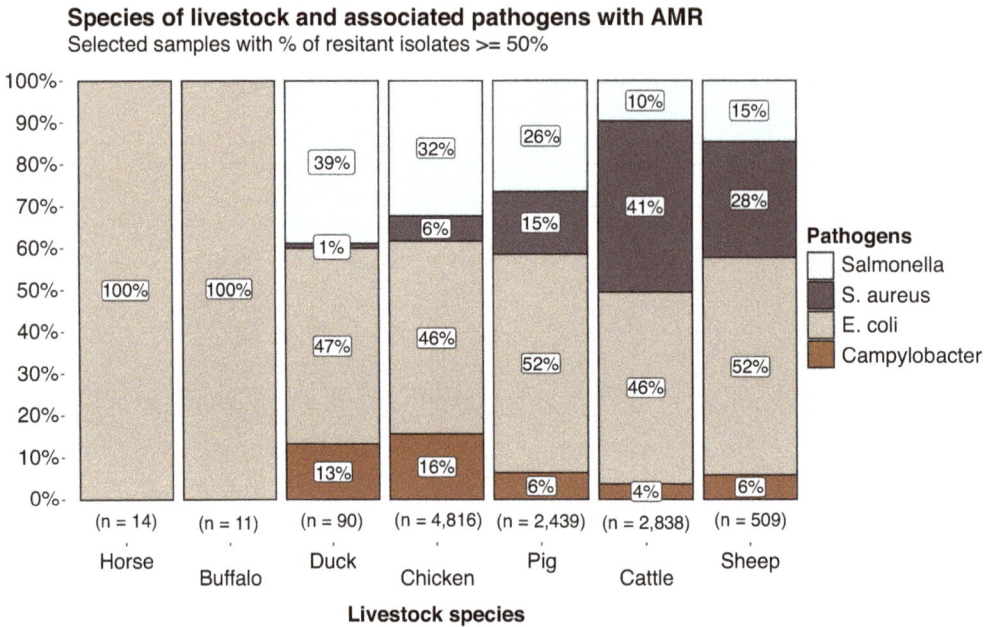

Fig. 6.7. The proportion of pathogens in livestock species as reported in studies worldwide. Depicted data is restricted to those sampling assays showing pathogen isolates with AMR frequency equal to or above 50% (data available from resistancebank.org). Figure author's own.

not only be technological but also require new ways in which societies relate to the environment and the living beings that sustain humanity.

AMR may be an unexpected consequence of our technological prowess as humans; yet, it is also a consequence of our biological condition. Food is ultimately meaningful to humans through the physiological mechanisms occurring in the digestive system, known as the gut. On a healthy human, with rich and propitious conditions of acidity, temperature, nutrient heterogeneity, and varying oxygen levels, the gut offers a rich landscape for microbial communities to colonize, flourish, and evolve (Lozupone *et al.*, 2012). As such, the human gut is a microbial ecosystem that comprises trillions of microorganisms, more than one thousand microbial species, and nearly 150 times more genes than the human genome, including bacteria, archaea, fungi, and viruses, with bacteria being the most abundant and diverse, playing a crucial role in transferring AMR genes to potentially harmful bacteria (Penders *et al.*, 2013; Tasnim *et al.*, 2017; Afzaal *et al.*, 2022). In such

a vast environment, bacteria and humans have coevolved for hundreds of thousands of years, interacting with each other and their hosts in intricate ways, forming a complex web of symbiotic relationships that contribute to the overall stability and function of the gut ecosystem. The evolutionary process in the human gut can be quite rapid and dramatic, with recorded examples spanning a few weeks (Venkatakrishnan *et al.*, 2021). Thus, the human gut, with its universe of microorganisms and interactions is a perfect scenario for the evolution of AMR and a central hub for the spread of ARGs (Lamberte and van Schaik, 2022).

Such enormous microbiological richness in the gut should inevitably carry within its collective genomes or 'hologenome' a set of ARG. As such, the collection of all ARGs in the gut microbiome is known as the resistome, and the human gut could be understood as a natural reservoir of ARGs and AMR microorganisms (Theophilus and Taft, 2023). Several factors, like antibiotic use, diet, high density of bacterial cells, and the immune system, further influence the structure of the microbial community and the selective

Resistant pathogen and its associated antibiotic class
Selected samples with % of resitant isolates >= 50

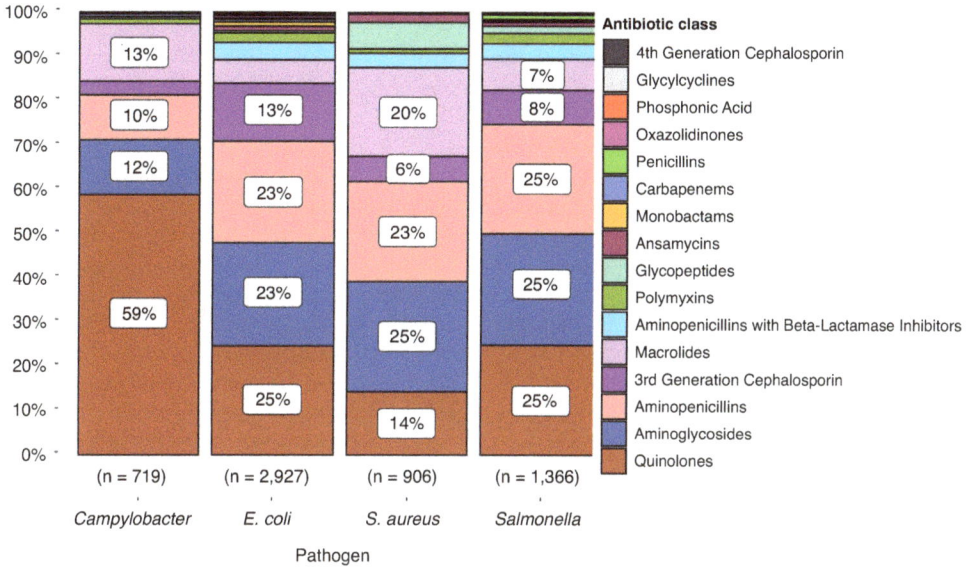

Fig. 6.8. The proportion of antibiotic classes in pathogens with AMR frequency equal to or above 50% (data available from resistancebank.org). Proportions are labeled for those above. Figure author's own.

pressures driving resistance (Oliver *et al.*, 2022; Stege *et al.*, 2022). Diet influences the structure of the gut microbiome, where changes in diet are mirrored in changes in the microbiome (David *et al.*, 2014). Clear patterns are not yet well understood, in part due to the complexity of the microbiome ecosystem; yet, there is evidence that more diverse diets and diets rich in fiber promote attenuated abundances of ARGs (Oliver *et al.*, 2022). The authors of this study propose that the consumption of fiber in favor of animal protein contributes to an anaerobic state of the gut ecosystem, which limits a propitious ecological niche for facultative anaerobes, known to harbor antibiotic resistance. Insights on the effect of diet on the resistome may eventually be part of health dietary guidelines in favor of better health overall and more efficient healthcare systems within the ethos of One Health (Oliver *et al.*, 2022). When antibiotics are used, whether for treatment or consumed indirectly through contaminated food or water, they disrupt the gut microbiota by selecting for resistant strains (Lamberte and van Schaik, 2022). This pressure promotes the growth of resistant

pathogens and encourages the transfer of resistance genes between gut microbes (van Schaik, 2015). Additionally, the growth of bacteria carrying AMR genes is influenced by both internal and external factors, such as probiotic use, inflammation, diseases that disrupt microbial balance, age, sex, and environmental exposure to elements like soil, water, and air (Fredriksen *et al.*, 2023).

6.4.3   The economic burden of AMR on the food industry

The effect of AMR on healthcare costs exceeds US$55 billion in the USA alone, with $20 billion in direct healthcare costs and $35 billion in lost productivity (Theophilus and Taft, 2023). The impact of AMR on human well-being and welfare has led some to christen it the 'silent pandemic' (Akram *et al.*, 2023). The development and widespread use of antibiotics have exerted selective pressure on bacterial populations, favoring the survival and proliferation of resistant strains. Pathogens are constantly evolving to combat the drugs used to treat potentially life-threatening diseases, rendering treatments ineffective. Furthermore, technological advancements in areas such as agriculture and animal husbandry have led to the extensive use of antimicrobials in livestock, further accelerating the development and spread of AMR. The world's increasing interconnectedness through global trade and travel has facilitated the rapid dissemination of resistant strains across geographical boundaries.

Each year, 20% of livestock production is lost to animal diseases, including a proportion caused by treatment-resistant infections (Adamie *et al.*, 2024). The economic and healthcare implications of AMR are projected to be substantial. The World Bank (Jonas *et al.*, 2017) estimates potential annual global GDP losses ranging from 1.1 to 3.8% by 2050, equivalent to between $1 trillion and $3.4 trillion. This projection aligns with the UK AMR Review's estimate of a cumulative cost of $100 trillion by 2050 if AMR remains unchecked (O'Neill, 2016).

Beyond macroeconomic impacts, AMR poses significant burdens on healthcare systems. In the USA alone, Thorpe *et al.* (2018) estimate that multi-drug-resistant bacterial infections result in between $2 billion and $20 billion in additional healthcare costs annually. Furthermore, the World Bank (Jonas *et al.*, 2017) projects that AMR could trigger a decline in livestock production, with low-income countries facing an 11% reduction and middle- and high-income countries experiencing a 6–9% reduction by 2050.

Through economic modeling and simulations of AMR-attributable effects on the productivity of the global livestock sector, Adamie *et al.* (2024), from the Economic Impact of Antimicrobial Resistance Project, found that AMR is significantly affecting the economy of livestock production systems, with negative effects and significant risks projected to grow over time. Losses in the livestock sector are attributable to bacterial infections in animals that fail to be effectively treated with antimicrobials, consequentially leading to the death or reduced output of livestock.

Various scenarios and their predictions were assessed by Adamie *et al.* (2024). Under the most optimistic scenario of a 5% resistance rate in modeled pathogens and sectorial diseases, across all world regions and productive sectors, losses due to AMR will equal the consumption needs of 746 million people. This most optimistic scenario predicts that by 2050 the estimated cumulative GDP loss due to AMR in livestock is $575 billion. Under a scenario where the AMR-attributable disease burden doubles in all regions of the world, the expected yearly production losses in livestock production will equal the consumption of two billion people and an estimated cumulative GPD loss of $953 billion. If the disease burden doubles and human health is severely compromised by spillover effects from livestock, global GDP losses could reach a staggering $5.2 trillion by 2050.

On the bright side of the work by Adamie *et al.* (2024), an effective reduction of AMU in livestock of around 30% is predicted to lead to a cumulative increase in the global GDP of $120 billion from the year 2025 to 2050. Potential losses to the economy may be among the most effective warnings to act on prevention. The quantification of the antibiotic resistance threat accentuates the need for establishing efficient control measures and developing coordinated approaches for food safety, as crucial steps to mitigate the risk of contamination and ensure the microbiological quality of postharvest raw materials. Thus, there is a generalized need for stronger food safety protocols and awareness in the agricultural and retail sectors to address the rising threat of antibiotic-resistant bacteria in food products. Slowing or stopping the evolutionary process that generates AMR involves curtailing the components involved and thus avoiding the expected economic losses to the food industry due to worldwide AMR intensification.

6.5 The Urgency of Global Action on AMR

Combating AMR involves global action plans promoting antimicrobial stewardship, surveillance, research, and public awareness campaigns. The 'One Health' approach is critical in tackling this multifaceted issue. The World Health Assembly of the WHO has gathered efforts in a Global Action Plan (GAP) on AMR;

published in 2015, it focuses on five strategic objectives: improve awareness, strengthen knowledge and scientific evidence, reduce the incidence of infection, optimize antimicrobial use, and increase investment in the development of new medical interventions (new antibiotics, diagnostic tools, medicines, and other technologies). The GAP has echoed in world regions and world institutions that have designed their action plans, such as the FAO Action Plan on Antimicrobial Resistance (FAO, 2021). However, the performance of the GAP has been considered poor in regions such as Africa with weak implementation of surveillance and antibiotic use (Iwu and Patrick, 2021).

A global assessment of 78 national action plans on AMR submitted to the WHO determined that fewer than half of the countries in the world have submitted a national action plan (NAP) on AMR to the WHO (Willemsen *et al.*, 2022). This study concluded that high-income countries generally reported greater progress in addressing AMR compared to low- and middle-income countries; the most common challenges to implementing AMR strategies were a lack of financial and human resources and insufficient laboratory capacity. Thus, greater AMR surveillance is urgently needed, and financial and resource support is required in developing countries to develop and implement national action plans. Countries vary widely in their strategic policy design, capacity to implement tools to reduce the burden of AMR, and efforts to monitor and evaluate these interventions (Patel *et al.*, 2023). Interestingly, Patel *et al.* (2023) found that the measure of responsible use of antimicrobials and facilitated access to antimicrobials was highly scored in low- and middle-income countries, while the Americas, Europe, and high-income countries scored lower.

Avoiding AMU increase can be possible through alternative technologies and practices that contribute to maintaining economically viable livestock production while avoiding disease in animals. Disease prevention through vaccination, antibiotic alternatives such as antimicrobial peptides, probiotics, prebiotics, and synbiotics, predatory bacteria, immunotherapeutics, and bacteriophages are potential alternatives and new prospective biotechnologies for tackling the risks of AMR (Allen, 2017). Mitigating the economic threat of AMR within global food production systems requires a multifaceted approach that integrates cultural and political strategies. This includes enhanced surveillance of both antibiotic usage and AMR prevalence in livestock, coupled with educational initiatives aimed at informing farmers about the complexities of AMR. Furthermore, veterinary training programs should emphasize responsible and judicious antibiotic stewardship in food-producing animals. Finally, incentivizing farmer compliance with established policies and regulations is crucial to ensure the long-term sustainability of the agricultural sector. By promoting responsible antibiotic use throughout the food production chain, these measures collectively contribute to safeguarding global food security and economic stability (Adamie *et al.*, 2024).

Success stories (Gomez *et al.*, 2021; Phu *et al.*, 2021) reported by Adamie *et al.* (2024) at mitigating the risks of AMR in food production systems include a compelling dairy farming case study from Ontario and a promising broiler chicken case study from Vietnam. In Ontario, where neonatal diarrhea in calves often leads to high AMU, a multidisciplinary intervention across ten small-scale dairy farms demonstrated the effectiveness of education and medicine management. By training farmers in calf health assessment, streamlining disease prevention protocols, and implementing a decision algorithm for AMU in diarrhea treatment, the intervention achieved an average 37% reduction in antimicrobial volume used, with some farms achieving reductions as high as 81%. Importantly, these reductions were achieved without negatively impacting calf mortality and even showed an 80% predicted reduction in the risk of inappropriate AMU. Meanwhile, in Vietnam, where small-scale poultry farming is vital for rural livelihoods and AMU in chickens is predominantly preventative, a randomized trial showcased the power of farmer education and advice. By focusing on antimicrobial alternatives, improved biosecurity, litter management, and vaccination, and employing a persuasive rather than restrictive approach, the intervention achieved a remarkable 66% reduction in AMU and a 40% reduction in mortality. These case studies highlight the potential of knowledge transfer, farmer empowerment, and proactive management strategies in mitigating AMR risks within diverse food production contexts, paving the way for a more sustainable and secure global food system.

Use of control, taxation, economic incentives and penalties, education and better culture, and continuous monitoring and strategies around AMU are fundamental elements to mitigate the problem with AMR. It has been suggested that *Aeromonas* spp. detected in aquatic species can enhance environmental surveillance of AMR (Grilo *et al.*, 2020).

Technological change in the food industry is another essential component to solving the problem of AMR. Bioprotective cultures and biopreservatives have been proposed as a sustainable alternative to antibiotics in livestock production. Some beneficial microorganisms, such as lactic acid bacteria (LAB), contribute to decreasing the pH in the food environment and are sources of antibacterial activity and compounds, such as bacteriocins (small peptides) and bacteriocin-like inhibitory substances (BLISs), which promise biotechnological applications to improve the means of food production

and reduce the risk of foodborne pathogens (Pérez-Rodríguez and Taban, 2019). To determine the value and biosafety of microorganisms and their products as biopreservatives, the international principle of 'Generally Recognized as Safe,' or GRASS, steers the selection and use of potentially valuable microorganisms in the food industry and requires thorough technical assessments of the genetic and biochemical properties of the microorganisms of interest, including the genetic potential for hosting or developing AMR (Pérez-Rodríguez and Taban, 2019). The development and implementation of preventive measures, alternative technologies, and responsible antibiotic stewardship programs are crucial to mitigate AMR-associated risks and ensure the sustainability of the food industry. The fight against AMR is a multifaceted endeavor requiring a concerted effort from all stakeholders. By embracing the One Health paradigm, promoting responsible antibiotic use, and investing in research and innovation, we can strive to safeguard human health, animal welfare, and the environment for future generations.

PART II: MICROBIAL HEALTH, BIOSECURITY FRAMEWORKS, AND REGULATORY POLICIES IN THE CONTEXT OF ONE HEALTH AND AQUACULTURE

Laura D. Urdes[1]*, Chris Walster[2]*, Iasmina Maria Moza[3] and Gratiela Gradisteanu[3]
[1]*The World Aquatic Veterinary Medical Association, Romania; Faculty of Veterinary Medicine, University Spiru Haret Bucharest, Romania; Faculty of Management and Rural Development, University of Agricultural Sciences and Veterinary Medicine, Bucharest, Romania;* [2]*The World Aquatic Veterinary Medical Association, UK;* [3]*The Faculty of Biology, University of Bucharest, Romania*

6.6 Microbiomes and One Health

The One Health approach addresses the interconnected health of humans, animals, and ecosystems, focusing on the impacts of microbial communities or microbiomes within each domain (Zinsstag *et al.*, 2023). We need to understand how the human, animal, and environmental microbiomes interact and influence each other and how these

relationships can be leveraged to improve health on a global scale.

Human exposure to diverse environmental microbiomes, particularly in urbanized areas, influences microbial diversity, immune health, and resistance to diseases (Matthews *et al.*, 2024).

Livestock and companion animals, as microbial reservoirs, have significant effects on the human microbiome, influencing zoonotic disease transmission. Resistant microorganisms are often present in humans, animals, food, and the environment, causing a very complex epidemiological issue, namely, AMR.

6.6.1 The human microbiota and One Health

Microbiota residing in the human gut, skin, and respiratory tract play a crucial role in maintaining overall health by contributing to essential physiological functions, such as nutrient absorption, immune modulation, and protection against pathogens (Lazar et al., 2018). These diverse microbial communities, consisting of bacteria, fungi, viruses, and archaea, coexist in a symbiotic relationship with the human body, influencing a broad range of processes vital for well-being.

In the gut, the microbiota assists in breaking down complex carbohydrates, fibers, and other nutrients that are otherwise indigestible by human enzymes. This process results in the production of short-chain fatty acids (SCFAs), such as butyrate, acetate, and propionate, which serve as an energy source for colon cells and contribute to the maintenance of the gut barrier. The gut microbiota also synthesizes essential vitamins, such as vitamin K and certain B vitamins, that are important for metabolic health. Beyond nutrient processing, the gut microbiota plays a pivotal role in shaping and regulating the immune system, helping the body distinguish between harmful pathogens and harmless antigens. Dysbiosis, or imbalance in the gut microbiome, has been linked to various health conditions, including inflammatory bowel disease (IBD), obesity, and even mental health disorders, highlighting the importance of microbial homeostasis for systemic health. Key members of the gut microbiota include bacteria such as *Bacteroides*, *Firmicutes*, *Lactobacillus*, and *Bifidobacterium*, which are known for their roles in digestion and immune support (Hrncir, 2022).

The skin microbiota acts as a first line of defense against environmental threats. These microbial communities help maintain skin integrity by preventing colonization by harmful pathogens through competitive exclusion and by producing antimicrobial compounds that inhibit the growth of opportunistic invaders (Byrd et al., 2018). The skin microbiota also interacts with the immune system, influencing both innate and adaptive responses to maintain immune tolerance and prevent chronic inflammation. For example, bacteria such as *Staphylococcus epidermidis* produce molecules that can modulate immune responses, offering protection against infections and contributing to overall skin health (Uberoi et al., 2024). Other important members of the skin microbiota include *Cutibacterium acnes*, which, despite its association with acne, also plays a role in maintaining skin health under normal conditions.

The respiratory tract microbiota plays a key role in protecting against respiratory pathogens and maintaining respiratory health. Microorganisms residing in the upper respiratory tract, such as the nasal passages and throat, contribute to the immune system's readiness by training immune cells and maintaining a balance that discourages pathogenic overgrowth. These microbes help prevent colonization by potential pathogens, such as *Streptococcus pneumoniae* or *Staphylococcus aureus*, by competing for nutrients and attachment sites. Common members of the respiratory microbiota include *Corynebacterium*, *Dolosigranulum*, and *Moraxella*, which contribute to maintaining respiratory balance and protecting against infections (Man et al., 2017). An imbalance in respiratory tract microbiota has been associated with respiratory diseases, such as asthma and chronic obstructive pulmonary disease (COPD), as well as recurrent respiratory infections.

Overall, the microbiota across different body sites are integral to human health. They help fine-tune metabolic processes, modulate immune responses, and provide a protective barrier against a multitude of pathogens. Advances in understanding the microbiome have opened new avenues for therapeutic interventions, including the use of prebiotics, probiotics, and microbiota transplantation to restore balance and promote health. Maintaining a diverse and healthy microbiota is essential for the optimal functioning of the human body, illustrating the intricate interplay between our microbial companions and physiological well-being.

Urbanization often leads to reduced exposure to a variety of beneficial environmental microbes, which can result in lower microbial diversity and dysregulated immune responses. This reduction in microbial diversity has been linked to an increase in autoimmune diseases, allergies, and other inflammatory conditions.

Livestock and companion animals serve as important reservoirs of microbial diversity, and their interaction with humans can have a significant effect on the human microbiome (Kuthyar and Reese, 2021). Pets, such as dogs and cats, bring a variety of environmental microbes into homes, contributing to increased microbial diversity in urban settings, which can support immune development and reduce the risk of allergies and autoimmune disorders. These animals, however, can also serve as vectors for zoonotic diseases, which are diseases that can be transmitted between animals and humans. Livestock, in particular, play a dual role as they contribute to human microbiome diversity through environmental exposure but also pose a risk for the transmission of zoonotic pathogens, such as *Salmonella* or *Campylobacter* (Yang *et al.*, 2024). The One Health approach emphasizes the importance of understanding these interactions to ensure that the benefits of microbial exposure are maximized while minimizing the risks associated with zoonotic disease transmission.

By adopting a One Health perspective, it becomes evident that maintaining biodiversity and healthy ecosystems is crucial for human well-being. Strategies such as promoting sustainable agricultural practices, reducing unnecessary antibiotic use in livestock, and encouraging responsible pet ownership can help foster a balance between beneficial microbial exposure and minimizing the risks of pathogenic infections. The One Health framework encourages collaboration among veterinarians, ecologists, medical professionals, and public health experts to address health challenges holistically, recognizing that the health of humans, animals, and the environment are interdependent.

The microbial communities in wildlife populations are pivotal in understanding zoonoses, as wildlife serves as both a reservoir and a bridge for pathogens that can spill over into human populations. Deforestation and habitat encroachment are key factors that disrupt these natural microbial communities, leading to increased contact between wildlife, humans, and domesticated animals (González-Barrio, 2022). Such disturbances alter the natural balance of wildlife microbiomes, often creating opportunities for pathogens to cross species barriers and adapt to new hosts. For example, habitat loss can force animals such as bats and primates into closer proximity to human settlements, increasing the risk of spillover events involving zoonotic pathogens like coronaviruses or *Ebola* virus (Koch *et al.*, 2020). The One Health approach emphasizes the need to address these environmental disruptions by promoting conservation efforts and sustainable land-use practices, which help maintain ecosystem integrity and reduce the likelihood of zoonotic disease emergence. By protecting wildlife habitats and minimizing human encroachment, we can decrease the potential for pathogen spillover, ultimately safeguarding both human and animal health. Collaboration across disciplines, including ecology, veterinary medicine, and public health, is essential to effectively mitigate these risks and ensure a balanced coexistence between humans, animals, and the environment.

Soil and aquatic microbiomes are vital for environmental sustainability, affecting plant health, soil fertility, and water quality (Hartmann and Six, 2023). These microbiomes interact with human and animal microbiomes through agriculture, water sources, and the food supply chain. The soil microbiome plays a crucial role in supporting plant health by facilitating nutrient cycling, enhancing soil fertility, and promoting plant growth. Bacteria such as *Rhizobium*, which forms symbiotic relationships with legumes to fix atmospheric nitrogen, and *Bacillus*, known for promoting plant growth and providing protection against pathogens, are key members of the soil microbiome. Other important soil bacteria include *Pseudomonas*, which helps suppress plant diseases, and *Streptomyces*, which produces antibiotics that protect plants from infections. The soil microbiome also influences the human microbiome through the consumption of food crops that are enriched with nutrients by these microbial communities (Poole *et al.*, 2018).

The aquatic microbiome is equally significant for maintaining water quality and ecosystem health. Aquatic bacteria such as *Nitrosomonas* and *Nitrobacter* are involved in nitrification, a critical process that converts ammonia into nitrites and nitrates, which are essential for aquatic life. Cyanobacteria, often referred to as blue-green algae, play a major role in primary production and oxygen generation in aquatic environments. Disturbances in aquatic ecosystems, however, such as nutrient runoff from agriculture, can lead to harmful algal blooms, which can negatively impact both aquatic life and human health (Lassudrie *et al.*, 2020). Aquatic microbiomes also interact with human populations through drinking water sources and recreational activities, underscoring the importance of maintaining balanced microbial communities for public health.

The misuse of antibiotics in human health care, veterinary medicine, and agriculture has led to AMR becoming a critical public health issue, with resistant microbes readily transferring across species and environments. Resistant strains, such as methicillin-resistant *Staphylococcus aureus* (MRSA) and ESBL-producing Enterobacteriaceae, have become increasingly prevalent, complicating the treatment of infections in both humans and animals (Brown *et al.*, 2021). Resistance to antivirals, such as those used to treat HIV, is also increasing (European Union, 2017).

As a result of this resistance, antibiotics and other antimicrobial drugs can become ineffective, and infections may become difficult or impossible to treat. The risk of spread of communicable diseases increases, which aggravates illness and disability, eventually causing death. AMR is a consequence of natural selection, genetic mutation, and vertical (intergenerational) transmission of AMR within microorganisms following exposure to antimicrobials such as antibiotics, antivirals, antifungals, and antiprotozoals. This process is enhanced by human intervention, by inappropriate use of antimicrobials in human and veterinary medicine and inadequate hygienic conditions and practices in health care or in the food chain, making it easier for resistant microorganisms to transmit.

In May 2015, the World Health Assembly adopted a Global Action Plan on antimicrobial resistance (AMR3) to improve awareness on the issue of AMR, to call for advanced knowledge and evidence base through surveillance and research, to reduce the frequency of infections through more effective prevention, to control the use of antimicrobial medicines in human and animal health, and to foster increased investment in new medicines, novel diagnostic tools, vaccines, and other interventions aimed at reducing the risk of AMR globally. The Action Plan was subsequently adopted by the World Organisation for Animal Health (WOAH) and the Food and Agriculture Organization (FAO). Within the EU, a One Health Action Plan against AMR with the One Health approach has been put in place since 2017 to encourage the development of new effective antimicrobials.

The One Health approach is crucial in addressing AMR, as it emphasizes the need for coordinated efforts across human and animal health in relation to the environment to reduce antibiotic use, promote responsible practices, and implement effective surveillance and stewardship programs (Fig. 6.9). Reducing antibiotic use in agriculture, especially for growth promotion in livestock, and ensuring the prudent use of antibiotics in healthcare settings are essential steps in mitigating the spread of AMR. Public awareness, education, and policy interventions are also vital in combating AMR and preserving the effectiveness of antibiotics for future generations.

By adopting a One Health perspective, it becomes evident that maintaining biodiversity and healthy ecosystems is crucial for human well-being. Strategies such as promoting sustainable agricultural practices, reducing unnecessary antibiotic use in livestock, and encouraging responsible pet ownership can help foster a balance between beneficial microbial exposure and minimizing the risks of pathogenic infections.

6.7   Aquaculture Biosecurity

To improve biosecurity, it is necessary to understand the concept, to appreciate the relative importance of the different actions it implies and the relationships between these actions. It requires a good understanding of the motivation and perceptions of farmers

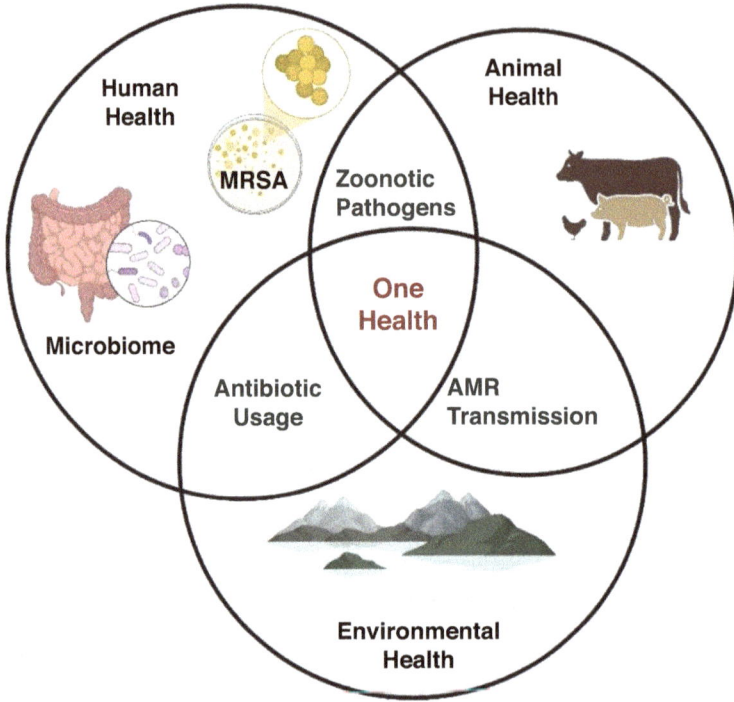

Fig. 6.9. The One Health approach: microbiomes, AMR, and public health. Figure author's own.

and veterinarians and a detailed comparative evaluation of investment alternatives, including an assessment of relative risk and financial viability/feasibility (DEFRA Research project SE4003[3]). Almost without exception, activities on one farm have consequences for the biosecurity of other farms. Communicable diseases spread onto and between farms when there are deficiencies in biosecurity measures on these farms.

6.7.1 Voluntary or regulated approaches

All aquaculture operations have some aspects of biosecurity in place, such as vaccination strategies, surveillance through mandatory schemes for listed diseases, and hygiene or welfare measures through quality assurance schemes. Governments may actively promote biosecurity through research funding or pump priming of industry schemes, but they are unlikely to regulate unless there is a perceived public good, such as economic or health benefits to society.

Currently, Responsibility and Cost Sharing (RCS)[4] is gaining traction in several countries. RCS is seen as providing agricultural industries with greater input into animal health and welfare policy while taking greater responsibility for disease prevention. Alternatively, some may look on it as a mechanism where the state's economic liability is linked to industries' biosecurity plans and approach to risk.

Given the above, any plan will be a mixture of voluntary and regulated approaches. Heed will need to be paid to national legislation, national strategies such as national aquatic health plans,[5] and the WOAH Code[6] and WOAH Manual,[7] along with industry codes of practice.[8] Voluntary approaches can be tailored to individual business needs and resources with greater flexibility than a fully regulated approach, and proposals such as RCS imply a biosecurity plan must be in place and that legislative or regulated approaches will continue to be targeted at ensuring the public good (i.e. an individual business's approach does not harm others).

6.7.2 Disease hazard identification— identifying diseases of significance

This could identify a short or extensive list. It may consist simply of diseases considered significant by the aquaculture operation (particularly if the plan is only for internal use) but, by default, it should include any listed diseases, diseases of regional concern, and diseases for which the business already operates some sort of control policy as elements of biosecurity will/should already be in place for these. Where the plan is to be used externally, such as to demonstrate freedom from disease to customers or the competent authority, it is recommended that a comprehensive list of potential disease hazards is drawn up as this will help demonstrate due diligence to others. In drawing up the list, consideration should be given to categorizing diseases into listed, non-listed, found in the wild, known to be on the EpiUnit, likely to be present at suppliers, emerging, zoonotic, and of concern to the public. Additionally, information on disease prevalence, geographic distribution, vectors, environmental influence on disease expression (i.e. facility type, water temperature), and potential economic cost of an outbreak should be recorded, for example. The primary purpose of recording such information is to assist decision making during the risk analysis stage but it can also provide a record of the decision process, aiding transparency in communication and assisting in auditing or certification.

Disease information websites such as the following are useful to consult and can help ensure data is up to date and accurate.

- WAHIS: World Animal Health Information System: https://wahis.woah.org/#/home
- WOAH Collaborating Centre for Emerging Aquatic Animal Disease: https://www.cefas.co.uk/icoe/aquatic-animal-health/designations/woah-collaborating-centre-for-emerging-aquatic-animal-disease/
- WOAH Aquatic Animal Health Code: https://www.woah.org/en/what-we-do/standards/codes-and-manuals/aquatic-code-online-access/
- Network of Aquaculture Centers Asia (NACA, n.d.) Quarterly Aquatic Animal Disease Report: https://enaca.org/?id=8
- Synopsis of Infectious Diseases and Parasites of Commercially Exploited Shellfish: https://www.dfo-mpo.gc.ca/science/aah-saa/diseases-maladies/index-eng.html

6.7.3 Disease risk analysis—evaluating an operation's risk from disease hazard

RISK = PROBABILITY × CONSEQUENCE

This is a simple equation for which it can be extremely difficult to decide what the values of the terms are. Probability is perhaps the easiest to determine a numerical value for (i.e. probability of introduction of disease from suppliers), but consequence can be extremely difficult (i.e. when loss of reputation is considered greater than economic loss or the disease is present on the EpiUnit). Further perceived difficulties in carrying out a quantitative risk analysis (cost, time, lack of data, which probability distribution to

Table 6.1. Example of a risk matrix.

Risk		Consequence				
Score		Very Low	Low	Medium	High	Very High
		1	3	5	7	9
Very Low	1	1	3	5	7	9
Low	3	3	9	15	21	27
Medium	5	5	15	25	35	45
High	7	7	21	35	49	63
Very High	9	9	27	45	63	81

use) make qualitative risk analysis more popular, but this can result in a risk matrix that is little better than chance. The perceived difficulties with quantitative analysis arise due to government projects such as Import Risk Analysis (IRA),[9] which involves assessment of multiple pathways and parameters. However, for a single EpiUnit, a robust quantitative risk analysis can be carried out fairly simply using epidemiological principles (Table 6.1). An alternative to a risk matrix would be the use of a decision tree. Whether qualitative or quantitative, all relevant stakeholders should be involved in what is an iterative process as people's risk perceptions vary (previous experience, reputation, pride, animal welfare, etc.) and should be accommodated to help acceptance of the final list.

Regardless of the methodology used for this step, the aim is to answer three questions of the disease hazards identified and create a prioritized list of diseases:

1. Are they a risk? This can be a yes or no answer, based on the information assessed or prioritized by using a risk matrix or decision tree or simply on that of most concern to the business. Remember, this risk is to business stock, people, or the wider environment. Diseases that are of no consequence (i.e. vector not present, incorrect water temperature, age of stock) can be removed and the reason why it was removed recorded. This information should not be discarded.
2. Are they on the operation (EpiUnit)? Again, this can be a yes or no answer. If there is doubt, then steps should be taken to confirm absence/presence. It is essential that an accurate baseline of disease presence is determined. Not only does this assist in determining appropriate actions to either control or eradicate the disease, but it also helps the accuracy of the economic assessment of the benefits of the biosecurity plan in the future.
3. What are the consequences? In general, this will be assessed on the economic impact of the risk. The consequence should evaluate not just the mortality and morbidity but also the likely degree of spread (within and outside the EpiUnit), as well as the cost of control or eradication. The assessment does not have to be quantified at this stage but can be expressed qualitatively as say high, medium, or low.

6.7.4 Identifying an aquaculture operation's critical control points—evaluating an operation's critical weak points for disease introduction and spread

A critical control point (CCP) is a point in a procedure or operational step that can be *controlled* to eliminate the hazard or minimize the likelihood of its occurrence. Critical limits (maximum and minimum) need to be set and a monitoring system established to ensure the control of CCPs.

Hand washing and disinfection of footwear prior to entering different areas of a site can decrease pathogen spread, but it is not a CCP since it would be difficult to set critical limits or monitor the effectiveness (due to variation in hand washing technique, avoidance of footbath, etc.), therefore it might be better to incorporate such procedures in 'good hygiene practice'. However, for a terrestrial site, it is possible for the water intake to be a CCP since it is possible to treat the water to decrease the pathogen load, set minimum and maximum log reductions in pathogens, and effectively monitor the process.

In essence, all potential disease entry points (water intake, contact with/presence of other aquatic operations, stock, feed delivery and storage, equipment, staff, etc.) need to be evaluated and, where possible, a CCP set or a mitigating measure incorporated into a standard operating procedure (SOP).

Specific disease prevention recommendations will depend on the aquatic species, the diseases of concern, routes of transmission, and the likelihood of exposure.

6.7.5 Disease risk management/mitigation options—approaches for correcting an operation's critical weak points for disease introduction and spread

The easiest way to prevent a disease outbreak at an aquaculture operation or EpiUnit is to prevent the pathogen from entering, although there will also be controls to prevent spread within the EpiUnit. This is achieved by erecting barriers, which might be physical (e.g. fencing), chemical (e.g. disinfection), biological (e.g. vaccines), behavioral (e.g. deterrence of contact), spatial (e.g. distance between units), social (e.g. staff not allowed to keep aquatic animals), or educational (e.g. staff training).

For a land-based closed system, these barriers can be thought of as solid and providing good (measurable) protection, since it is easy to erect a fence to prevent unwanted visitors and ensure water intake is free of pathogens. A fully closed system is able to not only control pathogen entry but also aim to eradicate any disease present. As the system becomes more open, then some of these barriers may become more porous and the ability to eradicate a disease may lessen since the business may not be able to isolate itself from disease reservoirs present in wild populations or in adjacent business. Although the same barriers are used in an open or closed system, the emphasis in an open system moves more toward control through methods such as testing and surveillance (for early detection) rather than eradication, unless there is cooperation between aquaculture operations in the adjacent areas or those linked epidemiologically. Cooperation between businesses has been shown to be effective for many disease scenarios.

While the aim of any biosecurity plan is to prevent introduction of disease from external sources, it should also address how to prevent spread of disease already present at the EpiUnit or disease spread if the biosecurity measures are breached.

The approach to correcting the critical weak spots can be summarized as follows:

- What are the biosecurity plan requirements (i.e. control or eradication, internal or external benefits)?
- What are the available resources (i.e. economic, laboratory, staff training)?
- What is epidemiologically achievable (i.e. testing, surveillance, isolation)?

All three considerations are interconnected. From these general concepts, more specific details that need to be addressed include:

- What is the site location and is there separation from adjacent EpiUnits? What biosecurity controls do they have? What

diseases are present on them and could be transferred?
- Are there other facilities close by that could act as sources of disease (slaughterhouses, rendering plants, feed mills, or other concentrations of animals)?
- What are the effective means of physical separation?
- What access control is there for people (visitor book, a requirement for changing areas and showers)?
- What vehicle access control is there? Do they need to enter the site? If so, are there facilities for washing and disinfection (on leaving or prior to entry if they have been at another aquaculture operation)?
- What is the water supply? Is there sharing, can wild populations be excluded, and what methods of pathogen reduction are there?
- What is the source of the feed supply and how is it stored?
- Is there prevention of exposure to vectors? What are the vectors and can contact be prevented?

In order to detect disease entry or spread, the following need to be considered:

- Are personnel aware of the signs and symptoms of disease?
- What might trigger an 'outbreak investigation'?
- What monitoring and surveillance activities are there?
- Are laboratories used accredited and are test results valid?
- Where, how, and to whom should disease occurrence be reported?
- If an outbreak occurs, will there be an effective response?

From these considerations, it will be apparent where to concentrate resources on risk mitigation and what the appropriate measures will be.

6.7.6 Disease epidemiology and surveillance information—important principles

As explained in the previous chapter, disease epidemiology and surveillance could be briefly described as follows:

- Epidemiology asks who has what (disease), when and where (time and space), why and how (risk factors and transmission). This process defines the disease outbreak.
- You cannot compare disease outbreaks without the above information and hence will not be able to correctly evaluate the benefits arising from the biosecurity plan.
- Terms that you need to understand and use include, probability, relative risk, odds ratio, prevalence, incidence, sensitivity, specificity (and the difference between test Se/Sp and laboratory Se/Sp), likelihood ratios, positive and negative predictive values, and the difference between monitoring and surveillance.
- The goals of surveillance for an aquaculture operation are:

  - the rapid detection of disease outbreaks (to minimize control costs);
  - the early identification of disease problems (including environmental causes);
  - assessment of the health status of the defined population (working toward certification or audit);
  - definition of priorities for disease control and prevention (from risk assessment and cost–benefit analysis);
  - identification of new and emerging diseases (as part of contingency planning);
  - evaluation of the disease control programs; and
  - confirmation of absence of a specific disease (certification, disease-free status).

The preconditions of an 'ideal' surveillance system are:

- a pre-agreed intervention threshold (e.g. percentage mortality that triggers an investigation);
- an agreed case definition (e.g. the signs/symptoms/pattern that constitute a case or episode of a specific disease);
- harmonized diagnostic procedures (e.g. to ensure you are comparing like with like);
- a defined population at risk (e.g. the EpiUnit);

- a method of recording cases (farm and veterinary records);
- a method to report data (within business, between businesses, veterinarian and official services);
- timely and competent data analysis (designed dependent on who/what will use the data—requires some foresight);
- means to assess the effect of an intervention (measures to assess prevalence);
- that it is a continuous process (iterative); and
- means to communicate findings in a timely fashion.

6.7.7 Determining disease status and freedom—qualitative and quantitative approaches

A qualitative approach uses multiple sources of evidence, such as laboratory records from farm diagnostic testing or structured surveys carried out by the official services, other information from farm and slaughter records, and disease information databases. Using a panel of experts, this type of approach has been used by WOAH and national authorities to assess claims of freedom from disease following disease eradication programs. While it would seem to be a common sense approach to use all of the available evidence, it tends to be a subjective assessment with problems of transparency and repeatability. Further, the outcome will be either 'likely to be free from infection' or 'not likely to be free from infection,' with the inherent uncertainty unmeasured. For a business, this might be an appropriate approach for internal use but may not be 'convincing' for external use.

A quantitative approach could be a structured representative survey of the relevant population. This involves determining the number of animals to test at the design prevalence (P*).[10] If the probability of disease presence is less than the agreed P*, then the population is considered free of disease. While the cost of these surveys can be reduced by targeting animals at risk rather than the whole population (which reduces the number to be tested), they can be expensive to carry out.

Other weaknesses include failing to include other relevant evidence and the fact that the surveys are ephemeral (e.g. only relevant at the time they are carried out, unless repeated).

A method which combines the advantages of both the above by using multiple sources of surveillance data (both random and non-random) is scenario tree modeling for disease freedom.

6.7.8 Disease diagnostics and monitoring

The clinical signs expressed by aquatic animals may not be pathognomonic, and additionally they may be subclinically infected (i.e. not showing signs). Therefore, there is a requirement for some form of diagnostic testing to ensure correct pathogen/disease identification and early identification of a disease incursion. Since an effective biosecurity plan requires knowledge of the disease status of stock already on the EpiUnit and any stock introduced, the implication is that these diagnostics will be carried out routinely over a specified period of time. By recording this information, we are monitoring the stock.

Diagnostics will include:

- Observation of stock behavior—this is an essential activity for any EpiUnit.
- Clinical examination.
- On-site diagnostic testing such as skin scrapes, gill snips, routine blood biochemistry[11] and post mortem.[12]
- Samples may be processed on site for culture and sensitivity or sent to an external laboratory.
- Currently there are a few pen or tank-side tests said to be available for viral or bacterial infections, with others under development. Issues in using such tests revolve around not only the test Se and Sp, which is needed to interpret the test correctly, but also variance in results due to operator technique.
- For external laboratory diagnostics, it is better to use an accredited laboratory (e.g. WOAH accredited).

- The frequency of testing should be decided and recorded within the biosecurity plan.
- All of this information should be recorded and form part of the farm and veterinary record.

Monitoring[13] will include:

- carrying out and recording all of the above;
- water quality parameters;
- stock movements and density;[14]
- recording of mortality;
- recording of morbidity;
- feed usage;
- recording movements both on and off site and also within the EpiUnit if it is divided into separate sections or zones;
- treatment records (often required under national legislation); and
- significant weather events.

6.7.9 Auditing and certifying biosecurity status

Auditing is verifying that procedures are in place and working correctly (checking what is supposed to have been done/occurred, has been done/occurred) and is often carried out by external third parties. The process can be summarized as:

- opening meeting;
- inspection;
- verification;
- documenting;
- closing meeting; and
- report.

The report should identify the business or EpiUnit and set out the biosecurity risks and the actions taken, along with the corrective actions required/suggested and a measure of the confidence the auditor has in these actions. The results could be summarized as in Table 6.2

Certifying is somewhat different to auditing and, for a veterinarian, is carried out within a set of strict rules which provides assurance of the validity. Whereas an audit evaluates the effectiveness of the biosecurity plan procedures, its implementation, and user

Table 6.2. Biosecurity audit report summary.

Criteria	Description
Excellent	• Effective procedures • Biosecurity principles applied satisfactorily • No action necessary
Acceptable	• Procedures requiring only minor correction • Biosecurity principles generally applied satisfactorily, only minor corrections necessary • Action advisable
Poor	• Procedures inadequate • Biosecurity principles inadequately applied • Potentially significant implications of biosecurity failure • Action required (and, if regulated, likely formal enforcement if no action taken)
Unacceptable	• Immediate serious risk of biosecurity failure • Animal health and welfare at risk • Biosecurity principles not applied • Immediate action necessary (and, if regulated, formal enforcement if no action taken)

compliance based on records and evidence supplied, a veterinary certificate could only state that an audit has taken place, since the attending veterinarian is unlikely to have personal knowledge that a procedure was carried out, even though the evidence may well imply this. A veterinary certificate can only attest to an event/occurrence/ test where the signing veterinarian is fully aware and conversant with all of the facts.

6.7.10 Contingency plans for accidental disease outbreaks

The WOAH code defines contingency plans as:

> a documented work plan designed to ensure that all needed actions, requirements and resources are provided in order to eradicate or bring under control outbreaks of specified diseases of aquatic animals. (WOAH, 2022)

A contingency plan is initiated from the results of monitoring (i.e. a response to increased mortality) and should consider the following:

1. What triggers an outbreak investigation (i.e. increase in mortality beyond expected levels)?

2. Who should this be reported to/who should be informed?
3. What diagnostic procedure will be used to confirm the suspected outbreak?
4. Specific instructions for all farm staff if a disease outbreak is confirmed and/or during confirmation. These should be general for any possible disease outbreak, but include:

   • how to handle affected animals;
   • how to dispose of any dead animals;
   • instructions to deal with affected animals last;
   • within-site movement restrictions; and
   • additional hygiene precautions.

5. If animals to be slaughtered, how is this to be carried out and how will they be disposed of hygienically?
6. Is emergency slaughter to market an option?
7. What disease control measures will be put in place?
8. Will the whole EpiUnit or business be quarantined and how will this be achieved?
9. Any additional surveillance of unaffected stock that might be necessary.
10. Details of any disinfection procedures necessary.
11. How to establish that the outbreak has been controlled or eradicated.

12. How will the EpiUnit be restocked?

13. Will there be a need for fallowing?

14. Any necessary training for staff.

A mind map is a potential way to record these decisions and has the potential advantage of ensuring all aspects are correctly considered.

The objective of the contingency plan is to ensure any outbreak is minimized and localized as much as possible. Following the precautionary principle, if there are any doubts, err on the side of caution to minimize any potential disease outbreak. Key parts of the contingency plan might include rapid culling of infected populations and any contacts, speedy diagnosis, and movement controls of animals, people, and equipment on the premises.

By being prepared for any disease incursions, the speed with which the outbreak can be controlled or eradicated is much improved. Quantitative modeling can assist in developing strategies in preparation for an outbreak and in predicting and evaluating the effectiveness of control policies during an outbreak.

6.8 Developing the Aquaculture Biosecurity Plan

6.8.1 Preparing, implementing, and verifying aquaculture operation biosecurity

Diseases cost money. Each business has a management decision to make, based on its risk assessment and risk policy, and has to choose to what extent it bears these costs. The options are either to prevent a disease from occurring or to deal with the consequences if it does enter the farm. There will inevitably be costs in either situation. The difference is in finding the path of least cost (OATA, 2023).

Disease is recognized as a business risk, which can be mitigated or controlled through novel vaccines and diagnostics, and by using models to assist in the prevention and control of disease emergencies. New concepts such as *compartmentalization* provide prevention and control options to deal with emergencies (Murray, 2008).

Emerging pathogens can occur at individual farm level and increase in virulence. Biosecurity measures can prevent the risk of transmission of these virulent pathogens to other populations. Tools such as physical separation of farms and fallowing between generations limit the risk.

6.8.2 The importance of biosecurity plans

Why they are needed:

- Aquaculture is required to increase production significantly to satisfy demand for aquatic products in the face of an increasing world population and maximized and stable capture fisheries.
- Intensification of fish production provides an ideal environment for disease-causing organisms to flourish and cause serious damage to productivity.
- Both live and dead aquatic animals are moved rapidly around the world, with transport times less than the incubation period of a disease.
- Aquatic animals are moved for purposes of restocking wild populations.
- There are other potential routes of disease transmission, such as people, angling equipment, and inappropriate stocking of non-native or even native species in non-confined waters.
- Medication and vaccination alone cannot prevent losses due to disease. Coupled with biosecurity, they can provide increased benefit.
- All of these issues mean that the threat from trans-boundary diseases has never been greater and nor has the need for a biosecurity plan.

For an aquaculture business, biosecurity should be an integral part of the management strategy to increase the marketing success of the business, and this applies equally to non-profit entities where marketing might be considered the success of a project. It provides reassurance to the trading partners that the product is healthy, aids in providing guarantees on the provenance of stock, and shows in a very transparent manner that the welfare of animals is taken seriously and that the business is earnest about its responsibilities to operate in a sustainable manner and minimize environmental impact.

For the EpiUnit, the importance of a written plan cannot be overstated as it is this that ensures that due consideration has been given to the process as outlined. For a unit or group of units, a written plan is essential, should they wish to be considered as a compartment, and the plan must be written in consultation with the competent authority. With governments moving closer toward 'public-private partnerships' or 'cost and responsibility sharing,' preparing some sort of biosecurity plan becomes a necessity.

Biosecurity implies the maintenance of a spatial separation between categories of biological things. It is a normal part of animal production and might be conceived as a modern way of thinking and talking about activities that have gone on before. It is a way of formalizing disease control strategies and communicating these strategies to those concerned. However, for the farmer, concepts are not useful or practical unless they can be shown to have a real impact on the economic viability and profitability of the enterprise. Also, remember that the degree of biosecurity required and how to achieve it is dependent on the type of aquaculture operation and the amount of risk present. Open operations are at more risk than closed ones. Commercial businesses' biosecurity needs might differ compared to hobby facilities.

Preliminary discussions with the veterinarian need to consider the following general questions:

- Is the business interested in implementing a biosecurity plan?
- What are the important diseases; do they have a financial, welfare, or regulatory impact?
- Can you actually control the disease or prevent its incursion?
- How are you going to achieve prevention, control, or possible eradication?
- Does the business or EpiUnit have the resources to achieve the plan?
- How do you ensure staff compliance and, if necessary, that of the suppliers?
- Is there a clear cost benefit or is it only probable? (Impinges on long-term compliance).

Answers to these questions can be obtained by completion of a 'Facility Preliminary Questionnaire,' which then allows the veterinarian to consider what in his professional opinion is required, potential costs/benefits, and likely success. This then becomes the basis for further detailed development. Copies of this completed questionnaire should be retained by both parties. The next step would be completion of the information required in the 'Facility Detailed Questionnaire.' You are now ready to develop a business- or EpiUnit-specific biosecurity plan.

The most critical element is that all stakeholders must understand the concept and be apprised of the financial viability and feasibility of the project. Adding to that, economic considerations and feasibility are important components of a successful biosecurity plan.

6.8.2.1 Stakeholder understanding

- This requires good communication between all parties.
- The producer is responsible for delivering a safe and high-quality product, prepared in accordance with national and international laws and regulations.
- It must be recognized there may be unknown consequences of the interaction between high-density populations and the environment, and there may be both knowledge gaps and fragmented responsibilities.
- Measures to decrease the possible introduction of disease or decrease pathogen load will inevitably inconvenience staff, and people naturally use the most convenient methods for daily activities. The more staff are inconvenienced, the less likely they are to follow biosecurity policies, unless they understand and believe that those policies are required and have value.
- Biosecurity may require allocating resources to low probability risks that may occur at some point in the future. People are not particularly good at making such decisions, placing too much weight on the recent past and overemphasizing these risks.

6.8.2.2 Economic considerations

- For the veterinarian working with a client, there are few if any publications that provide an economic assessment of biosecurity, mainly due to the numerous variables that need to be considered on a site-by-site or disease basis, but a description of the process for Bovine Viral Diarrhoea is given by Stott *et al.* (2003). It would seem that, at the business level, there will be a need to assess the economics on a case-by-case basis.
- At the international or national level, there are clear examples of the cost of breaches or lack of biosecurity, such as the billions of dollars lost in production of shrimp in the 1990s due to the global dissemination of the various shrimp viruses, the losses associated with infectious salmon anemia (ISA) in Scotland and, more recently, in Chile, and the introduction of *Bonamia* around the world.
- The European Food Safety Authority (EFSA) opinion on salmon welfare clearly links good welfare to biosecurity, and some governments are linking biosecurity to the availability of subsidies or compensation.
- There is also a social responsibility to implementing biosecurity, not only by providing a 'healthier' product, but also it would seem sensible to avoid harming your neighbors through accidental introduction of disease.

6.8.2.3 Feasibility

- Consider management priorities and concerns–short and long-term objectives.
- What is the farmer's expected return on investment (effort and cost).
- What is realistically achievable with the resources available?
- Will the plan be effective?
- Concentrating on key diseases may decrease the complexity of the biosecurity plan and will impact on prevention, control, or possible eradication of lower risk or lower cost diseases.

6.8.3 Steps required to prepare a biosecurity plan

See Fig. 6.10.

6.8.3.1 Development

- Collate information on the business/EpiUnit's current health status, production data, any current biosecurity procedures, and business goals.
- Agree hazard identification and prioritization (there may be variation between the business and the veterinarian's assessments).
- Carry out a risk assessment (is the business/EpiUnit at risk, to what extent and with what consequence). Use business's historical records, WAHIS data,[15] IAAD data,[16] proximity to other facilities, and facility design and build, etc. Best practice is for the veterinarian and the business to individually carry out an assessment and then reach mutual agreement.
- Evaluate CCPs and agree possible remediation.
- Decide presence/absence of diseases at the aquaculture operation through suitable clinical evaluation and diagnostic testing.
- Decide appropriate risk mitigation/management—what can be done to prevent or minimize disease incursion? (i.e. disease vector control, equipment disinfection, restriction of visitor access, provision of work clothing, washing/showering facilities, vehicle access control, quarantining or isolating new animals, improvements to business infrastructure, and restrictions on farmer/employee ownership of animals.)
- Design a suitable surveillance, monitoring, and testing program.
- Contingency planning—what to do if disease occurs, what triggers a disease outbreak investigation, possible options for treatment, isolation, or depopulation, disposal of clinical waste/cadavers.
- Agree necessary record keeping and responsibilities.
- Agree a suitable review process for the plan (monthly, annually, etc.).
- Consider veterinary auditing and certification.

Questions a Farmer Might Ask	Formal Biosecurity Process/Steps	Documentation & Records

Which diseases are serious potential hazards?

BIOSECURITY LEVEL I

Is my farm at risk?
How much risk?
Operational impact of disease?

Where can these hazardous diseases get in?

BIOSECURITY LEVEL II

What can be done to prevent disease entry or escape?

What should I do if disease gets in?

BIOSECURITY LEVEL III

Are any of these diseases on the farm?

How do I continue to monitor disease absence/presence?

BIOSECURITY LEVEL IV

How do I get third-party recognition of disease freedom?

BIOSECURITY LEVEL V

1. Hazard Identification & Prioritization → Prioritized Disease List

2. Risk-Impact Assessment → Evaluation of Disease Impacts

3. Critical Control Point (CCP) Evaluation → Identify the CCPs

4. Mitigation, Management & Remediation of CCP Risks → Implement CCP Corrective Actions

5. Contingency Planning → Isolation, Treatment Depopulation Plans

6. Clinical Evaluation & Diagnostic Testing → Farm, Lab & Vet Records Results

7. Ongoing Disease Surveillance & Monitoring → Farm, Lab & Vet Records Results

8. Veterinarian Auditing & Certification → Certificate of Veterinary Inspection (CVI)

9. Veterinary Authority (Gov't) Verification & Endorsement → Gov't Endorsed Certificate of Veterinary Inspection (CVI)

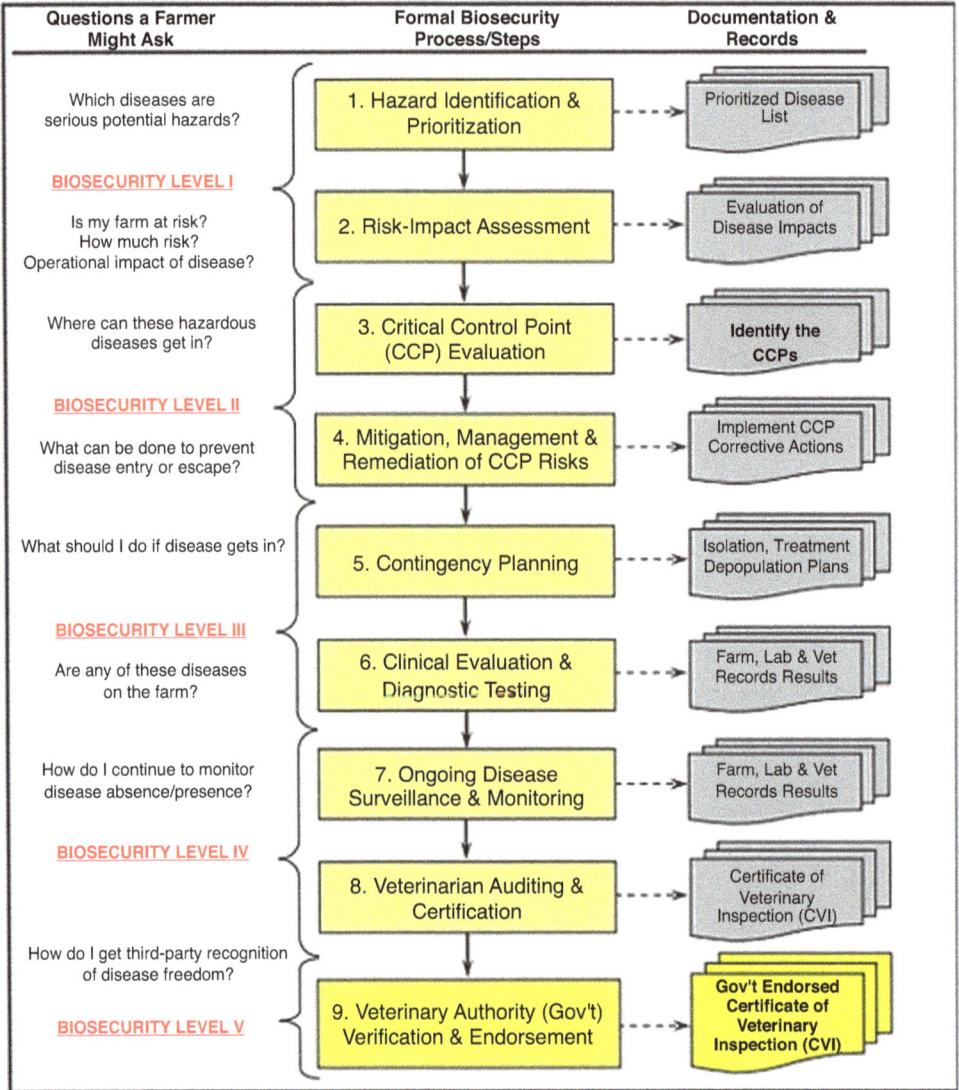

Fig. 6.10. Representation of a biosecurity program for an EpiUnit. Figure modified from Palić and Scarfe (2020). Permission granted by Dusan Palić.

Additional steps might include assessing whether and how the plan will integrate with other health and production issues and whether the plan will work in reality—in other words, *is it feasible?*

6.8.3.2 *Implementation*

- An accurate baseline of the disease status of the business/EpiUnit must be established prior to implementation.

- Stakeholders to have received training in all aspects of the plan. This may require some form of formal evaluation or assessment.
- Suitable plan documentation should be easily available to stakeholders to act as an aide-mémoire.
- Ensure correct record keeping is in place. Without records the plan is meaningless.
- Signage should be placed to remind staff and visitors that biosecurity is in place and what procedures are required before entering or exiting the facility.
- Monitor compliance, correct any mistakes, and review after a suitable time period.

6.8.3.3 Auditing

- The aim of auditing is to assist the business in achieving its objectives and, when carried out by an external assessor (i.e. veterinarian), provides an independent and objective opinion based on verifiable facts.
- Auditing should be carried out annually by a veterinarian, but it is advisable that the business carry out an internal audit on a monthly basis to ensure records are available and complete.

6.8.3.4 Certifying

- Certification is a procedure to ensure that a product, process, or service conforms to specified requirements.
- First Party certification is basically a self-declaration that the producer has met a certain standard.
- Second Party certification basically shows that the producer has met the requirements of a standard set by an external group (i.e. consumers, government, or NGOs). The producer is externally assessed, but often the same group that sets the standard also assesses it, thus there may be an issue as the standards may reflect the interests of the group that sets and assesses them.
- Third Party certification is based upon standards created by a multi-stakeholder process. Compliance is voluntary and is assessed by an accredited, independent third party holding no vested interest in the standards, certification, product, or stakeholder group. The International Organization for Standardization (ISO) defines third party certification as the highest order of proof of compliance.
- Veterinarians in the UK and Europe must follow the 10 principles of veterinary certification.[17]
- Following veterinary auditing, the EpiUnit may be issued a Certificate of Veterinary Inspection or receive government endorsement of disease freedom.

6.8.4 EU legislation, policy, and regulations on food safety measures and antimicrobial use

6.8.4.1 Food safety measures

Within the EU, products of animal origin intended for human consumption must meet specific requirements and the EU legal framework for health and food safety. These requirements are implemented by the national authorities responsible for official controls on residues of contaminants, pharmacologically active substances, and pesticides that may be present in products of animal origin destined for human consumption.

A third/non-EU country can export to the EU animals and products of animal origin for human consumption provided they meet three main requirements:

- animal health,
- public health, and
- control plan for 'monitoring of residues of contaminants, pharmacologically active substances and pesticides.'[18]

6.8.4.1.1 ANIMAL HEALTH REQUIREMENTS. Regulation (EU) 2021/404 provides 'the list of the non-EU countries which are authorized to export live animals, germinal products and products of animal origin.'

Regulation (EU) 2020/692, supplementing Regulation (EU) 2016/429, provides for 'rules for entry, movement and handling of consignments of certain animals, germinal products and products of animal origin.'

Animal health	Commission Implementing Regulation (EU) 2021/404[19]

6.8.4.1.2 PUBLIC HEALTH REQUIREMENTS. The establishments registered to export products of animal origin are listed in the European Commission's online platform for animal and plant health certification, TRACES.

TRACES	https://food.ec.europa.eu/horizontal-topics/traces_en
Public health (hygiene and residues)	Commission Implementing Regulation (EU) 2021/405[20]

6.8.4.1.3 RESIDUE MONITORING AND CONTROL PLAN (RCP). The requirements applicable to a non-EU country for controls on residues of pesticides, pharmacologically active substances (authorized in veterinary medicinal products or as animal feed additives), and contaminants used in animals and animal products destined for human consumption are provided in Regulation (EU) 2022/2292. The Regulation supplements Regulation (EU) 2016/429 of the European Parliament and of the Council on transmissible animal diseases (i.e. the Animal Health Law) and the Official Control Regulation (EU) 2017/625, which set out conditions for addition of non-EU countries to the list of those countries authorized to export animals and products for human consumption to the EU. Regulation (EU) 2022/2292 provides for 'additional food safety requirements for inclusion of non-EU countries on the list of countries authorized to export certain animals or goods to the European Union.'

Countries with an approved residue control plan are listed with an **x** in Annex-1 of Regulation (EU) 2021/405 (amended by Commission Implementing Regulation of the EU 2024/334 of January 19, 2024). The list is regularly updated by the Commission, and amended as needed upon positive opinion expressed by the EU member states.

The residue control plan refers to:

- use of pharmacologically active substances;
- residues of pharmacologically active substances (maximum limits) and residues of pesticides and contaminants (maximum levels); and
- other requirements on pharmacologically active substances and contaminants in products of animal origin and composite products.

Thus, third countries seeking access to the EU market, when preparing a residue control plan for pharmacologically active substances for submission to the Commission services, should follow the approach set out for EU member states in Regulation (EU) 2022/1644 and Regulation (EU) 2022/1646. Alternatives that are aligned with the EU approach can also be used to prepare residue monitoring plans during the listing request process, as for example the *Codex Alimentarius* Guidelines for the design and implementation of national regulatory food safety assurance programs associated with the use of veterinary drugs in food production animals (CAC/GL 71–2009).

Approved residue control plan	Delegated regulation - 2022/2292[21]
Public health (hygiene and residues)	Commission Implementing Regulation (EU) 2021/405[22]

6.8.4.2 *Composite products*

Composite products are food products containing mixed ingredients of processed products of animal origin and plant-based products. These are among the most commonly traded food commodities worldwide.

Three groups of composite products are recognized, based on their shelf stability: non-shelf stable (perishable), shelf-stable with colostrum-based or meat products (except gelatine, colostrum, and highly refined products[23]), and shelf-stable without colostrum-based or meat products (except gelatine, colostrum, and highly refined products derived from meat).

As a general rule, all ingredients of animal origin present in a composite product must originate from EU-authorized establishments. Regulation (EC) 853/2004 lists specific products of animal origin and processed products of animal origin. The non-EU countries authorized to export composite products to the EU are listed in TRACES NT.[24]

The rules for import of composite products are provided in Regulation (EU) 2022/2292, which supplements Regulation (EU) 2017/625 on official controls. The list of registered and approved establishments is available in TRACES NT.

Case Study

Approved residue control plan for composite products

For composite products, if a country is marked with an **x** in Annex-1 for certain categories of animal origin—products like eggs, milk, meat, or fish/aquaculture (but excluding honey)—they can request to have an **o** added to Annex-1 for the remaining categories. This allows them to use animal ingredients from other categories (without an **x**) from countries authorized by the EU (member state or third country).

In the case of both composite and non-composite products (raw materials), if a country uses raw materials from a member state or third country approved for entry into the Union and marked with an **x** for the included goods, they are marked with a **Δ**; this is known as 'triangular trade.'

For shelf-stable composite products containing animal-derived ingredients like eggs or dairy, a country not listed for these ingredients can still use them in the product if they purchase them from an EU member state or a third country authorized to export them. The third country must have a residue control plan in place for certain commodities (meat, fish/aquaculture, eggs, or milk) in order to be listed for exporting to the EU. Once listed, the third country must adhere to regulations regarding residue plans for pharmacologically active substances, pesticides, and contaminants for each ingredient present in the composite product.

Animal health requirements for composite products

The annexes to Regulation (EC) 2021/404 list the third countries authorized to export live animals, germinal products, and products of animal origin.

The 'rules for entry movement and handling of consignments of certain animals, germinal products and products of animal origin' are found in Regulation (EU) 2020/692 supplementing Regulation (EU) 2016/429.

EU animal health certificates are found in three EU Regulations:

1. Animals and animal products: Commission Implementing Regulation (EU) 2020/2235.[25]
2. Germinal products and certain terrestrial animals: Commission Implementing Regulation (EU) 2021/403.[26]
3. Certain animal by-products: Commission Regulation (EU) 142/2011.[27]

Public health requirements for composite products

The establishments registered to export products of animal origin are listed in TRACES.

When requests are received by the Commission, depending on the type of composite (whether shelf-stable/non-shelf-stable), the following aspects must be considered:

For non-shelf-stable composite products:

- The country must be authorized for each ingredient in the composite product (e.g. frozen cake with perishable ingredients).

For shelf-stable composite products:

- Compliance with the criteria for composite products:

  o For *shelf-stable* composites, countries authorized to export composite products with meat, fishery, egg, or dairy products can export to the EU products with ingredients based on one or more of the four products listed above. This doesn't apply to honey, which means countries authorized to export honey will only be able to export honey to the EU.

○ *Non-shelf-stable* and *Shelf-stable composites with colostrum-based or meat products* require official certificates (available in Chapter 50 – Annex III to Reg. (EU) 2020/2235).

○ *Shelf-stable composites without colostrum-based or meat products* (e.g. biscuits) require a private attestation (model available in Annex V to Reg. (EU) 2020/2235).

6.8.4.3 *Antimicrobial use*

Commission Delegated Regulation (EU) 2023/905 of February 27, 2023, regarding the application of the prohibition of use of certain antimicrobial medicinal products in animals or products of animal origin exported from non-EU countries ('third countries') to the Union, constitutes the basis for these specific rules and impacts livestock and animal products for human consumption with several exceptions: gelatine, collagen, highly refined products, composite products, wild animals, insects, frogs, snails and reptiles, animals and products of animal origin for human consumption not following to be placed on the EU market (i.e. dispatches transiting the EU, samples for analytical purposes), and products not intended for human consumption.

The EU's regulatory framework for veterinary medicinal products has been updated with the introduction of Regulation (EU) 2023/905, which builds upon Regulation (EU) 2019/6. The latter regulation replaced Directive 2001/82 and Regulation (EC) No 726/2004, which previously governed the authorization, manufacture, import, export, distribution, and use of veterinary medicines within the EU. Regulation (EU) 2023/905 introduces stricter rules on the use of certain antimicrobial medicines in animals that are reserved for human treatment. The criteria for designating these reserved antimicrobials are outlined in Commission Delegated Regulation (EU) 2021/1760, and the list of reserved antimicrobials is established in Commission Implementing Regulation (EU) 2022/1255. As a result, non-EU countries exporting animals

or animal products to the EU must comply with the new rules on antimicrobial use, as set out in Regulation (EU) 2023/905.

To ensure compliance with EU regulations, a system was implemented to control the importation of animals and animal products from non-EU countries. This led to the amendment of Regulation (EU) 2017/625 in 2021, which included verification of adherence to specific rules outlined in Article 118(1) of Regulation (EU) 2019/6. These rules govern official controls, including the conditions for exporting animals and animal products from third countries to the EU. Non-EU countries permitted to export to the EU are listed in Annex I of the Commission Implementing Regulation (EU) 2024/334. This list confirms that these countries have a validated control plan for the use of certain substances, and their national veterinary authorities guarantee compliance with EU requirements, as supplemented by Delegated Regulation (EU) 2023/905 regarding the use of antimicrobial medicines.

6.9 The Agreement on the Application of Sanitary and Phytosanitary Measures (The SPS Agreement of the WTO)

The Sanitary and Phytosanitary (SPS) Measures Agreement concerns the application of food safety and animal and plant health regulations within the World Trade Organization (WTO) member states, by setting out the science-based basic rules for sanitary and phytosanitary standards. The WTO itself does not develop such standards. Instead, most WTO countries participate in the development of these standards through representation in other international bodies, and with leading scientists and policy maker experts in the field.

The SPS standards ensure that consumers are supplied with safe food to eat, preventing the spread of pests or diseases among animals and plants, while eliminating unnecessary barriers to international trade and ensuring that strict health and safety regulations are not used to protect local producers. The standards are set out only to the extent required to protect human, animal, and plant health, and 'not to arbitrarily or unjustifiably discriminate between countries where identical or similar conditions prevail.'[28]

To provide the level of health protection deemed appropriate, countries are allowed to establish their own standards and use existing international standards or recommendations. Provided there is scientific justification, countries can also take measures which result in higher standards, or they can set different standards, but these standards 'must be based on appropriate assessment of risks' and countries must ensure this approach is not arbitrary (or, in other words, these measures are not used for protectionist purposes and do not result in unnecessary barriers to international trade) while encouraging consistent decision making (Marcuta *et al.*, 2023).

6.9.1 Key features of the SPS Agreement

6.9.1.1 *Conditional international trade*

To ensure that food is safe for consumers, and to prevent the spread of pests or diseases in member countries, products intended for international trade are required to meet several conditions, such as be produced in disease-free areas, undergo inspection and/or specific treatment or processing, conform to allowable maximum levels of residues, or use only certain additives in food. SPS measures can sometimes result in restrictions on trade, necessary to ensure food safety and animal and plant health protection. Sometimes, countries go beyond what is needed for sanitary and phytosanitary health protection, using these restrictions to preserve national (domestic) producers from international competition.

6.9.1.2 *Harmonization*

The SPS Agreement encourages governments to establish national SPS measures consistent with international standards, guidelines, and recommendations. This process is known as 'harmonization.'

6.9.1.3 *Science-based standards*

WTO countries may choose not to use the international standards, but should the national requirement lead to a greater restriction of trade, it may be required to scientifically demonstrate that the restrictions are necessary as the international standard would not result in the level of health protection the country considered appropriate.

6.9.1.4 *Adaptive measures*

Harmonizing the SPS requirements on food, animal, and plant products originating from different countries is not always possible, due to existing pests or pathogens, food safety conditions, differences in climate, and existence of disease-free areas which do not always correspond to national boundaries. These aspects are taken into account by the SPS Agreement, which provides for 'alternatives to the SPS measures,' given that these are not discriminative or do not favor local producers or foreign suppliers.

6.9.1.5 *Risk-based assessment*

Countries must develop SPS measures by conducting suitable risk assessments and, if needed, disclose the factors considered, assessment methods used, and acceptable risk level. The SPS Agreement promotes the adoption of systematic risk assessment by all WTO member countries.

6.9.1.6 *Alternatives for achieving acceptable risk levels*

Various options exist for achieving acceptable levels of risk, assuming these options are technically and economically viable, and offer the same level of food safety or animal and plant health as those in other countries. Countries are encouraged to choose measures that are no more trade restrictive than necessary to achieve their health goals. In some cases, countries can demonstrate that their standards provide the same level of health protection required by the SPS Agreement, making them equivalent. This approach ensures health protection while maximizing the availability of safe food options for consumers and fostering healthy economic competition.

6.9.1.7 *Transparency*

WTO governments must open to review the way the regulations on food safety and animal and plant health are implemented within the country. The WTO country members have established a systematic communication flow which provides the basis for national standards. At the same time, these countries are required to notify each other of any modification of existing or new sanitary and phytosanitary requirements affecting trade, and to set up offices (i.e. enquiry points) to respond to requests for more details about existing or new national SPS measures.

Notes

[1] https://www.cdcfoundation.org/what-public-health

[2] www.emro.who.int/health-topics/public-health-surveillance/index.html

[3] https://randd.defra.gov.uk/ProjectDetails?ProjectId=12738. DEFRA research projects can be found at https://randd.defra.gov.uk

[4] For further information on RCS see: *Responsibility and Cost Sharing for Animal Health and Welfare: Final Report*: https://www.gov.uk/government/publications/responsibility-and-cost-sharing-for-animal-health-and-welfare-final-report

[5] E.g. AQUAPLAN – Australia's National Strategic Plan for Aquatic Animal Health: https://www.agriculture.gov.au/agriculture-land/animal/aquatic/aquaplan

[6] WOAH *Aquatic Animal Health Code*: https://www.woah.org/en/what-we-do/standards/codes-and-manuals/aquatic-code-online-access/

[7] WOAH *Manual of Diagnostic Tests for Aquatic Animals*: https://www.woah.org/en/what-we-do/standards/codes-and-manuals/aquatic-manual-online-access/

[8] E.g. Code of Good Practice for Scottish Finfish Aquaculture: https://thecodeofgoodpractice.co.uk

[9] An example is: https://www.mpi.govt.nz/dmsdocument/2803-Salmonids-for-human-consumption-Import-risk-analysis-September-1997

[10] Design prevalence (P') = the disease prevalence we expect or know is there. In most cases, for disease freedom, a P' of <2% is used (usual WOAH level), but it can be varied, depending on the type of disease or what your trading partner requires.

[11] E.g. monitoring the presence/course of pancreas disease through the measuring of liver enzyme levels.

[12] Should be treated as clinical waste and disposed of accordingly.

[13] Monitoring is not the same as surveillance, as surveillance = monitoring + action.

[14] Since it is often difficult to observe aquatic animals fully or measure their biomass accurately, an estimate is acceptable.

[15] WAHIS — World Animal Health Information System: https://wahis.woah.org/#/home

[16] WOAH Collaborating Centre for Information on Aquatic Animal Diseases (IAAD): https://www.cefas.co.uk/icoe/aquatic-animal-health/designations/woah-collaborating-centre-for-emerging-aquatic-animal-disease/

[17] See https://www.rcvs.org.uk/setting-standards/advice-and-guidance/code-of-professional-conduct-for-veterinary-surgeons/supporting-guidance/certification/

[18] Commission Delegated Regulation (EU) 2022/2292 of 6 September 2022 supplementing Regulation (EU) 2017/625 of the European Parliament and of the Council with regard to requirements for the entry into the Union of consignments of food-producing animals and certain goods intended for human consumption; Annex I to Commission Implementing Regulation (EU) 2021/405 and Commission Implementing Regulation (EU) 2024/334.

[19] https://eur-lex.europa.eu/search.html?DTA=2021&SUBDOM_INIT=ALL_ALL&DB_TYPE_OF_ACT=regulation&DTS_SUBDOM=ALL_ALL&typeOfActStatus=REGULATION&DTS_DOM=ALL&type=advanced&excConsLeg=true&qid=1680595349369&DTN=0404

[20] https://eur-lex.europa.eu/search.html?DTA=2021&SUBDOM_INIT=ALL_ALL&DB_TYPE_OF_ACT=regulation&DTS_SUBDOM=ALL_ALL&typeOfActStatus=REGULATION&DTS_DOM=ALL&type=advanced&excConsLeg=true&qid=1680595417738&DTN=0405

[21] https://eur-lex.europa.eu/eli/reg_del/2022/2292/oj
[22] https://eur-lex.europa.eu/search.html?DTA=2021&SUBDOM_INIT=ALL_ALL&DB_TYPE_OF_ACT=regulation&DTS_SUBDOM=ALL_ALL&typeOfActStatus=REGULATION&DTS_DOM=ALL&type=advanced&excConsLeg=true&qid=1680595417738&DTN=0405
[23] Regulation (EC) No 853/2004.
[24] https://webgate.ec.europa.eu/IMSOC/tracesnt-help/Content/en/index.html
[25] https://eur-lex.europa.eu/search.html?DTA=2020&SUBDOM_INIT=ALL_ALL&DB_TYPE_OF_ACT=regulation&DTS_SUBDOM=ALL_ALL&typeOfActStatus=REGULATION&DTS_DOM=ALL&type=advanced&excConsLeg=true&qid=1680594835587&DTN=2235
[26] https://eur-lex.europa.eu/search.html?DTA=2021&SUBDOM_INIT=ALL_ALL&DB_TYPE_OF_ACT=regulation&DTS_SUBDOM=ALL_ALL&typeOfActStatus=REGULATION&DTS_DOM=ALL&type=advanced&excConsLeg=true&qid=1680594994725&DTN=0403
[27] https://eur-lex.europa.eu/search.html?DTA=2011&SUBDOM_INIT=ALL_ALL&DB_TYPE_OF_ACT=regulation&DTS_SUBDOM=ALL_ALL&typeOfActStatus=REGULATION&DTS_DOM=ALL&type=advanced&excConsLeg=true&qid=1680596847077&DTN=0142
[28] WTO Analytical Index: Guide to WTO Law and Practice, Article XX of the GATT 1994: https://www.wto.org/english/thewto_e/whatis_e/tif_e/agrm4_e.htm

References

AbuOun, M., Jones, H., Stubberfield, E., Gilson, D., Shaw, L.P. *et al.* (2021) A genomic epidemiological study shows that prevalence of antimicrobial resistance in Enterobacterales is associated with the livestock host, as well as antimicrobial usage. *Microbial Genomics* 7, 000630. DOI: 10.1099/mgen.0.000630.

Adamie, B.A., Akwar, H.T., Arroyo, M., Bayko, H., Hafner, M. *et al.* (2024) *Forecasting the Fallout from AMR: Economic Impacts of Antimicrobial Resistance in Food-Producing Animals. A Report from the EcoAMR Series*. World Organisation for Animal Health, Paris, and World Bank, Washington, DC. DOI: 10.20506/ecoAMR.3541.

Afzaal, M., Saeed, F., Shah, Y.A., Hussain, M., Rabail, R. *et al.* (2022) Human gut microbiota in health and disease: Unveiling the relationship. *Frontiers in Microbiology* 13, 999001. DOI: 10.3389/fmicb.2022.999001.

Ager, E.O., Carvalho, T., Silva, E.M., Ricke, S.C. and Hite, J.L. (2023) Global trends in antimicrobial resistance on organic and conventional farms. *Scientific Reports* 13, 22608. DOI: 10.1038/s41598-023-47862-7.

Agga, G.E. and Amenu, K. (2024) Editorial: Antimicrobial resistance in food-producing environments: A one health approach. *Frontiers in Antibiotics* 3, 1436987. DOI: 10.3389/frabi.2024.1436987.

Agga, G.E., Galloway, H.O., Netthisinghe, A.M.P., Schmidt, J.W. and Arthur, T.M. (2022) Tetracycline-resistant, third-generation cephalosporin–resistant, and extended-spectrum β-lactamase–producing *Escherichia coli* in a beef cow-calf production system. *Journal of Food Protection* 85(11), 1522–1530. DOI: 10.4315/jfp-22-178.

Aghamohammad, S. and Rohani, M. (2023) Antibiotic resistance and the alternatives to conventional antibiotics: The role of probiotics and microbiota in combating antimicrobial resistance. *Microbiological Research* 267, 127275. DOI: 10.1016/j.micres.2022.127275.

Ajulo, S. and Awosile, B. (2024) Global antimicrobial resistance and use surveillance system (GLASS 2022): Investigating the relationship between antimicrobial resistance and antimicrobial consumption data across the participating countries. *PLOS One* 19(2), e0297921. DOI: 10.1371/journal.pone.0297921.

Akram, F., Imtiaz, M. and Haq, I. (2023) Emergent crisis of antibiotic resistance: A silent pandemic threat to 21st century. *Microbial Pathogenesis* 174, 105923. DOI: 10.1016/j.micpath.2022.105923.

Allen, H.K. (2017) Alternatives to antibiotics: Why and how? *NAM Perspectives* 7(7). DOI: 10.31478/201707g.

Almansour, A.M., Alhadlaq, M.A., Alzahrani, K.O., Mukhtar, L.E., Alharbi, A.L. *et al.* (2023) The silent threat: Antimicrobial-resistant pathogens in food-producing animals and their impact on public health. *Microorganisms* 11(9), 2127. DOI: 10.3390/microorganisms11092127.

Angamarca, E., Castillejo, P. and Tenea, G.N. (2023) Microbiota and its antibiotic resistance profile in avocado Guatemalan fruits (*Persea nubigena* var. *guatemalensis*) sold at retail markets of Ibarra city, northern Ecuador. *Frontiers in Microbiology* 14, 1228079. DOI: 10.3389/fmicb.2023.1228079.

Argimón, S., Masim, M.A.L., Gayeta, J.M., Lagrada, M.L., Macaranas, P.K.V. *et al.* (2020) Integrating whole-genome sequencing within the National Antimicrobial Resistance Surveillance Program in the Philippines. *Nature Communications* 11(1), 2719. DOI: 10.1038/s41467-020-16322-5.

Batuman, O., Britt-Ugartemendia, K., Kunwar, S., Yilmaz, S., Fessler, L. *et al.* (2024) The use and impact of antibiotics in plant agriculture: A review. *Phytopathology* 114(5), 885–909. DOI: 10.1094/phyto-10-23-0357-ia.

Brown, N.M., Goodman, A.L., Horner, C., Jenkins, A. and Brown, E.M. (2021) Treatment of methicillin-resistant *Staphylococcus aureus* (MRSA): Updated guidelines from the UK. *JAC-Antimicrobial Resistance* 3(1), dlaa114. DOI: 10.1093/jacamr/dlaa114.

Browne, A.J., Chipeta, M.G., Haines-Woodhouse, G., Kumaran, E.P.A., Kashef Hamadani, B.H. *et al.* (2021) Global antibiotic consumption and usage in humans, 2000–18: A spatial modelling study. *The Lancet Planetary Health* 5(12), e893–e904. DOI: 10.1016/s2542-5196(21)00280-1.

Brunn, A., Kadri-Alabi, Z., Moodley, A., Guardabassi, L., Taylor, P. *et al.* (2022) Characteristics and global occurrence of human pathogens harboring antimicrobial resistance in food crops: A scoping review. *Frontiers in Sustainable Food Systems* 6. DOI: 10.3389/fsufs.2022.824714.

Byrd, A., Belkaid, Y. and Segre, J. (2018) The human skin microbiome. *Nature Reviews Microbiology* 16, 143–155. DOI: 10.1038/nrmicro.2017.157.

Caderhoussin, A., Couyin, D., Gruel, G., Quétel, L., Pot, M. *et al.* (2024) The fly route of extended-spectrum-β-lactamase-producing *Enterobacteriaceae* dissemination in a cattle farm: From the ecosystem to the molecular scale. *Frontiers in Antibiotics* 3, 1367936. DOI: 10.3389/frabi.2024.1367936.

Callahan, B.J., Fukami, T. and Fisher, D.S. (2014) Rapid evolution of adaptive niche construction in experimental microbial populations. *Evolution* 68(11), 3307–3316. DOI: 10.1111/evo.12512.

Choy, S.K., Neumann, E.-M., Romero-Barrios, P. and Tamber, S. (2024) Contribution of food to the human health burden of antimicrobial resistance. *Foodborne Pathogens and Disease* 21(2), 71–82. DOI: 10.1089/fpd.2023.0099.

Criscuolo, N.G., Pires, J., Zhao, C. and Boeckel, T.P. (2021) Resistancebank.org, an open-access repository for surveys of antimicrobial resistance in animals. *Scientific Data* 8(1), 189. DOI: 10.1038/s41597-021-00978-9.

Cycoń, M., Mrozik, A. and Piotrowska-Seget, Z. (2019) Antibiotics in the soil environment—degradation and their impact on microbial activity and diversity. *Frontiers in Microbiology* 10, 338. DOI: 10.3389/fmicb.2019.00338.

David, L.A., Maurice, C.F., Carmody, R.N., Gootenberg, D.B., Button, J.E. *et al.* (2014) Diet rapidly and reproducibly alters the human gut microbiome. *Nature* 505(7484), 559–563. DOI: 10.1038/nature12820.

Denamur, E. and Matic, I. (2006) Evolution of mutation rates in bacteria. *Molecular Microbiology* 60(4), 820–827. DOI: 10.1111/j.1365-2958.2006.05150.x.

Ding, D., Wang, B., Zhang, X., Zhang, J., Zhang, H. *et al.* (2023) The spread of antibiotic resistance to humans and potential protection strategies. *Ecotoxicology and Environmental Safety* 254, 114734. DOI: 10.1016/j.ecoenv.2023.114734.

Dolejska, M. and Papagiannitsis, C.C. (2018) Plasmid-mediated resistance is going wild. *Plasmid* 99, 99–111. DOI: 10.1016/j.plasmid.2018.09.010.

Dolejska, M. and Literak, I. (2019) Wildlife is overlooked in the epidemiology of medically important antibiotic-resistant bacteria. *Antimicrobial Agents and Chemotherapy* 63(8), e01167–19. DOI: 10.1128/AAC.01167-19.

Domingues, C.P.F., Rebelo, J.S., Dionisio, F. and Nogueira, T. (2023) Multi-drug resistance in bacterial genomes—a comprehensive bioinformatic analysis. *International Journal of Molecular Sciences* 24(14), 11438. DOI: 10.3390/ijms241411438.

European Union (2017) *A European One Health Action Plan against Antimicrobial Resistance (AMR)*. European Commission. Available at: https://health.ec.europa.eu/system/files/2020-01/amr_2017_action-plan_0.pdf (accessed 12 June 2025).

FAO (2021) The FAO Action Plan on Antimicrobial Resistance 2021–2025. FAO, Rome. Available at: https://openknowledge.fao.org/handle/20.500.14283/cb5545en (accessed 12 June 2025).

Founou, L.L., Founou, R.C. and Essack, S.Y. (2021) Antimicrobial resistance in the farm-to-plate continuum: More than a food safety issue. *Future Science OA* 7(5), FSO692. DOI: 10.2144/fsoa-2020-0189.

Fredriksen, S., Warle, S., Baarlen, P., Boekhorst, J. and Wells, J.M. (2023) Resistome expansion in disease-associated human gut microbiomes. *Microbiome* 11(1), 166. DOI: 10.1186/s40168-023-01610-1.

Gomez, D.E., Arroyo, L.G., Renaud, D.L., Viel, L. and Weese, J.S. (2021) A multidisciplinary approach to reduce and refine antimicrobial drugs use for diarrhoea in dairy calves. *The Veterinary Journal* 274, 105713. DOI: 10.1016/j.tvjl.2021.105713.

González-Barrio, D. (2022) Zoonoses and wildlife: One health approach. *Animals (Basel)* 12(4), 480. DOI: 10.3390/ani12040480.

Grilo, M.L., Sousa-Santos, C., Robalo, J. and Oliveira, M. (2020) The potential of *Aeromonas* spp. from wildlife as antimicrobial resistance indicators in aquatic environments. *Ecological Indicators* 115, 106396.

Guenther, S., Ewers, C. and Wieler, L.H. (2011) Extended-spectrum beta-lactamases producing *E. coli* in wildlife, yet another form of environmental pollution? *Frontiers in Microbiology* 2, 246. DOI: 10.3389/fmicb.2011.00246.

Han, B., Ma, L., Yu, Q., Yang, J., Su, W. *et al.* (2022) The source, fate and prospect of antibiotic resistance genes in soil: A review. *Frontiers in Microbiology* 13, 976657. DOI: 10.3389/fmicb.2022.976657.

Hartmann, M. and Six, J. (2023) Soil structure and microbiome functions in agroecosystems. *Nature Reviews Earth & Environment* 4, 4–18. DOI: 10.1038/s43017-022-00366-w.

Hrncir, T. (2022) Gut microbiota dysbiosis: Triggers, consequences, diagnostic and therapeutic options. *Microorganisms* 10(3), 578. DOI: 10.3390/microorganisms10030578.

Iwu, C.D. and Patrick, S.M. (2021) An insight into the implementation of the global action plan on antimicrobial resistance in the WHO African region: A roadmap for action. *International Journal of Antimicrobial Agents* 58(4), 106411. DOI: 10.1016/j.ijantimicag.2021.106411.

Jacobsen, A.B.J.E., Ogden, J. and Ekiri, A.B. (2023) Antimicrobial resistance interventions in the animal sector: Scoping review. *Frontiers in Antibiotics* 2, 1233698. DOI: 10.3389/frabi.2023.1233698.

Jia, W.-L., Song, C., He, L.-Y., Wang, B., Gao, F.-Z. *et al.* (2023) Antibiotics in soil and water: Occurrence, fate, and risk. *Current Opinion in Environmental Science & Health* 32, 100437. DOI: 10.1016/j.coesh.2022.100437.

Johnsen, P.J., Gama, J.A. and Harms, K. (2021) Bacterial evolution on demand. *eLife* 10, e68070.

Jonas, O.B., Irwin, A., Le Gall, F.G. and Marquez, P.V. (2017) *Drug-Resistant Infections: A Threat to Our Economic Future*. World Bank, Washington, DC. Available at: https://documents.worldbank.org/en/publication/documents-reports/documentdetail/323311493396993758/final-report (accessed 12 June 2025).

Koch, L.K., Cunze, S., Kochmann, J. and Klimpel, S. (2020) Bats as putative *Zaire ebolavirus* reservoir hosts and their habitat suitability in Africa. *Scientific Reports* 10, 14268. DOI: 10.1038/s41598-020-71226-0.

Kuthyar, S. and Reese, A.T. (2021) Variation in microbial exposure at the human-animal interface and the implications for microbiome-mediated health outcome. *mSystems* 6(4), e0056721. DOI: 10.1128/msystems.00567-21.

Lamberte, L.E. and van Schaik, W. (2022) Antibiotic resistance in the commensal human gut microbiota. *Current Opinion in Microbiology* 68, 102150. DOI: 10.1016/j.mib.2022.102150.

Lassudrie, M., Hégaret, H., Wikfors, G.H. and Silva, P.M. (2020) Effects of marine harmful algal blooms on bivalve cellular immunity and infectious diseases: A review. *Developmental and Comparative Immunology* 108, 103660. DOI: 10.1016/j.dci.2020.103660.

Lazar, V., Ditu, L.M., Pircalabioru, G.G., Gheorghe, I., Curutiu, C. *et al.* (2018) Aspects of gut microbiota and immune system interactions in infectious diseases, immunopathology, and cancer. *Frontiers in Immunology* 9, 1830. DOI: 10.3389/fimmu.2018.01830.

Lozupone, C.A., Stombaugh, J.I., Gordon, J.I., Jansson, J.K. and Knight, R. (2012) Diversity, stability and resilience of the human gut microbiota. *Nature* 489(7415), 220–230. DOI: 10.1038/nature11550.

Maciel, M.J., Machado, G. and Avancini, C.A.M. (2019) Investigation of resistance of *Salmonella* spp. isolated from products and raw material of animal origin (swine and poultry) to antibiotics and disinfectants. *Revista Brasileira de Saúde e Produção Animal* 20, e0162019. DOI: 10.1590/s1519-9940200162019.

Man, W., de Steenhuijsen Piters, W. and Bogaert, D. (2017) The microbiota of the respiratory tract: Gatekeeper to respiratory health. *Nature Reviews Microbiology* 15, 259–270. DOI: 10.1038/nrmicro2017.14.

Marcuta, L., Popescu, A., Tindeche, C., Fintineru, A., Smedescu, D. *et al.* (2023) Study on the evolution of fair trade and its role in sustainable development. *Scientific Papers Series "Management, Economic Engineering in Agriculture and Rural Development"* 23(2), 427–436.

Matthews, K., Cavagnaro, T., Weinstein, P. and Stanhope, J. (2024) Health by design; optimising our urban environmental microbiomes for human health. *Environmental Research* 257, 119226. DOI: 10.1016/j.envres.2024.119226.

Molina, D., Carrión-Olmedo, J.C., Jarrín, P. and Tenea, G.N. (2024) Genome characterization of a multi-drug resistant *Escherichia coli* strain, L1PEag1, isolated from commercial cape gooseberry fruits (*Physalis peruviana* l.). *Frontiers in Microbiology* 15, 1392333. DOI: 10.3389/fmicb.2024.1392333.

Mulchandani, R., Zhao, C., Tiseo, K., Pires, J. and Boeckel, T.P. (2024) Predictive mapping of antimicrobial resistance for *Escherichia coli*, *Salmonella*, and *Campylobacter* in food-producing animals, Europe, 2000–2021. *Emerging Infectious Diseases* 30(1), 96–104. DOI: 10.3201/eid3001.221450.

Murray, A.G. (2008) Existing and potential use of models in the control and prevention of disease emergencies affecting aquatic animals. *Revue Scientifique et Technique (International Office of Epizootics)* 27(1), 211–228.

Murray, C.J.L., Ikuta, K.S., Sharara, F., Swetschinski, L., Aguilar, G.R. *et al.* (2022) Global burden of bacterial antimicrobial resistance in 2019: A systematic analysis. *Lancet* 399(10325), 629–655.

NACA (n.d.) *Shrimp Health Management Extension Manual*. The Marine Products Export Development Authority, Cochin, India. Available at: https://enaca.org/?id=292 (accessed 17 June 2025).

Nwobodo, D.C., Chigozie Ugwu, M., Oliseloke Anie, C., Al-Ouqaili, M.T.S., Chinedu Ikem, J. *et al.* (2022) Antibiotic resistance: The challenges and some emerging strategies for tackling a global menace. *Journal of Clinical Laboratory Analysis* 36(9), e24655. DOI: 10.1002/jcla.24655.

OATA (2023) *Biosecurity and the Ornamental Fish Industry*. OATA, Westbury, Wiltshire, UK. Available at: https://ornamentalfish.org/wp-content/uploads/Biosecurity-and-the-ornamental-aquatic-industry-May-2023.pdf (accessed 17 June 2025).

Oliver, A., Xue, Z., Villanueva, Y.T., Durbin-Johnson, B. and Alkan, Z. (2022) Association of diet and antimicrobial resistance in healthy U.S. adults. *mBio* 13(3), e00101–22. DOI: 10.1128/mbio.00101-22.

O'Neill, J. (2016) *Tackling Drug-Resistant Infections Globally: Final Report and Recommendations*. The Review on Antimicrobial Resistance. Available at: https://amr-review.org/sites/default/files/160518_Finalpaper_withcover.pdf (accessed 17 June 2025).

Palić, D. and Scarfe, A.D. (2020) Biosecurity in aquaculture: Practical veterinary approaches for aquatic animal disease prevention, control, and possible eradication. In: Dewulf, J. and Immerseel, F. (eds) *Biosecurity in Animal Production and Veterinary Medicine: From Principles to Practice*. ACCO Publishing House, Leuven, Belgium and, pp. 497–523.

Patel, J., Harant, A., Fernandes, G., Mwamelo, A.J., Hein, W. *et al.* (2023) Measuring the global response to antimicrobial resistance, 2020–21: A systematic governance analysis of 114 countries. *The Lancet Infectious Diseases* 23(6), 706–718. DOI: 10.1016/s1473-3099(22)00796-4.

Penders, J., Stobberingh, E.E., Savelkoul, P.H.M. and Wolffs, P.F.G. (2013) The human microbiome as a reservoir of antimicrobial resistance. *Frontiers in Microbiology* 4, 87. DOI: 10.3389/fmicb.2013.00087.

Pérez-Rodríguez, F. and Taban, B.M. (2019) A state-of-art review on multi-drug resistant pathogens in foods of animal origin: Risk factors and mitigation strategies. *Frontiers in Microbiology* 10, 2091. DOI: 10.3389/fmicb.2019.02091.

Phu, D.H., Cuong, N.V., Truong, D.B., Kiet, B.T. and Hien, V.B. (2021) Reducing antimicrobial usage in small-scale chicken farms in Vietnam: A 3-year intervention study. *Frontiers in Veterinary Science* 7, 612993. DOI: 10.3389/fvets.2020.612993.

Poole, P., Ramachandran, V. and Terpolilli, J. (2018) Rhizobia: From saprophytes to endosymbionts. *Nature Reviews Microbiology* 16, 291–303. DOI: 10.1038/nrmicro.2017.171.

Prestinaci, F., Pezzotti, P. and Pantosti, A. (2015) Antimicrobial resistance: A global multifaceted phenomenon. *Pathogens and Global Health* 109, 309–318.

Ren, Y., Chakraborty, T., Doijad, S., Falgenhauer, L. and Falgenhauer, J. (2021) Prediction of antimicrobial resistance based on whole-genome sequencing and machine learning. *Bioinformatics* 38(2), 325–334. DOI: 10.1093/bioinformatics/btab681.

Ritchie, H. and Spooner, F. (2024) Large Amounts of Antibiotics Are Used in Livestock, but Several Countries Have Shown This Doesn't Have to Be the Case. OurWorldinData.org. Available at: https://ourworldindata.org/antibiotics-livestock (accessed 17 June 2025).

Rolain, J.-M. (2013) Food and human gut as reservoirs of transferable antibiotic resistance encoding genes. *Frontiers in Microbiology* 4, 173. DOI: 10.3389/fmicb.2013.00173.

Samtiya, M., Matthews, K.R., Dhewa, T. and Kumar Puniya, A. (2022) Antimicrobial resistance in the food chain: Trends, mechanisms, pathways, and possible regulation strategies. *Foods* 11(19), 2966. DOI: 10.3390/foods11192966.

Schaik, W. van (2015) The human gut resistome. *Philosophical Transactions of the Royal Society B: Biological Sciences* 370(1670), 20140087. DOI: 10.1098/rstb.2014.0087.

Schar, D., Zhao, C., Wang, Y., Larsson, D.G.J., Gilbert, M. *et al.* (2021) Twenty-year trends in antimicrobial resistance from aquaculture and fisheries in Asia. *Nature Communications* 12(1), 5384. DOI: 10.1038/s41467-021-25655-8.

Shi, A., Fan, F. and Broach, J.R. (2021) Microbial adaptive evolution. *Journal of Industrial Microbiology and Biotechnology* 49(2), kuab076. DOI: 10.1093/jimb/kuab076.

Souque, C., Escudero, J.A. and MacLean, R.C. (2021) Integron activity accelerates the evolution of antibiotic resistance. *eLife* 10, e62474. DOI: 10.7554/elife.62474.

Stall, B.E. and Thayer, P.L. (1962) Streptomycin resistance of the bacterial spot pathogen and control with streptomycin. *Plant Disease Reporter* 46, 389–392.

Stege, P.B., Hordijk, J., Shetty, S.A., Visser, M., Viveen, M.C. *et al.* (2022) Impact of long-term dietary habits on the human gut resistome in the Dutch population. *Scientific Reports* 12(1), 1892. DOI: 10.1038/s41598-022-05817-4.

Stott, A.W., Lloyd, J., Humphry, R.W. and Gunn, G.J. (2003) A linear programming approach to estimate the economic impact of bovine viral diarrhoea (BVD) at the whole-farm level in Scotland. *Preventive Veterinary Medicine* 59, 51–66.

Su, M., Satola, S.W. and Read, T.D. (2018) Genome-based prediction of bacterial antibiotic resistance. *Journal of Clinical Microbiology* 57(3), 10. DOI: 10.1128/jcm.01405-18.

Sweileh, W.M. (2021) Global research activity on antimicrobial resistance in food-producing animals. *Archives of Public Health* 79(1), 49. DOI: 10.1186/s13690-021-00572-w.

Tang, K.W.K., Millar, B.C. and Moore, J.E. (2023) Antimicrobial resistance (AMR). *British Journal of Biomedical Science* 80, 11387. DOI: 10.3389/bjbs.2023.11387.

Tasnim, N., Abulizi, N.A., Pither, J., Hart, M.M. and Gibson, D.L. (2017) Linking the gut microbial ecosystem with the environment: Does gut health depend on where we live? *Frontiers in Microbiology* 8. DOI: 10.3389/fmicb.2017.01935.

Tenea, G.N. and Ortega, C. (2021) Genome characterization of *Lactiplantibacillus plantarum* strain UTNGt2 originated from *Theobroma grandiflorum* (white cacao) of Ecuadorian Amazon: Antimicrobial peptides from safety to potential applications. *Antibiotics* 10(4), 383. DOI: 10.3390/antibiotics10040383.

Thayer, P.L. and Stall, R.E. (1962) The survey of *Xanthomonas vesicatoria* resistance to streptomycin. *Proceedings of the Florida State Horticultural Society* 75, 163–165.

Theophilus, R.J. and Taft, D.H. (2023) Antimicrobial resistance genes (ARGs), the gut microbiome, and infant nutrition. *Nutrients* 15(14), 3177. DOI: 10.3390/nu15143177.

Thorpe, K.E., Joski, P. and Johnston, K.J. (2018) Antibiotic-resistant infection treatment costs have doubled since 2002, now exceeding $2 billion annually. *Health Affairs* 37(4), 10.1377. DOI: 10.1377/hlthaff.2017.1153.

Tiseo, K., Huber, L., Gilbert, M., Robinson, T.P. and Van Boeckel, T.P. (2020) Global trends in antimicrobial use in food animals from 2017 to 2030. *Antibiotics* 9(12), 918. DOI: 10.3390/antibiotics9120918.

Uberoi, A., McCready-Vangi, A. and Grice, E.A. (2024) The wound microbiota: Microbial mechanisms of impaired wound healing and infection. *Nature Reviews Microbiology* 22, 507–521. DOI: 10.1038/s41579-024-01035-z.

Van Boeckel, T.P., Brower, C., Gilbert, M., Grenfell, B.T., Levin, S.A. *et al.* (2015) Global trends in antimicrobial use in food animals. *Proceedings of the National Academy of Sciences* 112(18), 5649–5654. DOI: 10.1073/pnas.1503141112.

Van Boeckel, T.P., Glennon, E.E., Chen, D., Gilbert, M., Robinson, T.P. *et al.* (2017) Reducing antimicrobial use in food animals. *Science* 357(6358), 1350–1352. DOI: 10.1126/science.aao1495.

Van Boeckel, T.P., Pires, J., Silvester, R., Zhao, C., Song, J. *et al.* (2019) Global trends in antimicrobial resistance in animals in low- and middle-income countries. *Science* 365, 6459. DOI: 10.1126/science.aaw1944.

Venkatakrishnan, A., Holzknecht, Z.E., Holzknecht, R., Bowles, D.E., Kotzé, S.H. *et al.* (2021) Evolution of bacteria in the human gut in response to changing environments: An invisible player in the game of health. *Computational and Structural Biotechnology Journal* 19, 752–758. DOI: 10.1016/j. csbj.2021.01.007.

Verraes, C., Boxstael, S., Meervenne, E., Coillie, E., Butaye, P. *et al.* (2013) Antimicrobial resistance in the food chain: A review. *International Journal of Environmental Research and Public Health* 10(7), 2643–2669. DOI: 10.3390/ijerph10072643.

Vezeau, N. and Kahn, L. (2024) Spread and mitigation of antimicrobial resistance at the wildlife-urban and wildlife-livestock interfaces. *Journal of the American Veterinary Medical Association* 262, 741–747. DOI: 10.2460/javma.24.02.0123.

Wainwright, M. and Kristiansen, J.E. (2011) On the 75th anniversary of prontosil. *Dyes and Pigments* 88(3), 231–234. DOI: 10.1016/j.dyepig.2010.08.012.

Willemsen, A., Reid, S. and Assefa, Y. (2022) A review of national action plans on antimicrobial resistance: Strengths and weaknesses. *Antimicrobial Resistance & Infection Control* 11(1), 90. DOI: 10.1186/s13756-022-01130-x.

WOAH (2022) *Aquatic Animal Health Code.* Glossary, p. xii. Available at: https://doc.woah.org/dyn/portal /index.xhtml (accessed 17 June 2025).

Xu, T., Dai, Y., Ge, A., Chen, X., Gong, Y. *et al.* (2024a) Ultrafast evolution of bacterial antimicrobial resistance by picoliter-scale centrifugal microfluidics. *Analytical Chemistry* 96(47), 18842–18851. DOI: 10.1021/acs.analchem.4c04482.

Xu, M., Wang, F., Stedtfeld, R.D., Fu, Y., Xiang, L. *et al.* (2024b) Transfer of antibiotic resistance genes from soil to rice in paddy field. *Environment International* 191, 108956. DOI: 10.1016/j.envint.2024.108956.

Yang, X., Li, C., Ouyang, D., Wu, B., Fang, T. *et al.* (2024) High microbiome diversity constricts the prevalence of human and animal pathogens in the plant rhizosphere worldwide. *One Earth* 7, 1301–1312. DOI: 10.1016/j.oneear.2024.06.005.

Zinsstag, J., Kaiser-Grolimund, A., Heitz-Tokpa, K., Sreedharan, R., Lubroth, J. *et al.* (2023) Advancing one human–animal–environment health for global health security: What does the evidence say? *The Lancet* 401(10376), 591–604.

Further Reading

Ackerman, P., Melville, S., MacWilliams, C. and MacKindley, D. (2007) *Tenderfoot Creek Salmon Hatchery: Fish Health Management Plan.* A fish health management plan example. Available at: https://waves -vagues.dfo-mpo.gc.ca/library-bibliotheque/347195.pdf (accessed 17 June 2025).

Arthur, J.R., Baldock, F.C., Subasinghe, R.P. and McGladdery, S.E. (2005) *Preparedness and Response to Aquatic Animal Health Emergencies in Asia: Guidelines.* FAO, Rome. Available at: https://www.fao. org/4/a0090e/a0090e00.htm (accessed 17 June 2025).

Corsin, F., Funge-Smith, S. and Clausen, J. (2007) *A Qualitative Assessment of Standards and Certification Schemes Applicable to Aquaculture in the Asia–Pacific Region.* FAO, Bangkok. Available at: https:// www.fao.org/4/ai388e/ai388e00.pdf (accessed 17 June 2025).

FAO (2011) *Technical Guidelines on Aquaculture Certification.* FAO, Rome. Available at: https://www.fao. org/4/i2296t/i2296t00.pdf (accessed 17 June 2025).

Hennessy, D.A. (2008) Economic aspects of agricultural and food and biodefence. *Biosecurity and Bioterrorism: Biodefense Strategy, Practice, and Science* 6, 66–77. DOI: 10.1089/bsp.2007.0016.

Magnusson, U., Sternberg Lewerin, S., Eklund, G. and Rozstalnyy, A. (2019) Prudent and efficient use of antimicrobials in pigs and poultry. In: *FAO Animal Production and Health / Manual 23.* FAO, Rome. Available at: https://openknowledge.fao.org/server/api/core/bitstreams/a8fbb826-a602-4bfc-86b2 -e3b840c0e0b7/content (accessed 13 June 2025).

Salmon Scotland (2023) Code of Good Practice for Scottish Finfish Aquaculture. Available at: www. thecodeofgoodpractice.co.uk/ (accessed 17 June 2025).

7 Human–Wildlife Interface

Laura D. Urdes[1]*, Devon R. Dublin[2], Julius Tepper[3] and Chris Walster[4]

[1]The World Aquatic Veterinary Medical Association, Romania; Faculty of Veterinary Medicine, University Spiru Haret Bucharest, Romania; Faculty of Management and Rural Development, University of Agricultural Sciences and Veterinary Medicine, Bucharest, Romania; [2]Ocean Policy Research Institute, Sasakawa Peace Foundation, 1-15-16 Toranomon, Minato-ku, Tokyo 105-8524, Japan; [3]The World Aquatic Veterinary Medical Association, USA; Long Island Fish Hospital, Manorville, New York, USA; [4]The World Aquatic Veterinary Medical Association, UK

Abstract

Wildlife and livestock interactions can lead to bidirectional disease transmission. Many diseases affecting both livestock and wildlife populations are a severe threat to endangered wildlife species and indigenous livestock breeds. Wildlife diseases may therefore have a serious impact on domestic animals and the public health, and can adversely affect wildlife conservation. The chapter describes the human–animal interface in its historical perspective, allowing one to realize its versatile and dynamic nature, as well as possible risks for cross-species transmission of pathogens while domestication, agriculture, urbanization, colonization, trade, and industrialization are underway. The text discusses emerging technologies used to detect and monitor diseases at the human–wildlife interface, particularly focusing on wildlife disease surveillance, including in aquatic species. It highlights the significance of amphibians as indicators of aquatic ecosystem health and notes the global impact of chytridiomycosis, a fungal disease caused by *Batrachochytrium dendrobatidis*, which has led to mass amphibian die-offs. The role of aquatic wildlife as a food source for other wild animals is also noted, and examples like avian influenza and fish mycobacteriosis are used to illustrate how diseases can emerge and spread at this interface.

7.1 Concept and Historical Context

Since prehistoric times, there has been an acknowledged intricate relationship between humans and animals (Shrestha *et al.*, 2018). The human–wildlife interface has been essentially present since the evolution of the human race and has been described as a significant facet of the One Health concept (Reperant and Osterhaus, 2013). Due to the nature of this interface, human–wildlife interactions can commonly occur. These interactions have been defined by the World Wildlife Federation as encounters between people and wildlife (neither positive nor negative) (Gross *et al.*, 2021) and, due to a variety of primarily human-derived changes, are not a simple topic (Narayan and Rana, 2023). For example, human–wildlife interactions, direct or indirect, can result in zoonotic pathogen transmission and reverse zoonoses, and can aid in the movement of pathogens of great concern to agricultural health.

*Corresponding author: urdeslaura@gmail.com

© CAB International 2025. *One Health Concepts and the Aquatic Ecosystem*
(eds L.D. Urdes *et al.*)
DOI: 10.1079/9781800623248.0007

The human–wildlife interface is largely driven either by humans making changes in land use, resulting in the utilization of areas normally inhabited by wildlife, or by changes in the distribution patterns of wildlife species, resulting in closer contact with humans. Some human–wildlife interactions can be viewed as negative, especially in instances of human–wildlife conflicts. These types of conflicts can occur when the wildlife adversely affects humans or when human objectives adversely affect wildlife (Madden, 2004). Although several other factors can make each human–wildlife conflict unique, it should be emphasized here that climate change is an important driver of this conflict, as shown in Fig. 7.1 (Gortazar *et al.*, 2014; Abrahms *et al.*, 2023; International Union for Conservation of Nature and Natural Resources, 2023). More market demands for wildlife products resulted in an increase in poaching and wildlife trafficking, which in turn resulted in a decrease in biodiversity, which compels animals to search for food in and around human dwellings, thus increasing human–wildlife conflicts (Debnath *et al.*, 2021).

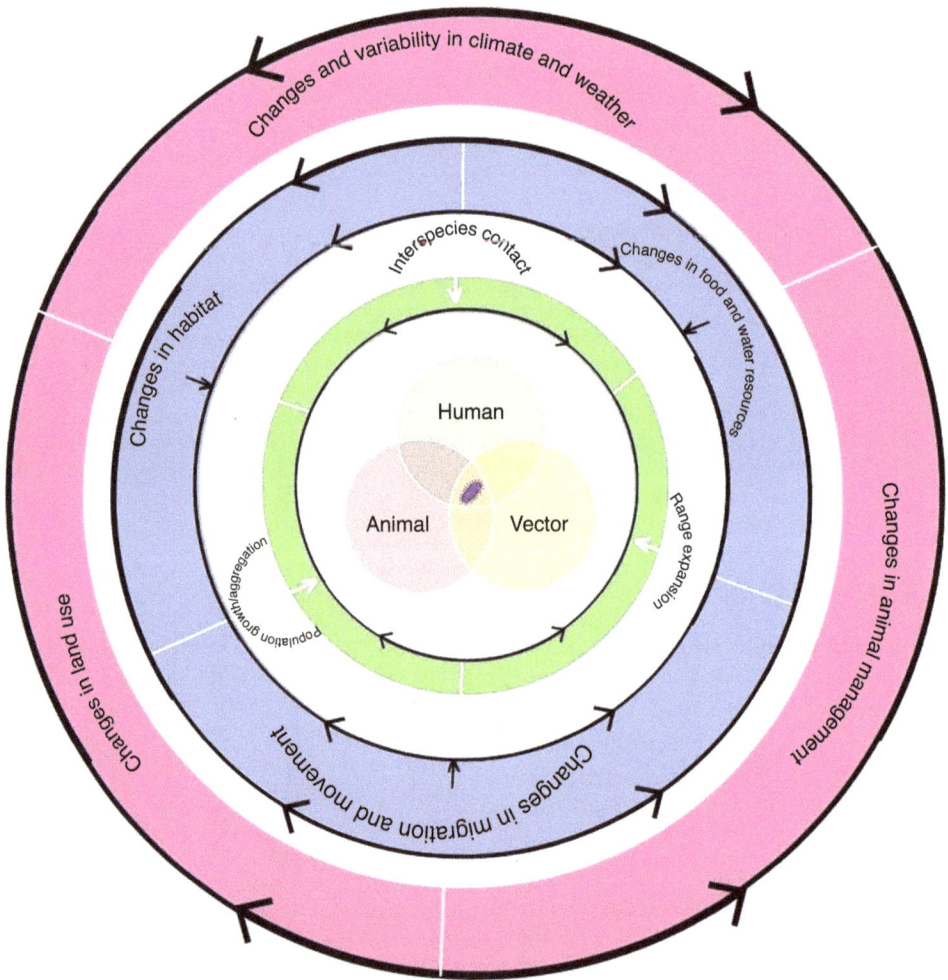

Fig. 7.1. Impacts of various drivers leading to increased human–wildlife interface and exposure to animal pathogens. Modified from Gortazar *et al.*, 2014.

Several issues are symptomatic of the human–wildlife interface. They include (i) damage to agricultural crops and pastures, (ii) loss of livestock and domestic animals, (iii) damage to infrastructure and equipment (e.g. houses, water storage facilities, etc.), and (iv) injury and death of humans caused by wild animals (Mardaraj and Sethy, 2015; World Wildlife Fund, 2022). Wild animals are killed by humans in retaliation for the damage or injury they cause, in defense when humans come under attack from animals, or to prevent future losses of agricultural or infrastructural assets (World Wildlife Fund, 2022). The other aspect of the human–animal interface that One Health is particularly concerned with is the spread of infectious diseases across species, as well as their emergence and eventual evolution (Reperant et al., 2012). Animal–human interactions, direct or indirect, are a key facet of this issue because zoonotic pathogen emergence requires spillover from animals to humans (Dobigny and Morand, 2022). The frequency and severity of zoonotic spillover events have been impacted by human activity throughout human history (Dobigny and Morand, 2022). Anthropogenic processes can highly influence the risks associated with pathogen transmission at the human–wildlife interface, as well (Hassell et al., 2017). Many elements that are thought to contribute to pathogen emergence, such as urbanization, globalization, climate change, dietary or habitat changes, domestication, agriculture, colonization, commerce, and industrialization, are triggered by humans (Lindahl and Grace, 2015; Dreyer et al., 2023). Indeed, human activity can contribute to increasing the spread of some pathogens, such as various fungal pathogens, through the alteration of natural regions, which can result in prospects for evolution of the pathogen in question (Fisher et al., 2012; Dreyer et al., 2023). Changes in these influences have been accompanied by evolving risks for cross-species transmission of pathogens (Reperant and Osterhaus, 2013). Furthermore, increasing populations and range expansions of invasive species can not only damage native ecosystems and negatively affect agriculture, but can also increase the transmission of the pathogens associated with these species (USDA, 2024a).

7.2 Cross-Species Transmission of Pathogens

In the field of veterinary medicine, the wildlife–livestock interface, in which wildlife serves as a potential source of pathogens of concern for agricultural health/production, has been the subject of probable pathogen transmission for millennia (Jori et al., 2021). This interface, of course, is not one-directional, as pathogens harbored by livestock could also have detrimental effects for wildlife. In various parts of Asia, losses in biodiversity, urbanization, and expansions of agricultural lands were suggested to have made past zoonotic viral spillover events in this region more probable, as these three factors have the potential to modify the human–wildlife interface (Goldstein et al., 2022).

About 60–75% of human infectious diseases are derived from pathogens that first appeared in non-human animal species (Ellwanger and Chies, 2021). Therefore, zoonotic spillover plays a fundamental role in the emergence of new human infectious diseases such as avian influenza and the recent COVID-19 pandemic (Bushman and Antia, 2019; Ellwanger and Chies, 2021). A common factor driving cross-species transmission can be phylogenetic similarity between host species (Bushman and Antia, 2019), although this is clearly not always the case for some pathogens (Porter et al., 2022, 2024). As demonstrated in Fig. 7.2, coronaviruses (CoVs) affect a wide variety of animals, including whales and dolphins (Li et al., 2023).

Similarly, climate change affects aquatic ecosystems by giving rise to melting polar ice caps and glaciers, increased ocean temperatures, and rising acidity and sea levels. Rising sea levels result in coastal flooding, thereby augmenting the risks of waterborne zoonoses (Naicker, 2011).

7.2.1 Avian influenza

While avian influenza viruses (AIVs) are primarily pathogens of aquatic birds, certain subtypes of these viruses (H5 and H7) can mutate into highly pathogenic (HP) forms when introduced into poultry flocks and can have

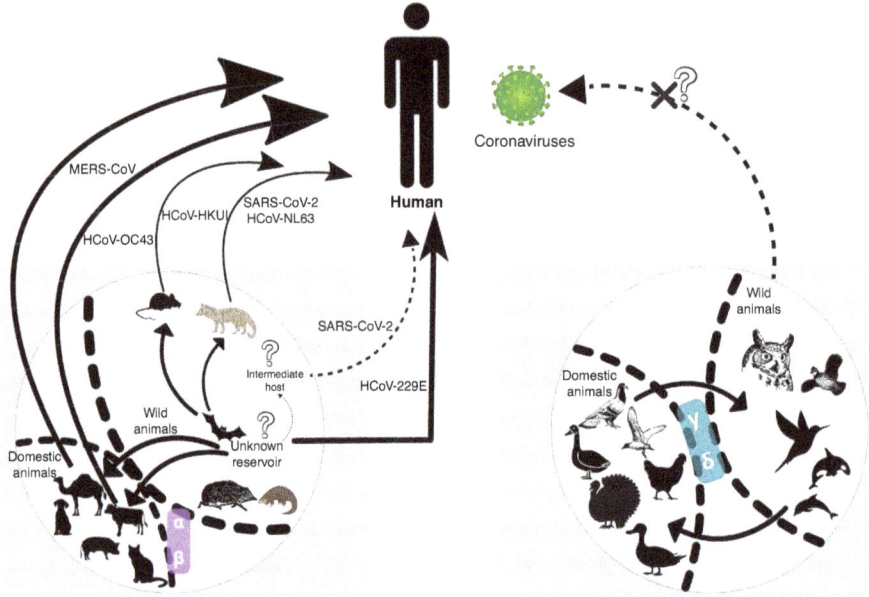

Fig. 7.2. Cross-species transmission of CoVs. Modified from Li *et al.*, 2023.

drastic consequences for poultry producers (Byrne *et al.*, 2021). For example, introduced and reassortant HP viruses in the USA affected more than 48 million poultry during 2015; initial estimates of losses were in the billions of US dollars (Shriner *et al.*, 2016). A second large outbreak of AIV occurred in the USA during 2022–2024 that had even greater consequences for the poultry industry and, more recently, the dairy industry. Unfortunately, this virus had the capacity to spill over into multiple wild avian and mammalian species, including marine mammals. For example, HP H5N1 was recently reported in harbor seals and gray seals in the north-eastern USA, concurrent with the same virus in wild birds in the same general area (Puryear *et al.*, 2023). Aside from marine mammals, this virus has also negatively affected many terrestrial wild mammals (USDA, 2024b), the vast majority of which are carnivorous or exhibit scavenging behavior. In addition to aquatic birds, these viruses have also been detected in several species of non-aquatic birds, many of which exhibit predatory and scavenging behavior (Ringenberg *et al.*, 2023). Thus, this virus, which is typically associated with aquatic ecosystems, has seemingly spilled into terrestrial mammals and birds and into

a new trophic level through the consumption of virus-laden waterbird carcasses. In the USA alone, AIVs have caused morbidity, mortality, or non-clinical positive cases in approximately 150 avian species, many of which are not considered as aquatic species (USDA, 2024b). Recent HP AIVs provide excellent examples of a virus of which reservoir hosts are wild aquatic birds that has spilled into non-aquatic wildlife and, occasionally, into humans. While AIVs are a good example of pathogens that can be associated with the wildlife–livestock interface, they can also be associated with the human–wildlife interface on occasion. For example, low pathogenic (LP) subtypes associated with the hemagglutinins H6, H7, H9, and H10 have been virologically confirmed to have infected people (CDC, 2024). While these are typically rare, an LP H7N9 virus in China was reported in over 1500 people during the years 2013–2017; this virus had a case fatality rate of approximately 40% for hospitalized patients. Regarding HP viruses, H5 viruses have been responsible for the bulk of the human case-patients. HP H5N1 was responsible for hundreds of human cases beginning in 1997; the human case fatality rate for this virus is roughly 50% (CDC, 2024). Reported human cases with HP H5N6 and H5N8 viruses have been minimal to date. Overall, human cases with HP H7 viruses (e.g. H7N3, H7N7) have been reported in a relatively small number of people (CDC, 2024).

The factors that determine if an avian influenza virus ultimately acquires the ability to spread efficiently among the human population are not well understood (Freidl *et al.*, 2014). Due to bird migrations, avian influenza virus outbreaks can occur, spreading to wild and domesticated mammals, and possibly humans (European Centre for Disease Prevention and Control, 2023). There is great concern over the infection of humans with highly pathogenic avian influenza virus (World Health Organization, 2023). While epidemiological and virological evidence suggest that avian influenza viruses have not acquired the ability of sustained transmission among humans, the human–animal interface increases the risk that a pandemic virus strain will emerge that can do so (Mumford *et al.*, 2007; World Health Organization, 2023).

Wild bird (primarily waterfowl) samples are routinely collected by the USDA/APHIS/WS National Wildlife Disease program for the purpose of national level surveillance of AIVs in the USA. The collected samples are subsequently tested for AIVs. The goal(s) of this surveillance program has evolved over the years, but at present the program has three primary goals: (i) to maximize the ability to detect and characterize the spatial locations of AIVs in wild waterfowl in the United States, (ii) to reveal the distribution of AIVs into new locations of interest, and (iii) to monitor wild waterfowl populations for incursions of novel viruses (e.g. Eurasian lineage H5 and H7 viruses) into the USA (USDA, 2023). Within this sampling framework, specific watersheds are selected for sample collection based on previous evidence of AIV infections in wild waterfowl and on locations known to have high levels of mixing of waterfowl populations (USDA, 2023). Surveillance and monitoring systems are presented in Chapter 5, Section 5.2.10, 'Disease surveillance and monitoring (MOSS).' In addition to this watershed-level surveillance, morbidity and mortality investigations are also conducted throughout the sampling year. While not possible throughout the entire sampling year, a large proportion of the samples collected annually for these surveillance efforts originate

from hunter harvested waterfowl. From January 2022 (which is the approximate origination of the current HP outbreak in the USA) to October 2024, this surveillance program documented over 10,000 HP AIV detections in wild birds throughout the USA (excluding Hawaii) (USDA, 2024b). For perspective, this same AIV surveillance system only detected a limited number of HP AIV cases in wild birds between December 2014 and June 2015 in the USA (USDA, 2016).

7.2.2 Domestication

An excellent example of problems associated with the domestication of a semi-aquatic animal can be found in the nutria (*Myocastor coypus*), which is a mammalian species that is tied to aquatic environments. This species, which is native to South America, was originally imported into the USA during the late 1880s for the fur trade (USDA, 2024a). Subsequently, wild populations of this invasive species have become established in multiple regions of the country. Nutria can cause several types of wildlife damage, through herbivory resulting in habitat damage, through burrowing, and through trafficking pathogens and parasites (CDFW, 2024). Crop damage is also an issue, as nutria are known to damage sugarcane, rice, and other crops (USDA, 2024a). In addition, zoonotic pathogens associated with nutria identified through serosurveys have included *Leptospira* spp., *Chlamydophila psittaci*, *Streptococcus equi zooepidemicus*, *Toxoplasma gondii*, and encephalomyocarditis virus in its native range (Martino *et al.*, 2014). These and other pathogens could be harbored in fur-farmed nutria, thereby providing ample opportunities for zoonotic pathogen spillover, as well as pathogen spillover to native wildlife.

A terrestrial example of the risks of cross-species transmission of pathogens at a wildlife–livestock interface caused by humans can be found in bighorn sheep (*Ovis canadensis*), where the arrival of a single domestic sheep (*Ovis aries*) on the winter range of bighorn sheep is thought to have resulted in the transmission of *Pasteurella* spp. bacteria, which significantly reduced survival and recruitment in the affected populations (George *et al.*, 2008). In addition, anecdotal reports of pneumonia outbreaks in bighorn sheep following interactions with domestic sheep in natural settings have been corroborated by several captive studies (Cassirer *et al.*, 2018).

7.3   Pathogen Surveillance in Wildlife

Because of the increased human–wildlife interface and the associated risks of emerging diseases, pathogen surveillance in wildlife populations is becoming increasingly important. Pathogen surveillance can be described as the ongoing detection and recording of pathogens in wildlife populations to support disease management (Artois *et al.*, 2009). Pathogen surveillance of wild animal populations will allow for the early detection of infectious and zoonotic diseases and lead to the adoption of swift countermeasures (Mörner *et al.*, 2002). It must be noted, however, that the detection of new pathogens is a challenging task (Artois *et al.*, 2009).

There are two strategies that may be employed: passive and active. Passive surveillance utilizes previously collected data and analyzes it, while active surveillance involves a proactive process of surveying for specific pathogens (Sleeman *et al.*, 2012). Passive surveillance is more likely to provide earlier detection of wildlife pathogens that are introduced into areas formerly free of the pathogen in question (Guberti *et al.*, 2014).

While emerging pathogens associated with the human–wildlife interface have been monitored for some time, some recent technologies have helped to expand our knowledge on this topic. High-throughput sequencing (HTS) methods have been described as having vast potential to help better understand the dynamics, transmission, and unforeseen ecological events at this interface (Titcomb *et al.*, 2019). Furthermore, HTS has the ability to conduct full-spectrum pathogen analysis in urban water systems (Zhao *et al.*, 2023). Thus, this technique may hold potential for natural aquatic ecosystems, as well. For example, HTS of environmental DNA (eDNA) collected from a marine ecosystem was used to describe potential pathogens, invasive pathogenic organisms, and other harmful agents in this environment (Ríos-Castro

et al., 2021). This method also holds promise for pathogen investigators to examine the effects of season, climate, and land-use changes on emerging pathogens (Titcomb *et al.*, 2019). Beyond pathogens, eDNA methods have proven to have applicability to detecting elusive/cryptic invasive species, thereby providing a useful method to monitor invasive species associated with aquatic ecosystems (Piaggio *et al.*, 2014).

7.3.1   Aquatic animal species

While many positive aspects to the human–wildlife interface exist, the expanding interface between aquatic animals and humans has been suggested as a possible threat to public health (Rinanda, 2015). There are numerous resources that provide guidance and capacity building in aquatic animal health surveillance. The Aquatic Animal Health Code (the Aquatic Code) of the World Organisation for Animal Health provides standards for aquatic animal health surveillance in Section 1 (World Organisation for Animal Health, 2021). Nevertheless, in comparison to terrestrial animal diseases, surveillance of aquatic animal diseases is deficient and requires expansion. The spread of fish diseases includes biological pathways that involve birds, as well as waterways that are used and altered in many ways by human activities (Oidtmann *et al.*, 2013). As a result, a landscape approach would be ideal in addressing this issue.

When considering aquatic species, surveillance in aquaculture is of importance, given the fact that the pathways identified can provide a link between farmed and wild aquatic species. A 12-point checklist developed based on available main references on surveillance for aquatic animal diseases was proposed to achieve this objective, as shown in Fig. 7.3.

7.3.2   Amphibians as bio-indicators of aquatic ecosystem health

Amphibians are good bioindicators of aquatic ecosystem health and provide a wide array of ecosystem services, including food sources (especially in South-east Asia), models in medical research, regulating of mosquitos and other pest

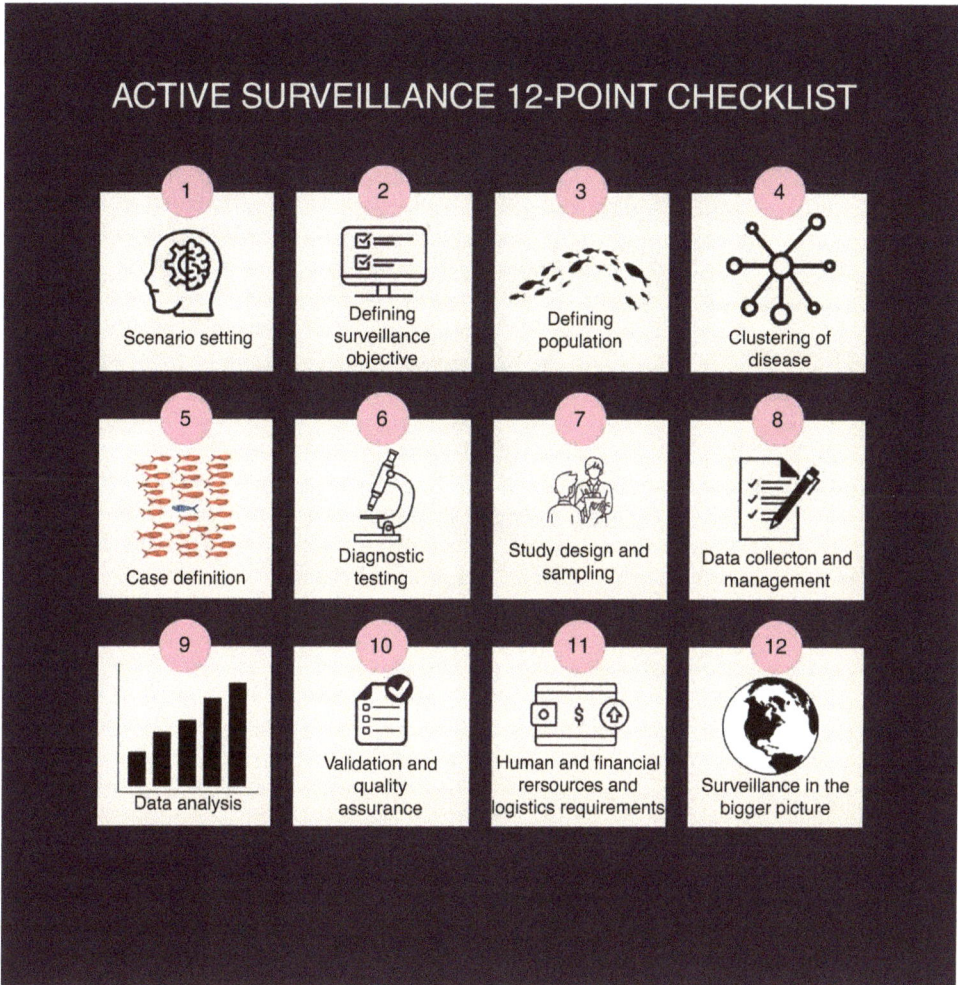

Fig. 7.3. Twelve-point checklist for aquatic disease surveillance. Modified from Bondad-Reantaso *et al.*, 2021.

species, structural supporting services (such as soil burrowing and aquatic bioturbation), and functional supporting services (such as decomposition and nutrient cycling) (Hocking and Babbitt, 2014). This is primarily because they utilize both aquatic and terrestrial habitats through their biphasic life cycle and are sensitive to environmental changes and pollutants since they absorb water and oxygen directly through their permeable skin (Blaustein *et al.*, 1994; US Environmental Protection Agency, 2015).

When compared to conventional measures, such as species richness, amphibians are superior in assessing environmental stresses. This

is because species richness combines species with differing responses to environmental stress (Sumanasekara *et al.*, 2015).

7.3.2.1 Chytridiomycosis (amphibian chytrid fungus disease)

Amphibians are sensitive to a range of environmental pollutants, including ultraviolet radiation, and chemical pollutants from industrial, mining, and agricultural waste, as well as the harmful effects of climate change (Figs 7.4 and 7.5) (Pounds, 2001; Blaustein *et al.*, 2003). Significant decreases in amphibian

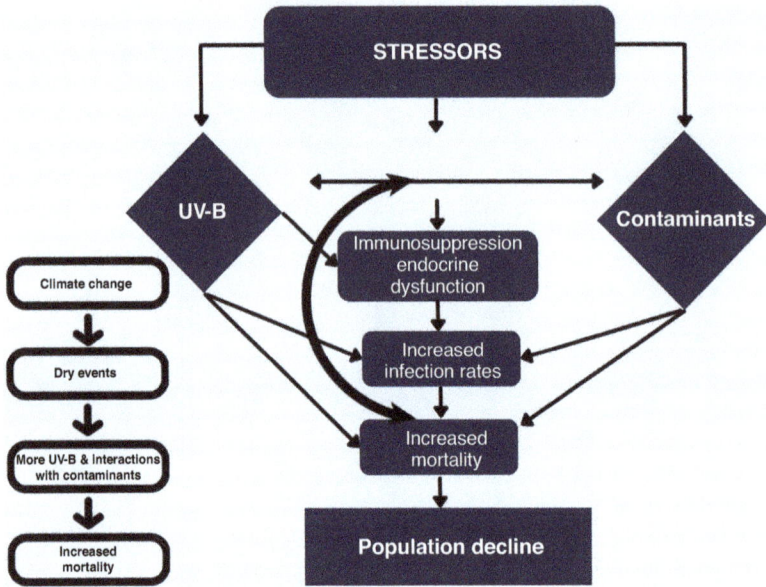

Fig. 7.4. Abiotic and biotic factors affecting amphibian populations. Modified from Blaustein *et al.*, 2003. Figure used with permission of Wiley.

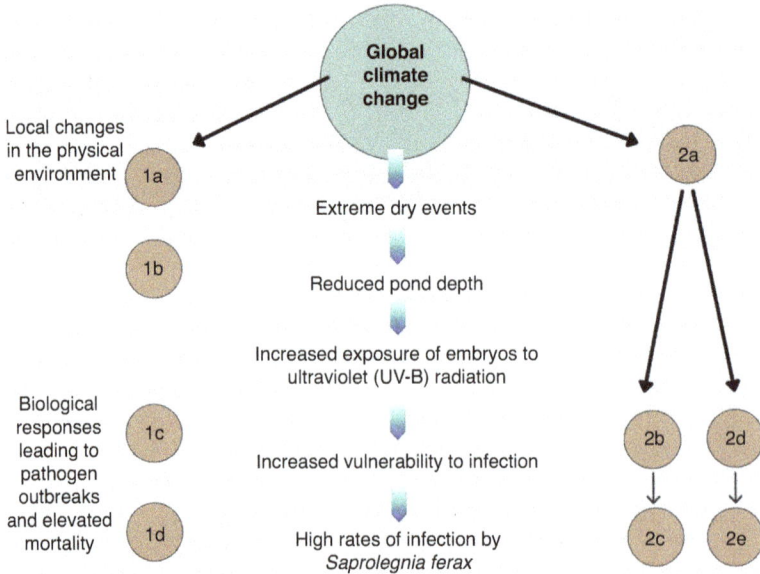

Fig. 7.5. Climate change and amphibian population decline. Modified from Pounds, 2001. Figure used with permission of Springer Nature.

populations may result in humans losing the associated ecosystem services (Hocking and Babbitt, 2014).

Over the past decade, one reason for amphibian population decline has been the emerging infectious disease chytridiomycosis, which causes mass mortality and significant population declines worldwide. The decline of at least 501 amphibian species over the past 50 years and 90 presumed extinctions are attributed to chytridiomycosis (Scheele *et al.*, 2019). Chytridiomycosis results from a sustained cutaneous infection by the fungi of the genus *Batrachochytrium* (e.g. *Batrachochytrium*

dendrobatidis) which emerged in the 1970s in Australia and the Americas (Berger *et al.*, 2016). The global trade of amphibians is presumed to have led to the introduction of this pathogen into new environments (Fisher and Garner, 2020). Using a model selection approach, research has suggested that anthropogenic activities, as described from a Human Footprint Index (based on information on infrastructure, land cover, and the access of humans to natural areas), were an important factor in the prediction of chytrid fungus' occurrence in the Atlantic Forest of Brazil (de Andrade Serrano *et al.*, 2022) (Figs 7.6 and 7.7).

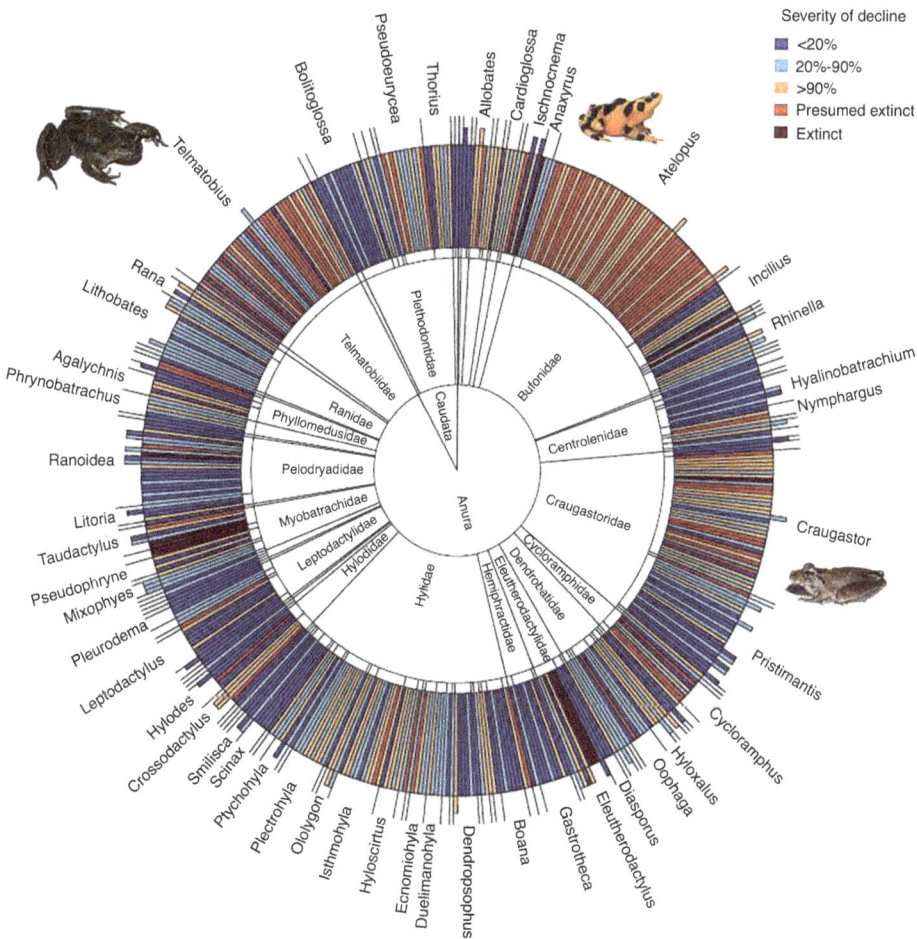

Fig. 7.6. Taxonomic distribution of chytridiomycosis-associated amphibian declines. (Reproduced from Scheele *et al.*, 2019).

Fig. 7.7. Andean frog (*Bryophryne cophites*) dead from chytrid fungus (*Batrachochytrium dendrobatidis*) Cusco region, Peru. (Photo by Emanuele Biggi, 2018. © Nature Picture Library, reproduced with permission.)

7.4 Technologies to Identify and Monitor Pathogenic Threats

Traditionally, disease surveillance tends to focus on the pathogen itself; however, when it comes to human–wildlife conflicts, it may be more useful to monitor the drivers that increase the human–animal interface, such as land-use change, global interconnectedness, and capital flows (Drewe *et al.*, 2023). Subsequent to the COVID-19 pandemic, which demonstrated that there are gaps in the world's ability to identify, track, and combat pathogenic threats in a timely manner, several proposals were developed to improve the response to emerging pathogens and future pandemics (Krofah *et al.*, 2021). To this end, the International Atomic Energy Agency (IAEA, 2021) launched the Zoonotic Disease Integrated Action (ZODIAC) initiative supporting a systematic and integrated approach to facilitate the early detection and characterization of pathogens (IAEA, 2021). What is needed, therefore, is a One Health approach, which essentially considers the human–wildlife interface, utilizing new technology. This approach begins at the systems level, ensuring that there are robust and resilient surveillance networks along with a highly skilled public health laboratory workforce, which is achieved through the identification of institutions and individuals, capacity building, and financial investment (Bird and Mazet, 2018; FAO, 2023). Operationalization of the One Health approach can be done using several existing tools. These include but are not limited to:

- States Parties Annual Reporting Tool (International Health Regulations, IHR);
- Performance of Veterinary Services (PVS) Pathway (World Organisation for Animal Health, WOAH);
- National Bridging Workshops (IHR/PVS);
- Laboratory Mapping Tool (Food and Agriculture Organization, FAO);
- Assessment Tool for Laboratories and Antimicrobial Resistance Surveillance Systems (FAO);

- Surveillance Evaluation Tool (FAO); and
- Monitoring and Evaluation Framework tools for After Action Reviews and Simulation Exercises (IHR).

What is needed, however, is to consider the local and regional realities to determine which would be most suitable for implementation and align the chosen ones for maximum impact when implemented (Pelican *et al.*, 2019). More details about making the One Health concept a practical and operational part of day-to-day health management and policy to improve health are provided in Chapter 9.

7.5 Threats to Wildlife Conservation

While many human–wildlife interactions can be positive, some can have negative effects (both direct and indirect) on wildlife conservation. Examples of this situation are as obvious as a mammal or bird being struck by a moving vehicle. However, several human–wildlife interactions that can cause threats to wildlife conservation are more obscure. Select examples of these less evident threats are detailed below.

7.5.1 Pathogen movement via trade

Aquatic animal trade can result in the inadvertent movement of pathogens (Hedrick, 1996), and the transboundary movement of pathogens can be facilitated further by the trade in live animals (Rodgers *et al.*, 2011). The latter can result in highly negative consequences, as this movement can enable pathogens to broaden their host range (Rodgers *et al.*, 2011). Trade in fish (i.e. eggs, larvae, or juveniles) also provides the potential for trafficking of pathogens to the same regions (Hedrick, 1996). Considering, for example, the numerous problems associated with the fish trade, one can understand the usefulness of the One Health approach in developing effective disease prevention and control to safeguard public health. A working example of a One Health assessment based on the ornamental fish trade is provided in Chapter 9. Bringing it all together.

7.5.1.1 Ornamental fish trade—facts and figures

Ornamental fish are selectively bred marine and freshwater animals, raised in various environments, including natural environments (mud ponds), concrete and rubber lined ponds and aquariums, as pets, or for other purposes (display and research). They consist of colorful fish species such as goldfish, guppies, angelfish, tetras, discus, koi, cichlids, rainbow fish, swordtails, *Corydoras*, plecos, loaches, killifish, etc. The ornamental fish trade has grown in importance worldwide, particularly as a source of foreign currency (Cognitive Market Research, 2025).

The worldwide trade in these aquatic species is facilitated by specialized breeders, wholesalers, retailers, and online platforms. The ornamental fish market has grown dramatically over time, involving over 120 exporting and/or importing countries all over the world, and is valued at over US$30 billion per year, including retail sales, associated materials, wages, and non-exported products. It is estimated that increasing numbers of new species are added to this trade every year. Over 2500 freshwater and marine fish species are involved in this commercial activity, with 60% of freshwater fish traded globally (Rhyne *et al.*, 2012, 2017; Dey, 2016).

Ornamental fish are either wild-caught (the majority of ornamental marine fish are captured from the wild), captive-bred, or captive-reared. Capturing fish can impact the local ecology of a coral reef or the catchment area of a river. Changes to the environment and the ecology of an aquatic ecosystem will impact the prevalence of pathogens. By taking a One Health approach, we can mitigate any negative impacts of these changes (Berthe *et al.*, 2018).

Based on recent estimations, there are between 15 and 150 million coral reef (marine) fishes traded annually, with the global value of this trade estimated for 1976 and 1999 at $28–40 million (Biondo and Burki, 2020). Among the major exporting countries are Singapore, Thailand, Malaysia, Indonesia, Israel, Brazil, Sri Lanka, and Colombia. The United States is the world's largest importing market of ornamental fish. Other importing countries holding stable shares in the market are the UK, Germany, Singapore, Japan, China and Hong Kong, France, the Netherlands, Italy, Malaysia,

Canada. and Belgium. Like Singapore, Germany, Hong Kong, Malaysia, and the Netherlands are important trading hubs, re-exporting a major portion of their imports (Dey, 2016).

7.5.1.2 Challenges of the trade

Specific environments can greatly affect the prevalence of pathogens and disease transmission. Almost all marine ornamental fish species are wild-caught. These fish would normally require special conditions and time to adapt to new environments and husbandry conditions. Also, transportation, handling, and improper feeding of wild caught fish sometimes cause high mortality rates throughout the supply chain, due to the lack of knowledge on the fish physiology and husbandry requirements.

Some countries practice over-intensive breeding, whereby breeders utilize selective breeding procedures to enhance desired qualities, often resulting in a wide spectrum of distinctive species with poor brood stock quality (lacking or reduced immune resistance) (Dey, 2016). In addition, the use of inappropriate fishing methods and overexploitation of wild fish populations has led to depletion of these populations with environmental consequences.

Humans have always been inveterate translocators of animals and disease agents. Accidental or deliberate introduction of exotic species in new habitats, accompanied by a lack of or poor disease monitoring of these stocks, leads to a greater risk of pathogen dissemination than in the past. Humans act as the main connecting interface between domestic and wild habitats, facilitating, through more or less direct activities, the introduction of pathogens existing in the wild but new to domestic animal populations, including pathogens with zoonotic potential. As stressed by many, we should be prepared to see in domestic habitats an increasing number of new diseases linked to the wild (Wobeser, 2006). Certainly, the opposite direction of pathogen transmission, from domestic to wild habitats, is also possible. However, by comparison, there are fewer records of these cases. Over the past decade, the need has been stressed for regulatory and monitoring systems along the ornamental fish trade chain, to allow collection of the data required to assess the health risks associated with this trade.

7.5.2 Fishing docks

Fishing docks, which are often locations at which waste (e.g. parts of harvested carcasses not used for human consumption or other purposes) is disposed of, can attract seabirds and marine mammals (Leguia *et al.*, 2023). Notably, some of the attracted species could be species of conservation concern. Some docks in select locations may also serve as tourist attractions, where deliberate feeding is used to attract wildlife for tourism purposes (Leguia *et al.*, 2023). Both activities, one unintentional, the other highly deliberate, increase potential human–wildlife interactions at artificially induced human–wildlife interfaces.

7.5.3 Urbanization

The effects of urbanization (positive or negative) on wildlife pathogens are debatable and may be influenced by a variety of factors (Bradley and Altizer, 2007). Nonetheless, interactions between humans and wildlife increase as human populations move into new areas that were previously not occupied by people; this has been primarily driven by population growth and agricultural expansion (Baker *et al.*, 2022). As an example, urban/suburban land use and housing density were noted as two of four important factors for forecasting the dissemination of West Nile virus in Georgia, USA (Gibbs *et al.*, 2006).

7.5.4 Disinfection

The use of disinfectants during viral disease outbreaks in humans is commonplace to help limit transmission. On occasion, these disinfectants can be released into the environment, thereby potentially impacting aquatic ecosystems. For example, during the 2014–2015 Ebola virus outbreak in Africa, the environment and humans were commonly sprayed with chlorine; this caused significant health effects for some individuals (Mehtar *et al.*, 2016). Similarly, chlorine disinfectants were commonly used in indoor and outdoor spaces to help control SARS-CoV-2 in various cities within China (Zhang et al., 2020). The chemicals and derivatives may eventually infiltrate natural water sources through direct runoff and sewage effluents, thereby posing a threat to aquatic plants and wildlife (Sedlak and von Gunten, 2011; Zhang et al., 2020).

7.5.5   Agriculture/industrialization

Anthropogenic factors (e.g. easy access to feed, abundant water, fewer predators, roosting/denning sites) associated with concentrated animal feeding operations (CAFO) and/or poultry production facilities can attract wildlife in large numbers. Due to plentiful wildlife of certain species, along with the large number of livestock/poultry that are typically housed at these types of facilities, a scenario that is conducive to pathogen transmission can be apparent at such facilities. For example, European starlings were suggested to have the capacity to mechanically move bacteria to feed and water sources at CAFOs, thereby providing a potential source of bacterial contamination within these types of facilities (Carlson et al., 2015). An additional example can be found in AIVs. Low pathogenic (LP) AIV H5 and H7 subtypes have the potential to mutate into HP forms of the viruses in some situations (Byrne et al., 2021). This can readily happen when a virus associated with the hemagglutinins mentioned above enter a poultry facility, as there are typically many permissive hosts/poultry present at these types of production facilities. Historically, these HP viruses were thought to be largely restricted to poultry; however, the A/goose/Guandong/1/1996 (GS/GD) H5N1 AIV lineage has challenged this line of thinking, as there have been many detections of the GS/GD lineage in wild birds (Ramey et al., 2022). Based on a variety of factors, including bidirectional exchange of viruses between poultry and wild birds, it has been suggested that HP AIVs may now pose a disease threat to wildlife in North America (Ramey et al., 2022). Data on species affected by HP AIVs in the USA support this school of thought, as over 150 species of wild birds have exhibited morbidity, mortality, or non-clinical positive cases in the USA alone (Ringenberg et al., 2023; USDA, 2024b). Further, detections of HP AIV in wild/

captive mammals have been recorded in over 20 species in the USA (USDA, 2024c).

7.6   Aquatic Wildlife as Sources of Food for Animals

7.6.1   General aquaculture

Various aquaculture facilities can present novel environments for wildlife populations that can create both attractants and repellents to some wildlife species (Barrett et al., 2019). It has been suggested that aquaculture facilities and practices can create novel pathways by which pathogens can be introduced to new locations (Rodgers et al., 2011). Fish farming, especially the aquaculture of farmed salmon, has caused concern regarding potential diseases shared between farmed and wild fish (Marty et al., 2010). Using a meta-analytic approach, a reduction in survival and/or abundance of multiple wild salmonid species was proposed to be associated with increased production of farmed salmon, thereby suggesting that this type of aquaculture production has resulted in decreased survival of wild trout and salmon populations in multiple regions (Ford and Myers, 2008).

7.6.1.1   Catfish culture ponds

As mentioned above, aquaculture practices and facilities can be wildlife attractants (Barrett et al., 2019). These facilities can attract several wildlife species, including piscivorous birds. Therefore, an unintended consequence of these facilities can be inter-pond transmission of pathogenic agents. For example, research has demonstrated that great egrets (Ardea alba), double-crested cormorants (Phalacrocorax auritus), American white pelicans (Pelecanus erythrorhynchos), and wood storks (Mycteria americana) can transport and shed viable hypervirulent Aeromonas hydrophila (vAh) following the consumption of channel catfish (Ictalurus punctatus) infected with that pathogen (Jubirt et al., 2015; Cunningham et al., 2018). Thus, it is not unreasonable to assume that the same birds could move this pathogen to natural areas following their visits to catfish ponds experiencing outbreaks.

7.7 Management and Control

The introduction of the scarecrow over 2500 years ago demonstrates the fact that humans have been searching for solutions to manage and control human–wildlife conflicts (World Wildlife Fund, 2022).

Indigenous people and local communities living alongside wildlife are important players in promoting sustainable interaction at the human–wildlife interface through the protection of biodiversity and the maintenance of harmonious coexistence (Sillero-Zubiri *et al.*, 2023). They are also important players in surveillance and mitigation of health risks at the human–wildlife interface.

Because management measures associated with the human–wildlife interface and the related zoonotic spillover are beyond the purview of the health system alone, mainstreaming policy in a transdisciplinary way is required, involving multiple stakeholders to ensure a comprehensive One Health approach is implemented (Debnath *et al.*, 2021). In this regard, the following are measures that must be considered in countries where deficiencies exist (Mardaraj and Sethy, 2015; Debnath *et al.*, 2021; Ellwanger and Chies, 2021):

- improve sanitary control of domestic animals and the environment, including waste management;
- increase surveillance of pathogens at human–animal interfaces;
- improve hunting regulations and trade of wild animals and wildlife products;
- improve land-use planning, maintenance of natural barriers, and the establishment of artificial barriers;
- reduce deforestation, habitat loss, and biodiversity loss;
- control all known vectors;
- effect wildlife translocation (if practical and suitable habitat with territorial vacancies exist and provided that the required screening for pathogens and quarantine protocols are put in place);
- increase investments in capacity building of technical staff;
- develop vaccination programs;
- identify and reduce biological and social factors of susceptibility to infections;
- invest in the mitigation of the emergence of infectious disease outbreaks;
- support conservation education and community empowerment for natural resource management;
- establish insurance programs and compensatory systems for losses which occur due to wildlife; and
- establish biosafety protocols for professionals working at human–animal interfaces.

The One Health approach could significantly assist in combating the emergence of infectious diseases at the human–wildlife interface. This is only possible, however, through national and international cooperation, the generation of and access to data, adequate funding, enactment and implementation of appropriate policies and legislation, and political will (Hassani and Khan, 2020).

References

Abrahms, B., Carter, N.H., Clark-Wolf, T.J., Gaynor, K.M., Johansson, E. *et al.* (2023) Climate change as a global amplifier of human–wildlife conflict. *Nature Climate Change* 13, 224–234.

Artois, M., Bengis, R., Delahay, R., Duchêne, M., Duff, J. *et al.* (2009) Wildlife disease surveillance and monitoring. In: Delahay, R.J., Smith, G.C. and Hutchings, M.R. (eds) *Management of Disease in Wild Mammals*. Springer Nature, Tokyo, pp. 187–213.

Baker, R.E., Mahmud, A.S., Miller, I.F., Rajeev, M., Rasambainarivo, F. *et al.* (2022) Infectious disease in an era of global change. *Nature Reviews Microbiology* 20, 193–205.

Barrett, L.T., Swearer, S.E. and Dempster, T. (2019) Impacts of marine and freshwater aquaculture on wildlife: A global meta-analysis. *Reviews in Aquaculture* 11, 1022–1044.

Berger, L., Roberts, A., Voyles, J., Longcore, J., Murray, K. *et al.* (2016) History and recent progress on chytridiomycosis in amphibians. *Fungal Ecology* 19, 89–99.

Berthe, F.C.J., Bouley, T., Karesh, W.B., Legall, I.C., Machalaba, C.C. *et al.* (2018) *One Health Operational Framework for Strengthening Human, Animal, and Environmental Public Health Systems at their Interface*. World Bank Group, Washington, DC. Available at: http://documents.worldbank.org/curate d/en/961101524657708673 (accessed 18 June 2025).

Biondo, M.V. and Burki, R.P. (2020) Systematic review of the ornamental fish trade with emphasis on coral reef fishes—an impossible task. *Animals* 10, 2014. DOI: 10.3390/ani10112014.

Bird, B.H. and Mazet, J.A.K. (2018) Detection of emerging zoonotic pathogens: An integrated One Health approach. *Annual Review of Animal Biosciences* 6, 121–139. DOI: 10.1146/annurev-animal-030117-014628.

Blaustein, A., Wake, D. and Sousa, W. (1994) Amphibian declines: Judging stability, persistence, and susceptibility of populations to local and global extinctions. *Conservation Biology* 8, 60–71.

Blaustein, A., Romansic, J., Kiesecker, J. and Hatch, A. (2003) Ultraviolet radiation, toxic chemicals and amphibian population declines. *Diversity and Distributions* 9, 123–140.

Bondad-Reantaso, M., Fejzic, N., MacKinnon, B., Huchzermeyer, D., Seric-Haracic, S. *et al.* (2021) A 12-point checklist for surveillance of diseases of aquatic organisms: A novel approach to assist multidisciplinary teams in developing countries. *Reviews in Aquaculture* 13(3), 1469–1487. DOI: 10.1111/raq.12530.

Bradley, C.A. and Altizer, S. (2007) Urbanization and the ecology of wildlife diseases. *Trends in Ecology and Evolution* 22, 95–102.

Bushman, M. and Antia, R. (2019) A general framework for modelling the impact of co-infections on pathogen evolution. *Journal of the Royal Society Interface* 16(155), 20190165. DOI: 10.1098/rsif.2019.0165.

Byrne, A.M.P., Reid, S.M., Seekings, A.H., Núñez, A., Obeso Prieto, A.B. *et al.* (2021) H7N7 avian influenza virus mutation from low to high pathogenicity on a layer chicken farm in the UK. *Viruses* 3(2), 259. DOI: 10.3390/v13020259.

Carlson, J.C., Hyatt, D.R., Ellis, J.W., Pipkin, D.R., Mangan, A.M. *et al.* (2015) Mechanisms of antimicrobial resistant *Salmonella enterica* transmission associated with starling–livestock interactions. *Veterinary Microbiology* 179, 60–68. DOI: 10.1016/j.vetmic.2015.04.009.

Cassirer, E.F., Manlove, K.R., Almberg, E.S., Kamath, P.L., Cox, M. *et al.* (2018) Pneumonia in bighorn sheep: Risk and resilience. *Journal of Wildlife Management* 82, 32–45.

CDC (2024) Reported Human Infections with Avian Influenza A Viruses. US Centers for Disease Control and Prevention. Atlanta, Georgia. Available at: https://www.cdc.gov/bird-flu/php/avian-flu-summary /reported-human-infections.html?CDC_AAref_Val=https://www.cdc.gov/flu/avianflu/reported-huma n-infections.htm (accessed 28 September 2024).

CDFW (2024) *California's Invaders: Nutria*. California Department of Fish and Wildlife, Sacramento, California.

Cognitive Market Research (2025) *Ornamental Fish Market Report 2025, Global Edition*, Report CMR136753. Cognitive Market Research Pune, Maharashtra, India. Available at: www.cognitivemar ketresearch.com (accessed 4 February 2025).

Cunningham, F.L., Jubirt, M.M., Hanson-Dorr, K.C., Ford, L., Fioranelli, P. *et al.* (2018) Potential of double-crested cormorants (*Phalacrocorax auritus*), American white pelicans (*Pelecanus erythrorhynchos*), and wood storks (*Mycteria americana*) to transmit a hypervirulent strain of *Aeromonas hydrophila* between channel catfish culture ponds. *Journal of Wildlife Diseases* 54, 548–552.

De Andrade Serrano, J., Toledo, L.F. and Sales, L.P. (2022) Human impact modulates chytrid fungus occurrence in amphibians in the Brazilian Atlantic Forest. *Perspectives in Ecology and Conservation* 20, 256–262.

Debnath, F., Chakraborty, D., Deb, A., Saha, M. and Dutta, S. (2021) Increased human-animal interface and emerging zoonotic diseases: An enigma requiring multi-sectoral efforts to address. *Indian Journal of Medical Research* 153(5–6), 577–584. DOI: 10.4103/ijmr.IJMR_2971_20.

Dey, V.K. (2016) The global trade in ornamental fish. *INFOFISH International* 4, 52–55. Available at: https: //www.bassleer.com/ornamentalfishexporters/wp-content/uploads/sites/3/2016/12/GLOBAL-TRAD E-IN-ORNAMENTAL-FISH.pdf

Dobigny, G. and Morand, S. (2022) Zoonotic emergence at the animal-environment-human interface: The forgotten urban socio-ecosystems. *Peer Community Journal* 2, e79. DOI: 10.24072/pcjournal.206.

Drewe, J., George, J. and Hasler, B. (2023) Reshaping surveillance for infectious diseases: Less chasing of pathogens and more monitoring of drivers. *Revue Scientifique et Technique* 42, 137–148. DOI: 10.20506/rst.42.3357.

Dreyer, S., Dreier, M. and Dietze, K. (2023) Demystifying a buzzword: Use of the term "human-animal-interface" in One Health oriented research based on a literature review and expert interviews. *One Health* 16, 100560.

Ellwanger, J. and Chies, J. (2021) Zoonotic spillover: Understanding basic aspects for better prevention. *Genetics and Molecular Biology* 44(1 Suppl. 1), e20200355. DOI: 10.1590/1678-4685-GMB-2020-0355.

European Centre for Disease Prevention and Control (2023) *Targeted Surveillance to Identify Human Infections with Avian Influenza Virus*. EU/EEA, Stockholm. Available at: https://www.ecdc.europa.eu /sites/default/files/documents/avian-influenza-virus-targeted-surveillance-to-identify-human%20inf ections-2023_1.pdf (accessed 16 June 2025).

FAO (2023) *The Emergency Prevention System for Animal Health: Enhancing the Prevention and Control of High-Impact Animal and Zoonotic Diseases through Biosecurity and One Health, Strategic Plan (2023–2026)*. Food and Agriculture Organization, Rome.

Fisher, M.C. and Garner, T.W.J. (2020) Chytrid fungi and global amphibian declines. *Nature Reviews Microbiology* 18, 332–343.

Fisher, M.C., Henk, D., Briggs, C., Brownstein, J.S., Madoff, L.C. *et al.* (2012) Emerging fungal threats to animal, plant and ecosystem health. *Nature* 484, 186–194.

Ford, J.S. and Myers, R.A. (2008) A global assessment of salmon aquaculture impacts on wild salmonids. *PLOS Biology* 6, 0411–0417.

Freidl, G., Meijer, A., de Bruin, E., de Nardi, M., Munoz, O. *et al.* (2014) Influenza at the animal–human interface: A review of the literature for virological evidence of human infection with swine or avian influenza viruses other than A(H5N1). *Eurosurveillance* 19(18), pii=20793. Available at: http://www.e urosurveillance.org/ViewArticle.aspx?ArticleId=20793

George, J.L., Martin, D.J., Lukacs, P.M. and Miller, M.W. (2008) Epidemic pasteurellosis in a bighorn sheep population coinciding with the appearance of a domestic sheep. *Journal of Wildlife Diseases* 44, 388–403. DOI: 10.7589/0090-3558-44.2.388.

Gibbs, S.E.J., Wimberly, M.C., Madden, M., Masour, J., Yabsley, M.J. *et al.* (2006) Factors affecting the geographic distribution of West Nile virus in Georgia, USA: 2002–2004. *Vector-Borne and Zoonotic Diseases* 6, 73–82.

Goldstein, J.E., Budiman, I., Canny, A. and Dwipartidrisa, D. (2022) Pandemics and the human-wildlife interface in Asia: Land use change as a driver of zoonotic viral outbreaks. *Environmental Research Letters* 17, 063009.

Gortazar, C., Reperant, L., Kuiken, T., de la Fuente, J., Boadella, M. *et al.* (2014) Crossing the interspecies barrier: Opening the door to zoonotic pathogens. *PLOS Pathogens* 10(6), e1004129. DOI: 10.1371/ journal.ppat.1004129.

Gross, E., Jayasinghe, N., Brooks, A., Polet, G., Wadhwa, R. *et al.* (2021) *A Future for All: The Need for Human-Wildlife Coexistence*. WWF, Gland, Switzerland.

Guberti, V., Stancampiano, L. and Ferrari, N. (2014) Surveillance, monitoring and surveys of wildlife diseases: A public health and conservation approach. *Hystrix Italian Journal of Mammalogy* 25(1), 3–8. DOI: 10.4404/hystrix-25.1-10114.

Hassani, A. and Khan, G. (2020) Human-animal interaction and the emergence of SARS-CoV-2. *JMIR Public Health and Surveillance* 6(4), e22117. DOI: 10.2196/22117.

Hassell, J.M., Begon, M., Ward, M.J. and Fèvre, E.M. (2017) Urbanization and disease emergence: Dynamics at the wildlife–livestock–human interface. *Trends in Ecology and Evolution* 32, 55–67.

Hedrick, R.P. (1996) Movements of pathogens with the international trade of live fish: Problems and solutions. *OIE Revue Scientifique et Technique* 15, 523–531.

Hocking, D. and Babbitt, K. (2014) Amphibian contributions to ecosystem services. *Herpetological Conservation and Biology* 9(1), 1–17. Available at: http://www.herpconbio.org/Volume_9/Issue_1/H ocking_Babbitt_2014.pdf

IAEA (2021) Zoonotic Disease Integrated Action Initiative (ZODIAC). International Atomic Energy Agency, Vienna. Available at: https://nucleus.iaea.org/sites/zodiac (accessed 16 June 2025).

International Union for Conservation of Nature and Natural Resources (2023) *IUCN SSC Guidelines on Human-Wildlife Conflict and Coexistence*, 1st edn. IUCN, Gland, Switzerland.

Jori, F., Hernandez-Jover, M., Magouras, I., Dürr, S. and Brookes, V.J. (2021) Wildlife-livestock interactions in animal production systems: What are the biosecurity and health implications? *Animal Frontiers* 11, 8–19.

Jubirt, M.M., Hanson, L.A., Hanson-Dorr, K.C., Ford, L., Lemmons, S. *et al.* (2015) Potential for great egrets (*Ardea alba*) to transmit a virulent strain of *Aeromonas hydrophila* among channel catfish (*Ictalurus punctatus*) culture ponds. *Journal of Wildlife Diseases* 51, 634–639.

Krofah, E., Gasca, C. and DeGarmo, A. (2021) *A Global Early Warning System for Pandemics: Mobilizing Surveillance for Emerging Pathogens*. Milken Institute, Santa Monica, California.

Leguia, M., Garcia-Glaessner, A., Muñoz-Saavedra, B., Juarez, D., Barrera, P. *et al.* (2023) Highly pathogenic avian influenza A (H5N1) in marine mammals and seabirds in Peru. *Nature Communications* 14, 5489.

Li, Q., Shah, T., Wang, B., Qu, L., Wang, R. *et al.* (2023) Cross-species transmission, evolution and zoonotic potential of coronaviruses. *Frontiers in Cellular and Infection Microbiology* 12, 1081370. DOI: 10.3389/fcimb.2022.1081370.

Lindahl, J.F. and Grace, D. (2015) The consequences of human actions on risks for infectious diseases: A review. *Infection Ecology and Epidemiology* 5, 30048.

Madden, F. (2004) Creating coexistence between humans and wildlife: Global perspectives on local efforts to address human–wildlife conflict. *Human Dimensions of Wildlife* 9, 247–257.

Mardaraj, P. and Sethy, J. (2015) Human-wildlife conflict: Issues and managements. In: Sahu, H.K. and Mishra, S.R.K. (eds) *Biodiversity Conservation, Research, and Management*, 1st edn. Himalaya Publishing House, Mumbai, India, pp. 158–173.

Martino, P.E., Stanchi, N., Pia, S.M. and Brihuega, B. (2014) Seroprevalence for selected pathogens of zoonotic importance in wild nutria (*Myocastor coypus*). *European Journal of Wildlife Research* 60, 551–554.

Marty, G.D., Saksida, S.M. and Quinn, T.J. (2010) Relationship of farm salmon, sea lice, and wild salmon populations. *Proceedings of the National Academy of Sciences of the United States of America* 107, 22599–22604.

Mehtar, S., Bulabula, A.N.H., Nyandemoh, H. and Jambawai, S. (2016) Deliberate exposure of humans to chlorine-the aftermath of Ebola in West Africa. *Antimicrobial Resistance and Infection Control* 5, 45.

Mörner, T., Obendorf, D., Artois, M. and Woodford, M. (2002) Surveillance and monitoring of wildlife diseases. *Revue Scientifique et Technique* 21(1), 67–76. DOI: 10.20506/rst.21.1.1321.

Mumford, E., Bishop, J., Hendrickx, S., Ben Embarek, P. and Perdue, M. (2007) Avian influenza H5N1: Risks at the human–animal interface. *Food and Nutrition Bulletin* 28(2 Suppl), S357–363.

Naicker, P. (2011) The impact of climate change and other factors on zoonotic diseases. *Archives of Clinical Microbiology* 2(2), 1–6. DOI: 10:3823/226.

Narayan, E. and Rana, N. (2023) Human-wildlife interaction: Past, present, and future. *BMC Zoology* 8, 5.

Oidtmann, B., Peeler, E., Lyngstad, T., Brun, E., Jensen, B. *et al.* (2013) Risk-based methods for fish and terrestrial animal disease surveillance. *Preventive Veterinary Medicine* 112(1–2), 13–26. DOI: 10.1016/j.prevetmed.2013.07.008.

Pelican, K., Salyer, S., Barton, B., Belot, G., Carron, M. *et al.* (2019) Synergising tools for capacity assessment and one health operationalization. *Revue Scientifique et Technique* 38(1), 71–89. DOI: 10.20506/rst.38.1.2942.

Piaggio, A.J., Engeman, R.M., Hopken, M.W., Humphrey, J.S., Keacher, K.L. *et al.* (2014) Detecting an elusive invasive species: A diagnostic PCR to detect Burmese python in Florida waters and an assessment of persistence of environmental DNA. *Molecular Ecology Resources* 14, 374–380.

Porter, S.M., Hartwig, A.E., Bielefeldt-Ohmann, H., Bosco-Lauth, A.M. and Root, J.J. (2022) Susceptibility of wild canids to SARS-CoV-2. *Emerging Infectious Diseases* 28, 1852–1855.

Porter, S.M., Hartwig, A.E., Bielefeldt-Ohmann, H., Root, J.J. and Bosco-Lauth, A.M. (2024) Experimental SARS-CoV-2 infection of elk and mule deer. *Emerging Infectious Diseases* 30, 354–357.

Pounds, J. (2001) Climate and amphibian declines. *Nature* 410, 639–640. DOI: 10.1038/35070683.

Puryear, W., Sawatzki, K., Hill, N., Foss, A., Stone, J.J. *et al.* (2023) Highly pathogenic avian influenza A(H5N1) virus outbreak in New England seals, United States. *Emerging Infectious Diseases* 29, 786–791.

Ramey, A.M., Hill, N.J., DeLiberto, T.J., Gibbs, S.E.J., Hopkins, M.C. *et al.* (2022) Highly pathogenic avian influenza is an emerging disease threat to wild birds in North America. *Journal of Wildlife Management* 86(2), e22171.

Reperant, L. and Osterhaus, A. (2013) The human-animal interface. *Microbiology Spectrum* 1(1), OH–0013. DOI: 10.1128/microbiolspec.OH-0013-2012.

Reperant, L., Cornaglia, G. and Osterhaus, A. (2012) The importance of understanding the human–animal interface. *Current Topics in Microbiology and Immunology* 365, 49–81.

Rhyne, A.L., Tlusty, M.F., Schofield, P.J., Kaufman, L., Morris, J.A. *et al.* (2012) Revealing the appetite of the marine aquarium fish trade: The volume and biodiversity of fish imported into the United States. *PLOS One* 7, e35808.

Rhyne, A.L., Tlusty, M.F., Szczebak, J. and Holmberg, R.J. (2017) Expanding our understanding of the trade in marine aquarium animals. *PeerJ* 5, e2949.

Rinanda, T. (2015) Aquatic animals and their threats to public health at human-animal-ecosystem interface: A review. *AACL Bioflux* 8, 784–789.

Ringenberg, J.M., Weir, K., Humberg, L., Voglewede, C., Oswald, M. *et al.* (2023) Prevalence of avian influenza virus in synanthropic birds associated with an outbreak of highly pathogenic strain EA/AM. *bioRxiv* (8), 565892.

Rodgers, C.J., Mohan, C.V. and Peeler, E.J. (2011) The spread of pathogens through trade in aquatic animals and their products. *OIE Revue Scientifique et Technique* 30, 241–256.

Ríos-Castro, R., Romero, A., Aranguren, R., Pallavicini, A., Banchi, E. *et al.* (2021) High-throughput sequencing of environmental DNA as a tool for monitoring eukaryotic communities and potential pathogens in a coastal upwelling ecosystem. *Frontiers in Veterinary Science* 8, 765606.

Scheele, B., Pasmans, F., Skerratt, L., Berger, L., Martel, A. *et al.* (2019) Amphibian fungal panzootic causes catastrophic and ongoing loss of biodiversity. *Science* 363(6434), 1459–1463. DOI: 10.1126/science.aav0379.

Sedlak, D.L. and von Gunten, U. (2011) The chlorine dilemma. *Science* 331, 42–43.

Shrestha, K., Acharya, K. and Shrestha, S. (2018) One health: The interface between veterinary and human health. *International Journal of One Health* 4, 8–14. DOI: 10.14202/IJOH.2018.8-14.

Shriner, S.A., Root, J.J., Lutman, M.W., Kloft, J.M., VanDalen, K.K. *et al.* (2016) Surveillance for highly pathogenic H5 avian influenza virus in synanthropic wildlife associated with poultry farms during an acute outbreak. *Scientific Reports* 6, 36237.

Sillero-Zubiri, C., Ardiantiono, F., Ying, C., Dimitra, C., Girma, E. *et al.* (2023) From conflict to coexistence: The challenges of the expanding human–wildlife interface. *Oryx* 57(4), 409–410. DOI: 10.1017/S0030605323000698.

Sleeman, J., Brand, C. and Wright, S. (2012) Strategies for wildlife disease surveillance. In: Aguirre, A.A., Ostfield, R.S. and Daszak, P. (eds) *New Directions in Conservation Medicine: Applied Cases in Ecological Health*. Oxford University Press, New York, pp. 539–551. Available at: http://digitalcommons.unl.edu/usgsstaffpub/971 (accessed 10 January 2024).

Sumanasekara, V., Dissanayake, D. and Seneviratne, H. (2015) Review on use of amphibian taxa as a bio-indicator for watershed health and stresses. In: *NBRO Symposium Proceedings*. Available at: https://www.nbro.gov.lk/images/content_image/pdf/symposia/32.pdf (accessed 10 January 2024).

Titcomb, G.C., Jerde, C.L. and Young, H.S. (2019) High-throughput sequencing for understanding the ecology of emerging infectious diseases at the wildlife-human interface. *Frontiers in Ecology and Evolution* 7, 126.

USDA (2024a) Nutria, An Invasive Rodent. Fact Sheet. APHIS 11-15-005. USDA, Washington, DC. Available at: https://www.aphis.usda.gov/sites/default/files/fsc-nutria-invasive-rodent.pdf (accessed 16 June 2025).

USDA (2016) *Final Report for the 2014-2015 Outbreak of Highly Pathogenic Avian Influenza (HPAI) in the United States*. USDA, Washington, DC.

USDA (2023) *Implementation Plan for Avian Influenza Surveillance in Waterfowl in the United States, Summer 2023–Spring 2024*. Washington, DC, USDA. Available at: https://www.aphis.usda.gov/animal_health/downloads/animal_diseases/ai/2023-24-wild-bird-ai-surveillance-implementation-plan.pdf (accessed 16 June 2025).

USDA (2024b) *2022-2024 Detections of Highly Pathogenic Avian Influenza in Wild Birds*. USDA, Washington, DC. Available at: https://www.aphis.usda.gov/aphis/ourfocus/animalhealth/animal-disease-information/avian/avian-influenza/hpai-2022/2022-hpai-wild-birds (accessed 16 June 2025).

USDA (2024c) 2022-2024 Detections of Highly Pathogenic Avian Influenza in Mammals. USDA, Animal and Plant Health Inspection Service, Washington, DC. Available at: https://www.aphis.usda.gov/livestock-poultry-disease/avian/avian-influenza/hpai-detections/mammals (accessed 29 September 2024).

US Environmental Protection Agency (2015) Mean Amphibian Species Richness: Southeast. Fact Sheet. EnviroAtlas.

Wobeser, G.A. (2006) *Essentials of Disease in Wild Animals*. Blackwell, Ames, Iowa.

World Health Organization (2023) *Influenza at the Human-Animal Interface*. WHO, Geneva. Available at: ht tps://www.who.int/publications/m/item/influenza-at-the-human-animal-interface-summary-and-ass essment-21-december-2023 (accessed 20 June 2025).

World Organisation for Animal Health (2021) *Aquatic Animal Health Surveillance*. Aquatic Animal Health Code, World Organisation for Animal Health, Paris.

World Wildlife Fund (2022) Human-wildlife conflict in the global biodiversity framework. Position paper. Available at: https://wwfint.awsassets.panda.org/downloads/wwf_position_paper_human_wildlife_c onflict_in_the_cbd_nov22.pdf (accessed 16 June 2025).

Zhang, H., Tang, W., Chen, Y. and Yin, W. (2020) Disinfection threatens aquatic ecosystems. *Science* 368, 146–147.

Zhao, Y., Huang, F., Wang, W., Gao, R., Fan, L. *et al.* (2023) Application of high-throughput sequencing technologies and analytical tools for pathogen detection in urban water systems: Progress and future perspectives. *Science of the Total Environment* 900, 165867.

8 Food Security and Innovative Food Production

Abstract

The chapter describes food availability, food access and utilization, and food stability from the perspective of Sustainable Development Goal 2: Zero hunger (SDG 2). The main topics addressed in this chapter are: efficiency of use of world food resources; food waste; biofuel; food technology; and insect, algae, and seaweed food production.

PART I: CURRENT PROBLEMS AFFECTING FOOD AVAILABILITY

Patricia Aguirre Mejía[1] and Ricardo Muñoz Cisternas[2]*

[1]Facultad de Posgrado, Universidad Técnica del Norte, Ecuador; [2]Observatorio de Políticas Públicas del Territorio, FARAC, Universidad de Santiago de Chile, Chile

8.1 Introduction

The concept of sustainability along with the term sustainable development hold different meanings for different individuals (White, 2013). The most widely recognized definition, 'development that meets the needs of the present without compromising the ability of future generations to meet their own needs,' first appeared in the report of the World Commission on Environment and Development, *Our Common Future* (WCED, 1987). However, numerous other definitions and interpretations exist, which fall beyond the scope of this section. Thus, sustainability can be understood as the ability to meet present needs without compromising the capacity of future generations to meet their own. It is a fundamental principle for addressing issues such as inequality, poverty, and food insecurity, and involves seeking a balance between social, economic, and environmental dimensions (Carballo, 2005; FAO *et al.*, 2018).

In the context of food security, sustainability is linked to agricultural practices that protect natural resources, promote biological diversity, and reduce the use of agrochemicals (Urdes *et al.*, 2022). It also involves the promotion of local and regional food systems that reduce dependency on imports and enhance access to fresh and nutritious food (UNDP and FAO, 2016; Wezel *et al.*, 2020). The Sustainable Development Goals (SDGs), adopted by the United Nations General Assembly in 2015, consist of 17 interconnected global objectives aimed at eradicating poverty, protecting the planet, and ensuring prosperity for all. Several of the SDGs are directly related to inequality, poverty, and food security. They promote a holistic approach, emphasizing the importance of policies that strengthen economic and social resilience, specifically, the need to eliminate hunger and extreme poverty, reduce inequality, and promote sustainable food systems (FAO *et al.*, 2018). These objectives are particularly

*Corresponding author (part 1): ricardo.munoz@usach.cl; Corresponding author (part 2): jkprado@utn. edu.ec

reflected in those SDGs that focus on eradicating poverty and hunger, reducing inequalities, and promoting sustainable production, as follows:

- SDG 1, No Poverty: Aims to eradicate poverty in all its forms and dimensions, including extreme poverty. This requires increasing the income of the poorest individuals as well as ensuring their access to essential services such as health and education.
- SDG 2, Zero Hunger: Seeks to end hunger, achieve food security and improved nutrition, and promote sustainable agriculture. This goal involves improving food production and distribution, reducing food waste, and ensuring access to healthy diets for all.
- SDG 10, Reduced Inequalities: Aims to reduce inequality within and among countries. This entails addressing economic, gender, ethnic, and opportunity inequalities, as well as promoting more progressive fiscal and social policies.
- SDG 12, Responsible Consumption and Production: Promotes more sustainable patterns of consumption and production, including the reduction of food waste and the encouragement of responsible agricultural practices.

The challenge of eradicating food insecurity requires addressing the structural causes of poverty and economic inequality, which not only hinder access to food but also delay progress toward sustainable development (Stewart, 2015). Several studies have shown that, in many countries, current efforts are insufficient to ensure universal access to nutritious food, particularly in contexts of high inequality found in undeveloped and developing countries of the Global South (Haini *et al.*, 2023).

8.2   Climate Change

Climate change is a critical issue across all sectors, with agriculture and food production also being significantly impacted by this phenomenon (Fig. 8.1). It is expected to contribute to the increase in hunger and malnutrition, as well as exacerbating food insecurity and degrading the living conditions of farmers. The most vulnerable populations and those who rely on agriculture, forests, and fisheries can be disproportionately affected (FAO *et al.*, 2022).

Rural communities, particularly those living in fragile environments, face an immediate and growing risk of crop and livestock losses, as well as reduced availability of marine, forest, and aquaculture products. Hunger and malnutrition can increase, as frequent and intense climate-related events negatively affect food availability, access, stability, and utilization as well as assets and livelihood opportunities in both rural and urban areas (FAO, 2017). Poor populations will be at risk of food insecurity due to the loss of their assets and the lack of adequate insurance coverage. The ability of rural populations to cope with and respond to the impacts of climate change depends on the cultural context, existing policies, and socio-economic factors that enable them to be resilient to the increasingly extreme and unpredictable effects of climate change (FAO, 2017).

Poverty, economic inequality, and food security are interconnected concepts that are crucial for understanding sustainable development. From a multidimensional perspective, poverty not only refers to insufficient economic income but also encompasses deficiencies in health, education, and access to basic services (Alkire *et al.*, 2021). Economic inequality can be defined as the disparity in the distribution of income and economic resources within a local, regional, or national community, or between countries. It is a variable closely linked to structural and systemic barriers (Stewart, 2015), serving as a determinant of poverty. Both poverty and economic inequality influence food security. According to the definition established by the World Food Summit of 1996 (FAO, 1996), food security exists when all people, permanently, have physical and economic access to sufficient, safe, and nutritious food to meet their dietary needs and preferences, to lead a healthy life.

8.3   Adaptation and Resilience

The disruption or decline in global and local food supply due to climate change can be

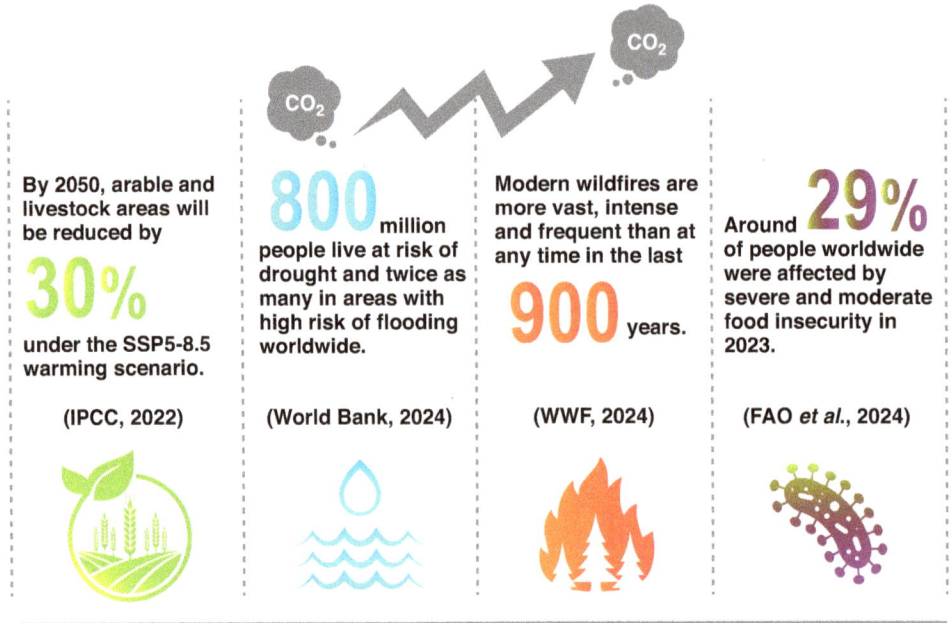

Fig. 8.1. Effects derived from the increase in global temperature. Figure author's own.

mitigated with proper irrigation infrastructure, efficient and well-managed watershed practices, advancements in land cultivation, sustainable agricultural and livestock management, and the development of crop varieties and forages adapted to changing climate conditions. Early warning systems used in the analysis of climate change impacts on agricultural production and the food supply chain can be a modern and effective use of climate data and weather forecasting. Additionally, a well-organized and prepared response system is essential for reacting to negative climate events. Water is crucial for increasing productivity, considering that improvements in agricultural water resource management will be key to protecting against the anticipated volatility in rainfed production. It is projected that the critical ecological flow in 42–79% of the world's watersheds will be impacted by 2050 (IPCC, 2022). Managing production risks in the face of increasing aridity and precipitation variability requires agricultural systems, both rainfed and irrigated, to be much more flexible and adaptable in their approach. The progressive adjustment of large-scale irrigation schemes would be essential for maintaining and

increasing production in line with the demand, while enhancing local water resource management practices can enable vulnerable groups to adapt their farming techniques.

The effects of climate change, along with the resulting variability in water availability, are already visible in Latin America and the Caribbean (LAC). Central and South America are characterized by the delayed onset of the rainy season and an increase in the frequency and intensity of droughts, as well as excessive rainfall, severe floods, and landslides. Rural communities, especially subsistence farmers, are impacted by water scarcity and soil erosion, which reduce food production and lead to food insecurity, with periods in which they lose their ability to meet basic dietary and nutritional needs. This further exacerbates poverty and the capacity to adapt to changes in climate patterns, increasing the pressure on these populations. Climate change adaptation and mitigation, with a focus on food security, involves, among other priorities, reducing greenhouse gas emissions in all agricultural and livestock activities. This goal is not often emphasized. It also entails reducing risks associated with the occurrence of

Fig. 8.2. Fires from agricultural sources in the Ecuadorian Andes, summer 2024. Figure author's own.

extreme events. All of this requires substantial financial investment in infrastructure to combat soil erosion, as well as the implementation of resource-efficient irrigation systems (WFP, 2024). These challenges are linked to extreme climate events, such as fires resulting from prolonged drought periods, as illustrated in Fig. 8.2.

The climate challenge facing communities and local governments is directly linked to the alarming rate of permanent migration from areas affected by climate variability and change. In the Caribbean, small island states are experiencing more intense hurricanes, the threat of rising sea levels in low-lying coastal areas, and losses in ecosystem services that impact people's livelihoods.

8.4 Agriculture and Climate Change

One of the factors that has intensified food insecurity by affecting agricultural production is climate change. According to the IPCC (2021), climate change affects food insecurity in several ways, including the following: impacts on agricultural production, affecting the yield of essential crops (such as cereals, legumes, etc.) critical for food security; the spread of agricultural pests and diseases that can expand into new regions, threatening essential crops; extreme weather events (such as heavy rains, hurricanes, heatwaves, etc.) that damage agricultural infrastructure, reducing food availability and increasing production and distribution costs; the degradation of ecosystems and biodiversity loss, which

affects the availability of key ecosystem services (e.g. pollination of crops; ocean warming threatens fish populations as a protein source for coastal populations); and inequalities in food security, as more vulnerable regions are exposed to higher risks due to their lower adaptive capacity and dependence on climate-sensitive agricultural systems.

Between 2000 and 2018, nearly 90% of deforestation was linked to agriculture, both for crop cultivation and livestock farming. At the same time, agriculture has been affected by climate change, particularly by alterations in precipitation patterns, soil quality, and biodiversity (Rainforest Alliance, 2023). Avocado cultivation serves as a key example of the challenges in sustainable development. The deforestation required for avocado plantations not only threatens high-altitude ecosystems, such as the Andean moors (*páramos*) and forested areas, but also intensifies the demand for irrigation water. Furthermore, avocado cultivation has become an increasingly significant source of rural employment, in contrast to short-cycle crops which, due to climate variations, pose a high risk to farmers, who often fail to cover production costs, thus contributing to poverty.

Agriculture contributes to climate change but is also part of the solution. Greenhouse gas emissions from the agricultural and forestry sectors currently represent more than 30% of annual emissions (deforestation and forest degradation account for 17.4%, agriculture for 13.5%). Sustainable agriculture, however, can help reduce greenhouse gas emissions and their impact through the management of ecosystem services, reduction in land-use change and deforestation linked to it, the use of more efficient crop varieties, better control of accidental fires, improved nutrition for ruminant livestock, more effective livestock waste management, organic soil management, conservation agriculture, and agroforestry systems. In addition to reducing greenhouse gas emissions, well-managed pasture and croplands can sequester significant amounts of carbon.

Forty percent of the land's biomass, and with it the biological carbon, is directly or indirectly managed by farmers, foresters, or pastoralists. It is their responsibility to adopt management systems that combine mitigation and adaptation, thereby enhancing the resilience of agricultural systems (FAO et al., 2024). Climate-smart agriculture programs are being implemented in several countries, funded and supported by the FAO. However, this is not very different from sustainable agriculture, which, in its principles, involves the protection of biodiversity, soil conservation, and the proper use of water (Rainforest Alliance, 2023). In the context of food security, the concept of sustainability is associated with agricultural practices that protect natural resources, promote biological diversity, and reduce the use of agrochemicals (An et al., 2022). The promotion of local and regional food systems is also important, as they reduce reliance on imports and support access to fresh, nutritious food (UNDP and FAO, 2016; Wezel et al., 2020).

All inhabited regions of the world are experiencing the effects of climate change, which is one of the main drivers of global hunger. Over the past decade, 1.7 billion people have been affected by extreme weather events and climate-related disasters (IPCC, 2022). Additionally, the communities that contribute the least to the climate crisis are the most affected by its impacts, with limited resources to address them. Responses to climate change are closely linked to sustainable food production practices. Climate change adaptation and mitigation strategies mutually reinforce each other through responsible approaches in agriculture and fisheries. Many of the climate and weather risk management strategies align with sustainable practices in these sectors, enabling their promotion through programs and policies focused on environmentally responsible production. The integration of these strategies is crucial for both fostering sustainable food production and developing effective climate change adaptation policies.

Family farming plays a fundamental role in addressing the challenges posed by climate change, as it accounts for 80% of global food production. Family farmers are uniquely positioned to spearhead the transition to more inclusive, effective, sustainable, and climate-resilient food systems. The FAO's Globally Important Agricultural Heritage Systems (GIAHS) program promotes the conservation of sustainable agricultural systems that represent global heritage, with the aim of ensuring the reproduction of responsible agricultural techniques, preserving local cultures, and securing food security. This is achieved through the conservation of environments that protect natural ecosystems and safeguard agrobiodiversity, as well as food sovereignty, which is essential for communities. Examples of these important sustainable agricultural systems include the small cornfield, the so called *milpas* in Mexico and Central America, and the Chakra and Aja systems in the Ecuadorian Amazon (Moreno and Flores, 2011). Fig. 8.3 presents the map of the Ecuadorian cacao Chakra corridor, where families generate income through cacao cultivation, while also having access to a variety of foods for both subsistence and sale. This system reduces pressure on forests and ensures family food security through more than 90 plant species (Torres et al., 2014).

8.5 Conflicts, Political Instability, and Food Security

Political instability and conflicts are directly linked to food scarcity, as producing food in conflict zones becomes extremely difficult. This makes farmers vulnerable to attack. Political instability often leads to insecurity on rural roads, which are the main routes for transporting food. Examples of such conflicts include attacks on avocado trucks, particularly when prices are favorable for farmers. Such a situation is observed in Ecuador, where both avocados, known as 'green gold,' and onions face such threats. Political instability weakens

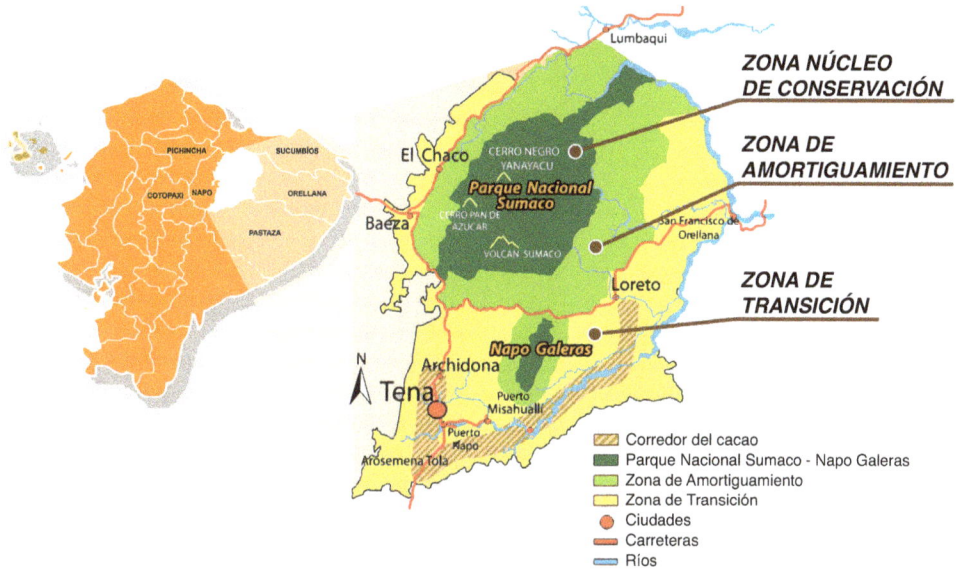

Fig. 8.3. Map of the cocoa farms, the so-called 'home gardens' (Chakras) corridor in Ecuador. Figure author's own.

food distribution systems, resulting in instability in the marketing of agricultural products and low prices due to the inability to distribute food on time. Thus, food security becomes a crucial guarantee for peace and sustainable development, as it forms the foundation for achieving a dignified, healthy, and fulfilling life—ultimately ensuring the quality of life of communities.

Hunger creates social instability when people lack access to food and are forced to make decisions out of desperation. This manifests in various ways, sometimes in social unrest, protests, and a lack of trust in the state, and other times in violent behaviors such as looting, theft, or support for violent movements or armed groups that offer short-term security and food relief in exchange for loyalty to their ideologies—situations that are becoming increasingly common in developing countries.

Eradicating hunger, or achieving zero hunger, should be considered a primary objective for all states within the international community, as it is a global issue with complex, interdependent risks and threats that do not recognize borders. This is why a global food security governance system has been established. The term 'governance system' refers to the set of

institutions, objectives, policies, agreements, and norms intended to regulate, manage, and coordinate matters related to food security. As Pérez de Armiño (2013) notes, however, this system suffers from various issues, such as fragmentation and lack of coordination, the absence of binding commitments from governments, and a lack of coherence.

8.6 Economic Inequality, Poverty, and Food Insecurity

Economic inequality acts as a determining factor for poverty and food insecurity (Casas, 2020). An unequal distribution of wealth and income limits access to food for low-income individuals, as well as to other basic goods and services. This situation is exacerbated in contexts of extreme poverty, where individuals lack the economic resources to obtain adequate food (Vilar-Compte et al., 2021), thus perpetuating a cycle of poverty and food insecurity. Poverty restricts families' ability to access sufficient and diverse food, leading to food insecurity. Studies have shown that poor families face higher rates of hunger and malnutrition due to economic constraints that affect both the quality and quantity of the food they can acquire (Mahadevan and Hoang, 2016). Furthermore, income inequality deepens these disparities by limiting the ability of certain groups to access food, even in economies where the overall food supply is adequate (Grzelak, 2017).

It is important to note that poverty influences the issue of accessing adequate food and can lead to malnutrition, particularly among pregnant women and children, with negative effects on health and development (Neves *et al.*, 2022). Poverty also limits access to education, participation, and social well-being, which in turn reduces people's opportunities to improve their living conditions. Therefore, food insecurity can create a vicious cycle that affects health, academic performance, productivity, and the ability of individuals to escape poverty (Guardiola and González, 2010; FAO *et al.*, 2013). In this regard, food insecurity is not only a consequence of poverty and inequality but also one of its causes, creating complex scenarios that require integrated solutions (Woodhill *et al.*, 2022). From an economic perspective, poverty reduces purchasing power, limiting access to food markets, while from a social perspective, inequalities perpetuate generations of poverty, reducing access not only to food but also to other essential services, such as education and health care, which are fundamental for long-term food security (Siddiqui *et al.*, 2020). Politically, the lack of investment in rural infrastructure and social protection systems exacerbates poverty and maintains scenarios of inequity, inequality, and food insecurity, leading to the marginalization and exclusion of individuals and their communities (Karmakar and Sarkar, 2014).

Market dynamics, such as dependence on agricultural exports, income inequality, and economic growth, amplify these issues in developing countries (Batool and Sheikh, 2024). Economic factors, such as financial barriers, low income, and informal employment, prevent access to nutritious food. Inequality restricts the economic growth necessary to alleviate poverty (Rahmanto *et al.*, 2021). Social factors, such as marginalization and exclusion of vulnerable groups, intensify food insecurity. These inequalities perpetuate cycles of poverty in both rural and urban communities (Ouma, 2017). Political

factors, such as inadequate policies and lack of regulation, exacerbate inequality and food insecurity, particularly affecting developing countries (Pollard and Booth, 2019).

8.7   Food Security and Economic Crisis

An important threat to food security is the emergence of an economic crisis, which is characterized by a significant decline in economic activity (Cárcamo *et al.*, 2022). This may manifest in the form of recessions, high unemployment, and a decrease in per capita income, all of which affect the growth and development of a country or region. Crises can be classified into different types, such as financial crises, production crises, and food crises, each with their own causes and effects. These crises are often caused by factors such as inadequate fiscal policies, external causes, or changes in global markets that impact the economic stability of a country or region (MacNabb and Fletcher, 2021).

It is understood, then, that there is a bidirectional relationship between economic crises and poverty. On one hand, a crisis leads to unemployment and reduced wages, which exacerbates poverty and reduces food consumption. On the other hand, an impoverished population, in addition to suffering from poor nutrition, may limit the possibilities of economic recovery due to the negative effect on labor productivity. Food insecurity refers to the lack of access to sufficient nutritious food for leading an active and healthy life. Economic crises can affect access to food in terms of both quality and quantity, limiting people's ability to maintain an active and healthy quality of life by impacting their purchasing power, thus limiting access to food resources. As an example, during the global financial crisis of 2008, many countries experienced a significant rise in food prices, leading to an increase in levels of hunger and malnutrition (Junior and Marzábal, 2012; Miyashiro, 2022).

A recent example is the crisis caused by the COVID-19 pandemic, which led to disruptions in global food supply chains and an increase in food insecurity. According to FAO *et al.* (2021), pandemic-related restrictions, such as border closures, travel limitations, quarantines, and disruptions in supply chains and trade, affected the physical ability to access diverse and nutritious foods. This included the loss of perishable, high-value products due to interruptions in labor and transportation. Additionally, access to food was further eroded by losses in income and livelihood, which increased the demand for food assistance programs such as food banks. In this context, in 2020, the number of people facing severe food insecurity increased by 12% of the global population. COVID-19, therefore, not only played a role in exacerbating pre-existing vulnerabilities but also contributed to inequalities in food access, pushing millions of people into conditions of food insecurity.

Another example of a crisis is the one caused by the war in Ukraine. This conflict has had significant repercussions on the production and export of cereals, particularly wheat and corn. According to FAO *et al.* (2022), global food prices reached record levels due to supply disruptions, leading to an increase in food insecurity in regions dependent on these imports. In this context, the war in Ukraine has had major effects on the global food supply, exacerbating food insecurity due to disruptions in the exports of grains, sunflower oil, and fertilizers, particularly from two of the world's leading producers: Ukraine and Russia. These disruptions have led to increases in international food and energy prices, raising costs for consumers and making access to an adequate diet more difficult. Additionally, the number of people facing food insecurity remained high in 2023, with the global prevalence of moderate or severe food insecurity affecting 28.9% of the world's population (2.33 billion people). Since the onset of the pandemic in 2019, there has been an increase of more than 65 million people in this condition, with a total of 864 million people facing severe food insecurity (FAO *et al.*, 2024).

8.8 Comprehensive Strategic Bases for Food Security

Understanding poverty, economic inequality, and these factors as causal drivers of food insecurity forms the foundation for designing comprehensive strategies to address the root causes of poverty and strengthen food systems.

Governments must implement public policies that tackle both economic inequality and food insecurity, addressing their structural causes (Cruz *et al.*, 2019). Essential to this is the application of integrated policies for income redistribution, social protection programs, investment in education and health care, and the promotion of decent employment (Castillo, 2016; FAO *et al.*, 2018). Social protection policies are critical for improving the incomes of vulnerable families, protecting their consumption of basic goods, and strengthening their capacity to respond to crises. In this regard, the design of income transfer programs, such as scholarships or assistance for fostering entrepreneurship or the development of small businesses, serves as an effective strategy to reduce poverty, generate employment, and improve access to food (FAO *et al.*, 2018).

Strengthening family farming or small-scale agricultural producers is crucial for improving food security and reducing economic inequality in rural areas (Fonseca-Carreño and Bossa-Pabon, 2022). To achieve this, it is essential to ensure their access to credit, technology, markets, and training, which not only increases their productivity and income but also promotes food diversification and environmental sustainability (León *et al.*, 2004; Stewart *et al.*, 2015; FAO *et al.*, 2018). Additionally, supporting local agricultural production contributes to the availability of and access to nutritious food, while encouraging crop diversification and the adoption of sustainable agricultural practices (León *et al.*, 2004). In this regard, strengthening family farming allows for an increase in the supply of fresh and healthy food in rural and marginalized communities, thus reinforcing the local economy and reducing dependence on external markets (Touch *et al.*, 2024).

Strengthening local and regional food systems is crucial for improving access to fresh and nutritious food, supporting small producers, and creating employment. To achieve this, promoting local markets, producer cooperatives, and public procurement programs for local food is key (FAO, 2017). Additionally, these systems can reduce reliance on imports by producing and marketing food locally, promoting short food supply chains that benefit both producers and consumers (Pasquier, 2019). Furthermore, local food systems tend to be more resilient to external shocks and environmental issues, which enhances their long-term sustainability (León *et al.*, 2004; FAO *et al.*, 2018; Pasquier, 2019).

Nutritional education plays a crucial role in improving eating habits and reducing malnutrition (Juela-Tiban and Chileno-Camacho, 2024). It should include promoting healthy diets, encouraging the consumption of fresh foods, and reducing processed food intake (FAO *et al.*, 2018). Additionally, it is essential for nutritional education strategies to target all population groups, with a special focus on women, children, and low-income families, who often face greater barriers to accessing healthy food (FAO *et al.*, 2018; Menasche and Machado, 2019). Strengthening nutritional education, along with implementing food guidelines based on locally available foods and policies that encourage the consumption of healthy products, significantly contributes to improving the quality of a population's diet (ALADI and FAO, 2015). These actions create a positive impact on public health and help prevent diet-related diseases.

Ensuring universal access to basic services, such as clean drinking water, sanitation, health care, and education, is essential for improving the food and nutrition security of the population (UNDP and FAO, 2016). These elements not only affect quality of life but also influence the availability and proper consumption of food. Investment in infrastructure is crucial to strengthening food systems

(Rodríguez-Peñaguirre and González-Arellano, 2022). Improving transportation, markets, and sanitation facilitates physical access to food and its proper preparation. Additionally, the availability of clean water and electricity is vital to ensure optimal conditions for food hygiene and preservation, thus contributing to the food security of communities (CAF, 2024).

The conducting of research and development on food insecurity and its relationship with poverty is essential to understand its complexity and to design effective intervention strategies. Furthermore, the continuous monitoring of progress and setbacks in achieving food security goals allows for the assessment of the effectiveness of implemented policies (Mundo-Rosas *et al.*, 2019). In this context, the evaluation of policies and programs is crucial to ensure that they meet their objectives. To this end, the collection and systematization of statistical data and the development of food security indicators are key in measuring progress and supporting decisions for adjustments when necessary (Mundo-Rosas *et al.*, 2019; Pasquier, 2019). These tools not only facilitate impact analysis but also contribute to the formulation of evidence-based strategies to improve food security. Table 8.1 presents a synoptic identification and description of global comprehensive strategies and the scope of action to which they would preferably be directed, based on the findings obtained from the literature review discussed earlier.

Table 8.1. Synopsis of comprehensive strategies aimed at poverty, economic inequality, and food security. Table author's own.

Global strategy	Description of the global strategy	Scope of action	Scope description
Public policies and social protection	Measures to comprehensively address economic inequality and food insecurity, considering redistributive policies and the strengthening of social welfare.	Income redistribution	Implementation of social protection policies, investment in education, health, and decent employment.
		Social protection	Monetary transfer programs and subsidies to improve access to food.
		Employment policies	Generation of well-paid job opportunities to strengthen access to food.

Continued

Table 8.1. Continued

Global strategy	Description of the global strategy	Scope of action	Scope description
Strengthening local and regional food systems	The development of resilient and sustainable food systems is key to ensuring access to nutritious food and promoting employment in vulnerable communities.	Local and regional food systems	Promotion of local markets, cooperatives, and public procurement to improve access to food.
		Infrastructure and basic services	Investment in transportation, markets, potable water, and electricity to enhance food distribution and ensure access to food.
		Food aid programs	Short-term hunger mitigation measures, linked to long-term strategies that address the structural causes of poverty and food insecurity.
Support for peasant family agriculture (PFA) and local production	The development and strengthening of family-based peasant agriculture is a key component for improving food security and reducing economic inequality in rural areas.	Access to resources	PFA and small agricultural producers need access to credit, technology, markets, and training to increase productivity and income.
		Agricultural diversification	Promotion of diverse crops and sustainable agricultural practices to improve food availability and climate resilience.
		Production and supply in local markets	Increase in fresh food production in rural and marginalized areas, strengthening the local economy and reducing dependence on external markets.
Nutritional education and promotion of healthy habits	The development and dissemination of knowledge on food and nutrition are essential to improving diet quality and reducing malnutrition.	Nutrition education	Promotion of healthy diets and reduction of malnutrition (e.g. consumption of ultra-processed foods).
		Priority groups	Targeting primarily women, children, and low-income families.
		Education and awareness programs	Training on food preparation and preservation to maximize utilization and minimize losses.

Continued

Table 8.1. Continued

Global strategy	Description of the global strategy	Scope of action	Scope description
Research and development	Continuous monitoring of policies and strategies is essential to ensure their effectiveness and make necessary adjustments to the strategies.	Food safety indicators	Development and use of statistical metrics to assess progress in food security or otherwise reduce food insecurity.
		Research and monitoring	Analysis of the issues generating food insecurity to design more effective strategies. .

PART II: HOW INSECT FOOD SUPPORTS SDG 2

Julia Prado-Beltran*

Faculty of Engineering in Agricultural and Environmental Sciences, Universidad Técnica del Norte, Ecuador

8.9 Using Insects as Tools to Enhance Food Security

One-third of agricultural land is used for growing food, while two-thirds is allocated for grazing (Ritchie and Roser, 2019; Ritchie *et al.*, 2023). Most of the world's food supply comes from cereals, which represent the largest food group. The second largest group includes root crops, followed by legumes, which are third. Globally, the production of these food groups totals approximately 2 billion tonnes of cereals, 600 million tonnes of root crops, and 60 million tonnes of legumes each year. Additionally, around 85 million tonnes of fats and oils and 180 million tonnes of sugars are produced annually worldwide (Gahukar, 2011; Latham, 2002).

Developing countries account for a larger share of production in these food categories compared to industrialized nations. In contrast, industrialized countries produce more animal-based foods, such as meat, milk, and eggs, than developing countries. Overall, 40% of dietary protein comes from animal products, while 60% is derived from crops (Mikulec *et al.*, 2024).

Livestock production, however, contributes to greenhouse gas emissions and the depletion of natural resources. Furthermore, the production of protein-rich foods is associated with issues such as acidification, eutrophication, and the overuse of water resources (Poore and Nemecek,

2018). The demand for grains and protein-rich foods is closely linked to meat consumption; specifically, for every kilogram of high-quality animal protein produced, cattle require about 6 kg of plant protein as feed (Trostle, 2008). Moreover, the rise in global crop prices is projected to increase the prices of beef, pork, and poultry by over 30% by 2050 compared to the year 2000 (Nelson *et al.*, 2009).

In 2021, the average cost of a healthy diet worldwide was US$3.66 per person per day. The highest cost was reported in Latin America and the Caribbean, where it reached $4.08 (FAO *et al.*, 2023). Between 2017 and 2021, food prices consistently rose, and this trend continued into 2022. As a result, the average cost of a healthy diet increased globally and across all regions. The FAO Food Price Index surged by 52% from 2019 to 2022. The increase in global demand for meat consumption, limited available land, and rising food prices in the future drive the search for alternative protein sources (van Huis, 2016; Macdiarmid and Whybrow, 2019).

Consequently, the growing demand for protein alternatives positions insects as a viable option for ensuring food security (Nowakowski *et al.*, 2022). Research indicates that various insect species are rich in protein, energy, vitamins, and micronutrients (van Huis, 2003; Akullo *et al.*, 2017; Ruzengwe *et al.*, 2022). Entomophagy, the practice of consuming insects or insect-based foods, is being recognized as a solution to promote diet diversification and combat malnutrition, especially in tropical and subtropical regions of Africa, Asia, and Latin America, where diets often include several types of insects (Park *et al.*, 2022).

In Mexico, people consume more than 500 species of insects, including caterpillars, bed bugs, ants, flies, grasshoppers, dragonflies, and beetles (Ramos-Elorduy, 2008; Cerritos, 2009). Brazil primarily enjoys a variety of insects native to the Amazon, such as beetles, bees, wasps, ants, butterflies, and termites (Paoletti *et al.*, 2000; Tang *et al.*, 2019). In Ecuador, Costa Rica, Panama, and Colombia, several species of Hymenoptera (bees, wasps, and ants) and Coleoptera (beetles) are consumed (Abril *et al.*, 2022).

In Africa, common edible insects include caterpillars, termites, locusts, grasshoppers, ants, bees, bed bugs, beetles, and soldier flies (van Huis, 2005; Ishara *et al.*, 2023). In India, ants are particularly popular, along with a range of aquatic insects, bed bugs, beetles, bees, wasps, grasshoppers, locusts, termites, and dragonflies. Approximately 300 species are consumed in China (Chen *et al.*, 2009). In Thailand, around 150 species of wild insects are consumed, totaling 7500t each year, highlighting their significance in the local diet (Hanboonsong *et al.*, 2013; Krongdang *et al.*, 2023).

Approximately 2300 species of insects are consumed as food worldwide. The most consumed insect orders are as follows: Coleoptera (beetles) account for 31.2% of total consumption, followed by Lepidoptera (moths and butterflies) at 17.1%. Hymenoptera (bees, wasps, and ants) contribute 15.2%, while Orthoptera (grasshoppers, locusts, and crickets) make up 13.2%. Hemiptera (which includes stink bugs, giant water bugs, cicadas, leafhoppers, planthoppers, true bugs, and scales) represent 11.2%, Isoptera (termites) account for 3%, and Diptera (flies) make up 2% (Cerritos, 2009; van Huis *et al.*, 2013, 2021b; Cámara *et al.*, 2018; Poshadri *et al.*, 2018; Govorushko, 2019; Hawkey *et al.*, 2021; Park *et al.*, 2022; Omuse *et al.*, 2024). Each edible insect order contains a variety of diverse families, as shown in Table 8.2.

Among the insects with significant potential for consumption, the European Union (EU) has authorized six species for both human and animal consumption. These include the mealworm (*Tenebrio molitor*) (Fig. 8.4), black soldier fly (*Hermetia illucens*), house cricket (*Acheta domesticus*), striped cricket (*Gryllodes sigillatus*), field cricket (*Gryllus assimilis*), and lesser mealworm (*Alphitobius diaperinus*). In various Asian and European countries, different production systems with varying levels of technology have been developed for several cricket species, including *Acheta domesticus*, *Gryllodes sigillatus*, *Gryllus bimaculatus*, and *Gryllus assimilis* (Cruz-Fagua *et al.*, 2021; Omuse *et al.*, 2024). The main advantages of practicing entomophagy include significant nutritional value as well as bioactive, hypolipidemic, and antimicrobial

Table 8.2. Predominant families of edible insect orders. Data from Omuse *et al.* 2024. Table author's own.

Coleoptera	Heteroptera	Hymenoptera	Isoptera	Lepidoptera	Orthoptera
Scarabaeidae	Cicadidae	Apidae	Termitidae	Saturniidae	Acrididae
Cerambycidae	Pentatomldae	Formicidae		Hepialidae	Tettigoniidae
Dytiscinae	Coreidae	Vespidae		Sphingidae	Gryllidae
Curculionidae	Belostomatidae			Cossidae	Romaleidae
Passalidae	Tessaratomidae			Noctuidae	
Lucanidae	Nepidae				
Buprestidae					
Hydrophilidae					
Tenebrionidae					
Elateridae					
Chrysomelidae					

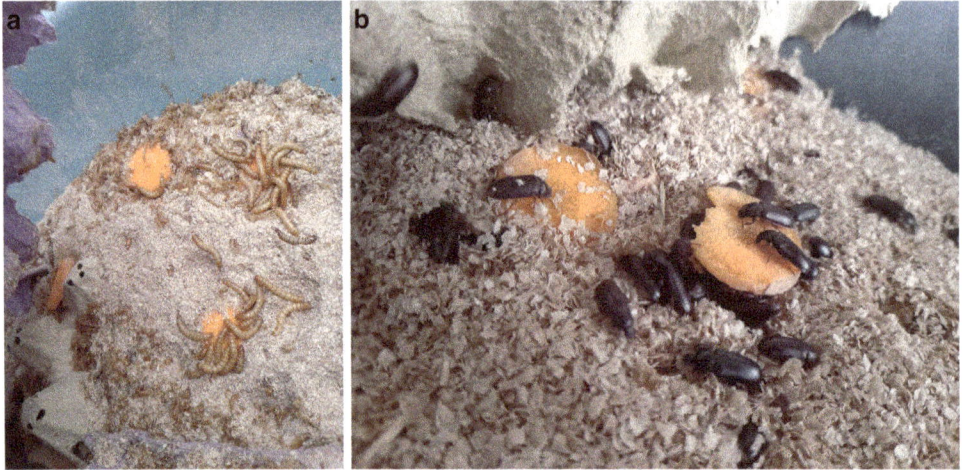

Fig. 8.4. Farming edible insects *Tenebrio molitor*: a. larvae, b. adults. Figure author's own.

properties. On the downside, there is also a significant risk of allergenic reactions to insect proteins.

8.9.1 Nutritional value

8.9.1.1 Protein and amino acids

Insects are primarily composed of protein, which can range from 23% to 77%. This protein content varies among species and can also be influenced by the insect's age and specific body parts (Rumpold and Schlüter, 2013b). For example, adults and larvae of Coleoptera (beetles) typically contain between 23% and 66% protein. In contrast, the pupae and larvae of Lepidoptera have a protein content ranging from 14% to 68%. Eggs, nymphs, and adults of Hemiptera (true bugs) contain between 42% and 74% protein, while the protein content in e. g.gs, larvae, pupae, and adults of Hymenoptera range from 13% to 77%. Additionally, adults and nymphs of Odonata have a protein content of between 46% and 65%. Finally, adults and nymphs of Orthoptera comprise between 23% and 65% protein (Kourimská and Adámková, 2016).

The nutritional quality of insect protein is largely determined by its amino acid content, especially the essential amino acids that our

bodies cannot produce and must obtain from our diet. Furthermore, insect protein is more easily digested than traditional meat while still effectively promoting a sense of fullness. This characteristic is important for managing hunger (Omuse *et al.*, 2024).

Feed conversion efficiency refers to the amount of feed required to produce 1 kg of animal protein, such as meat or milk. For example, the feed conversion ratio for cattle is approximately 7:1, meaning that 7 kg of feed are needed to produce 1 kg of milk or meat. Pigs have a better conversion ratio of 5:1, while chickens have an even more efficient ratio of 2.7:1. In comparison, crickets require only 1.7 kg of feed to produce 1 kg of body weight. This indicates that crickets are nearly twice as efficient as chickens in converting feed into biomass, at least three times more efficient than pigs, and four times more efficient than cattle (Smil, 2002; Hanboonsong *et al.*, 2013; van Huis *et al.*, 2013; Arévalo-Arévalo *et al.*, 2022).

Insects are a good source of high-quality amino acids, particularly essential ones. The main amino acids found in edible insects vary by order. In the Lepidoptera, Orthoptera, Coleoptera, and Diptera orders, the primary amino acids include glutamic acid, aspartic acid, phenylalanine, and alanine. In the Hemiptera order, the most notable amino acids are proline, leucine, tyrosine, alanine, valine, and methionine. For the Hymenoptera order, glutamic acid, leucine, and alanine are prominent (Sánchez-Muros *et al.*, 2014; Rumpold and Schluter, 2015; Omuse *et al.*, 2024).

When comparing the amino acid profiles of edible insects to the amino acid requirements for adults, published by the World Health Organization (2007), it is evident that many edible insects meet the requirements for methionine, methionine+cysteine, phenylalanine, and tyrosine. Additionally, when the amino acid profiles of insects are compared to those of animal feeds, they are found to be like meat meals. However, none of the insect species studied have lysine levels that exceed those found in fish meal (Alamu *et al.*, 2013; Avendaño *et al.*, 2020).

8.9.1.2   Fat content

The total fat content in insects can vary widely, ranging from 2% to 62% (van Broekhoven *et al.*, 2015; Lange and Nakamura, 2021). The fatty acid profile of insects closely resembles that of both animal fat and vegetable oils. However, insects tend to have a significantly higher proportion of unsaturated fatty acids compared to cattle and pigs, with some species exceeding 75% (Fontaneto *et al.*, 2011; Tzompa-Sosa *et al.*, 2014; Zielińska *et al.*, 2015; Lehtovaara *et al.*, 2017).

The ether extract is the second most important component, with the following content ranges: Orthoptera (4–22%), Hemiptera (6–46%), Isoptera (21–46%), Blattodea (27–34%), and some Lepidoptera (6–77%). Notably, the extract content is higher in the larval and pupal stages compared to the adult stage (Womeni *et al.*, 2009; Sánchez-Muros *et al.*, 2014; Avendaño *et al.*, 2020). The average amount of saturated fatty acids (SFA) in edible insects ranges from 31% to 42%, with palmitic acid (C16:0) and stearic acid (C18:0) being the most prevalent. The fraction of monounsaturated fatty acids (MUFA) varies between 22% and 49%, with oleic acid (C18:1) being the dominant type present in all insects (Chen *et al.*, 2022).

Polyunsaturated fatty acids (PUFA) account for 16% to 40% of the fat content, primarily consisting of linoleic acid (C18:2), linolenic acid (C18:3), and arachidonic acid (C20:4n6). A key distinction between terrestrial and aquatic insects is the presence of eicosapentaenoic acid (EPA, C20:5n3) and docosahexaenoic acid (DHA, C22:6n3) in aquatic insects, while C20:4n6 is absent (St-Hilaire *et al.*, 2007; Dobermann *et al.*, 2017). Insects cannot synthesize cholesterol *de novo*; instead, they contain approximately 0.1% cholesterol, which they obtain through their diet. This cholesterol content can vary, depending on the life cycle stage of the insect (Ramos-Elorduy, 2008; Payne *et al.*, 2016).

8.9.1.3   Fibre

The amount of crude fiber in insects can vary significantly, ranging from 0.12% to 29% (Rumpold and Schlüter, 2013a). Insects with harder exoskeletons typically have a higher fiber content). One of the key components of the insect exoskeleton is chitin, which is a type of polysaccharide. Chitin is classified as fiber because it is indigestible by humans and non-ruminant animals (Muzzarelli *et al.*, 2001; Paoletti *et al.*, 2007; Osimani *et al.*, 2018). The chitin content in insects can range from 11.6 mg to 137.2 mg per kg of dry matter. This variation in chitin is associated with defense against parasitic infections and allergic reactions, as well as the enhancement of the immune system (Finke, 2007; Mark, 2015; Tripathi and Singh, 2018).

8.9.1.4   Minerals

Insects are a rich source of iron, zinc, magnesium, and copper. Including iron in the human diet could help prevent anemia. Zinc is essential for supporting the immune system, promoting bone maturation, and preventing growth retardation (Durst and Hanboonsong, 2015; Montowska *et al.*, 2019).

For comparison, meat contains approximately 6 mg of iron per 100 g of dry matter, while insects have a higher iron content, ranging from 8 to 31 mg, depending on their diet. In terms of zinc, meat provides about 12.5 mg per 100 g of dry matter, whereas some insect species can contain as much as 26.5 mg of zinc (Christensen *et al.*, 2006; van Huis *et al.*, 2013; Oonincx *et al.*, 2015; Latunde-Dada *et al.*, 2016).

8.9.2 Bioactive compounds

Insects can convert their food into phenolic compounds more efficiently than plants (Panzella *et al.*, 2020). Bees, flies, wasps, and grasshoppers contribute to these phenolic compounds, which are found in more accessible forms. The proportions of these compounds depend on the insects' diet and life cycle (Liu *et al.*, 2012; Jantzen da Silva Lucas *et al.*, 2020; Ssepuuya *et al.*, 2021). Additionally, some edible insects are rich in peptides that can serve as alternatives to antibiotics. Certain species of Coleoptera (beetles) produce substances such as benzoquinones and pentacene, which exhibit anti-inflammatory effects. Notably, benzoquinones are active agents in reducing tumor cells (Wahrendorf and Wink, 2006; Tonk and Vilcinskas, 2017; Nino *et al.*, 2021).

D-glucosamine, commonly derived from cockroaches, is another beneficial compound. It is used to treat arthritis and back pain, support joint and bone health, and address respiratory issues (Bertuzzi *et al.*, 2018; Melgar-Lalanne *et al.*, 2019). Cantharidin, a defensive agent secreted mainly by beetles, hemipterans (true bugs), and dipterans (flies), is utilized to treat conditions such as rheumatism, anemia, carcinoma, and skin diseases (Costa-Neto, 2005; Zimian *et al.*, 2010; Zhang *et al.*, 2023). On the other hand, termites are rich in various pharmaceutical agents, including polysaccharides, sesquiterpenoids, cerebrosides, heteroglycans, and isoflavonoids, which are known for their neuroprotective and immune-enhancing effects (Hsieh and Ju, 2018; Zhao *et al.*, 2019). Edible insects also play a significant role in regulating blood pressure (Yamada *et al.*, 2002) because some species of Lepidoptera (butterflies and moths), Coleoptera (beetles), and Orthoptera (grasshoppers and crickets) contain ACE inhibitors (angiotensin-converting enzyme), which can be beneficial in treating cardiovascular diseases (Cito *et al.*, 2017).

8.9.3 Economic opportunities

Recently, the importance of diversifying protein sources has been highlighted, particularly to include those that are produced sustainably to minimize environmental impact. In this regard, insects are being recognized as a viable alternative protein source, as they are associated with environmentally sustainable production practices and possess high nutritional value. Additionally, raising insects for human consumption aligns with the principles of a circular economy, as they can be cultivated using organic waste (Rumpold and Schlüter, 2013b; Willett *et al.*, 2019; Barragán–Fonseca *et al.*, 2020).

Beyond the consumption of whole insects, there is increasing interest in high-protein powders derived from them. These powders can be incorporated into a variety of food products, including canned goods, energy bars, snacks, pasta, specialty food ingredients, and even chocolate (Fleta-Zaragozano, 2018; Reverberi, 2020; Entomo Farms, 2024). For example, the flour made from the insect species *Gryllus sigillatus* is high in protein and its lipid content provides energy and essential fatty acids, thereby contributing to the objective of achieving zero hunger, as outlined in the United Nations Sustainable Development Goals (SDG) (Zielińska *et al.*, 2015; Ribeiro *et al.*, 2018; Govorushko, 2019; Dion-Poulin *et al.*, 2021).

By 2019, the global market for edible insects surpassed $112 million, and it is projected to grow at a compound annual growth rate (CAGR) of over 47% from 2023 to 2032. By 2027, the estimated market value is expected to reach approximately $1.18 billion (Global Markets Insights Inc, 2023). In 2003, the Food and Agriculture Organization (FAO) identified the use of insects in human and animal nutrition as a promising alternative source due to their nutritional advantages and low environmental impact (Shelomi, 2016; van Huis *et al.*, 2021a).

Worldwide, many farms are dedicated to the production of edible insects. Thailand leads in this area, with the highest production of crickets globally. The country has over 20,000 medium and small farms, collectively producing 7500 t of crickets annually, while the import value of insects for food is around $1.14 million each year (Hanboonsong *et al.*, 2013). In South Korea, more than 2500 insect farms produce products for both human and animal consumption, as well as for medicinal purposes. In Africa, there are currently over 850 insect farms in operation (Verner *et al.*, 2021).

The development of the edible insect industry is underscored by the formation of producer associations that promote the sector, educate the public, and coordinate efforts for the legalization of the use of insects (Hanboonsong *et al.*, 2013). The economic benefits associated with breeding edible insects primarily relate to the trade of these products; with the expansion of the market for edible insects, production can provide farmers with a stable source of income (Dobermann *et al.*, 2017).

The annual profit for a cricket farmer in Thailand is approximately $2000. This figure is based on a farm consisting of eight insect-rearing cages, with an average of 8.5 growth cycles per year and a sales price of $0.44 per kilogram of fresh crickets. Profits per harvest have been reported to range from $1650 to $2500, with a profit margin of 50%. In South Korea, the estimated annual profit for a cricket farm is $20,500, based on a productive capacity of 3000 kg per year. For mealworms, the profit is estimated to be $100,000 for an annual production of 3500 kg (Reverberi, 2020; Verner *et al.*, 2021).

Thus, raising edible insects could provide a viable production alternative for rural areas. This initiative has the potential to improve the lives and livelihoods of farmers. In rural areas of developing countries, the consumption and trade of edible insects—especially through street vending—contribute to the socio-economic empowerment of rural communities, particularly women. These efforts support several goals established in the SDGs, including End poverty; Zero hunger; Gender equality; Decent work and economic growth; Responsible production and consumption; and Climate action (Halloran *et al.*, 2017a; Roos and van Huis, 2017; Arévalo-Arévalo *et al.*, 2022).

Currently, in Colombia, there is no established culture recognizing crickets as human food. However, in the municipality of La Mesa, Cundinamarca, there has been a shift in the perception of the cricket species *Gryllodes sigillatus*. This change is attributed to initiatives aimed at presenting crickets as a viable agricultural production alternative that can help mitigate the effects of climate change while enhancing food security and income for rural women in vulnerable situations. Empowerment and entrepreneurship workshops, along with the development of meals using cricket flour as an ingredient, have contributed to this positive change in perception (Bermúdez-Serrano, 2020; Vernot, 2021; Bermúdez-Serrano *et al.*, 2023).

One of the largest organizations in this field is the International Platform of Insects for Food and Feed (IPIFF), which was established in the EU in 2012 and currently has 75 members. IPIFF is a non-profit organization that advocates for the interests of the insect production sector. It focuses on relevant market policies within the EU, as well as the needs of producers and citizens. IPIFF promotes the utilization of insects for human consumption and encourages the use of insect products as a high-quality nutrient source for animal feed (IPIFF, 2024).

Other regional associations include the Asian Asian Food and Feed Insect Association (AFFIA), the North American Coalition for Insect Agriculture (NACIA), the Insect Protein Association of Australia (IPAA), and the Brazilian Association of Edible Insect Breeders (ASBRACI). In 2021, the Latin American Association of Insect Producers (APICAL) was created, boasting over fifty members (Bermúdez-Serrano *et al.*, 2023). Some notable companies involved in the production of insects for human consumption worldwide include Ynsect, Entomo Farm, and Jimmini's (France); Aspire Food Group, All Things Bugs, Bug Muscle, Crunchy Critter Farms, and Insectios (United States); Thailand Unique (Thailand); Entocube (Finland); Gryllies and Totem Nutrition (Canada); Mophagy, Edible Unique, and BugGrub (UK); Italibugs and Microvita (Italy); SensFood (Czech Republic and Germany); BenetoFoods and Happy Cricket Food (Germany); EnGrillo, Jacuna, Entomovit, Chapa farms, Orbitaverde, and Gricha (Mexico); Nutrinsecta and Startup Hakkuna (Brazil); Good Bug Food Shop and Insect Systems (Netherlands); Unique Bio Technology (Nigeria); Cricket One (Vietnam); EntoFitFood (Spain); Wikiri Sapoparque and Crick Superfoods (Ecuador); BioFly (Colombia); InProtin (Guatemala); Illucens, Arthrofood, and Mosca Soldado Negro (Colombia); Illucenscience and Kawat (Perú); Chepulines Cocina (Argentina); among others (Bug Burger, 2022).

The price of edible insects varies based on several factors, including the product's composition, the insects' diet, the breeding method, their country of origin, the retailer's margin, packaging, processing rates, the length of the distribution circuit, and, of course, the specific species. As a result, you will find insects priced at a wide range—sometimes differing by two to three times the amount. It's advisable to compare different offers to find the best value for your money (Halloran *et al.*, 2017b; Ambrosio *et al.*, 2021; Niyonsaba *et al.*, 2021). Here are some average prices for popular natural edible insects (Next Food, 2025):

- A 25 g bag of crickets or locusts typically costs between €5 and €7, while a 1 kg bag ranges from €50 to €100.
- Silkworms are priced between €4.50 and €9 for a 25 g box.
- Mealworms sell for around €4.50 to €7.50 for 25 g, with a 1 kg bag costing approximately €80.
- Grasshoppers have an average price of €5 for a 25 g can, with kilo prices starting at around €90.
- Black ants tend to be more expensive, ranging from €6 to €8 for a 10 g portion.

8.9.4 Opportunities in farming edible insects

By 2050, it is expected that the consumption of animal-based products will increase by between 60 and 70%. This surge in demand will require significant resources. Traditional sources of protein are not only the most expensive but also overexploited and harmful to the environment. Therefore, it has become essential for the agricultural sector to explore new sources of animal protein.

Utilizing insects as a source of animal protein can significantly reduce greenhouse gas emissions, minimize water waste, and decrease the demand for arable land. Additionally, producing and consuming insect-based products can improve the nutritional status of over 805 million people worldwide who suffer from nutritional deficiencies, including 191 million people in Latin America and the Caribbean (Oonincx

et al., 2010; Barennes *et al.*, 2015; Cámara *et al.*, 2018; Bermúdez-Serrano, 2020; FAO *et al.*, 2024).

Producing 1 kg of insect protein requires significantly less water compared to beef, as insects can extract moisture from their food. Estimates suggest that the water footprint of insect production in small farms is 5–10 times lower than that of traditional livestock farming. Additionally, insect cultivation requires 11 times less space for each kilogram produced. Furthermore, the production of 1 kg of insect protein contributes nearly zero greenhouse gas emissions, while traditional protein sources like beef, pork, and poultry generate between 10 and 100 times more greenhouse gases per kilogram of meat produced (van Huis, 2019; Arévalo-Arévalo *et al.*, 2022).

In various regions worldwide, the consumption of insects is becoming an attraction for the tourism sector. This trend is being integrated into gastronomic routes for both local and international tourists. For example, certain states in Mexico and countries in South-east Asia, such as Indonesia and Thailand, promote insect consumption as part of their culinary experiences (Ambrosio *et al.*, 2021). Additionally, government organizations play a role in regulating the production and consumption of insects. In 2014, China's Ministry of Health recognized silkworm pupae as a functional food. Similarly, in 2016, the Food and Drug Administration of South Korea approved mealworms and crickets as viable food sources (Hanboonsong and Durst, 2020; Ravagli and Carolina, 2021; Arroyo-Marlés, 2023).

8.9.5 Challenges faced in insect farming

There are many concerns regarding the consumption of insects as food (EFSA Scientific Committee, 2015). Negative effects associated with eating insects include microbiological, parasitological, and allergic issues (Soares de Castro *et al.*, 2018). Research indicates that gram-positive bacteria are commonly found in insects, so effective preparation methods such as curing, steaming, freezing, boiling, roasting, smoking, fermentation, and frying must be employed (Klunder *et al.*, 2012; Alamu *et al.*,

2013; Feng *et al.*, 2018; Kewuyemi *et al.*, 2020). Most edible insects can be processed, and it is important to know how to maintain their nutritional value while reducing toxic components (Fernández-García *et al.*, 2009; Kinyuru *et al.*, 2013; Mutungi *et al.*, 2019).

Consumer rejection poses a significant challenge to edible insect adoption, influenced by visual elements, preparation, sensory experience, and information (Bucea-Manea-Tonis *et al.*, 2023). In regions where entomophagy (the practice of eating insects) is common, insects are valued as a food source (Bermúdez-Serrano *et al.*, 2023). However, in countries where this practice is unfamiliar, consumers often react negatively. These adverse reactions stem from cultural factors, expectations of unpleasant taste, lack of knowledge about texture, and uncertainty regarding the origins of the food (Barsics *et al.*, 2017; Dobermann *et al.*, 2017; Hartmann and Siegrist, 2017; Bisconsin-Junior *et al.*, 2019).

A key strategy to overcome this rejection is to incorporate insect powder into other food products. When doing so, it's important to consider the percentage of insect powder added to the formulation and its potential effects on consumer acceptance. To increase acceptability in regions where consumption remains a challenge, it is also essential to implement communication programs and innovate in processing and incorporating insects into various food products, gastronomic activities, edible insect cookbooks, and restaurant snack and menu options (Simon *et al.*, 2006; Ruby *et al.*, 2015; Caparros Megido *et al.*, 2016; Shelomi, 2016; Osimani *et al.*, 2018).

Additionally, it is crucial to develop or unify regulations for the cultivation, transformation, and commercialization of insects to create a framework for preparing other food products. In several parts of Europe, insect consumption is restricted. However, in the first quarter of 2021, the European Food Safety Authority (EFSA) approved the consumption of mealworms (larvae of *Tenebrio molitor*) for human use. This decision was based on the nutritional advantages of including mealworms as an ingredient in flour for snacks. Furthermore, it was confirmed that food stability issues were addressed while maintaining controlled production limits, as assessed by the EFSA Panel on Nutrition,

Novel Foods and Food Allergens (NDA) (Arroyo-Marlés, 2023).

In the United States, the Federal Food and Drug Administration (FDA) plays a crucial role in protecting and promoting the health of both humans and animals by overseeing most of the food supply. The FDA enforces regulations, inspects manufacturing and processing facilities to ensure good manufacturing practices, and collaborates closely with local food safety agencies. However, the FDA has not yet established clear guidelines for the production and use of insects as food for humans and animals. As a result, businesses involved in the production and distribution of these insects have experienced significant delays (Lähteenmäki–Uutela *et al.*, 2017; Arévalo-Arévalo *et al.*, 2022).

In Canada, insects intended for human and animal consumption are categorized as novel foods. Their production and marketing are regulated by Health Canada under the Food and Drug Act. Before these products can be sold, they require premarket notification and must undergo safety, suitability, and nutritional assessments to ensure they meet safety standards in the country (Lähteenmäki–Uutela *et al.*, 2017; Arévalo-Arévalo *et al.*, 2022). Currently, there is no clear legislation in Mexico, Colombia, and Ecuador regarding the production and marketing of insects for animal and human consumption (Cartay, 2018; Arévalo-Arévalo *et al.*, 2022).

Research on the profitability of insect farming is limited. The variability in production costs and selling prices for different species complicates financial analysis (van Huis, 2020; Niyonsaba *et al.*, 2021). In the production of edible insects, it is important to consider the need for skilled labor, as well as the materials and equipment necessary for processing them for human or animal consumption (Rumpold and Schlüter, 2013b; Morales-Ramos *et al.*, 2020; Ites *et al.*, 2020).

8.10 Seaweed and Algae Farming

Seaweeds are photosynthetic eukaryotic organisms that inhabit oceans, rivers, lakes, and other bodies of water. They include various species of marine plants and algae with significant potential for use in animal and human nutrition (Gutiérrez Cuesta *et al.*, 2017; Sawarkar *et al.*, 2024). Seaweed and coastal ecosystems can absorb rising levels of greenhouse gases (CO_2) through a 'blue carbon' fixation process involving photosynthesis. Furthermore, algae have environmental advantages, as they capture CO_2 from the atmosphere, helping to mitigate climate change. The energy stored in these ecosystems can be transformed into economically valuable products for various industries, including food, animal feed, fuels, and fertilizers (Vigani, 2020; Choudhary *et al.*, 2021).

The growth of macroalgae is influenced by light and photosynthesis, but factors such as nutrients, temperature, herbivores, diseases, irradiation, competition, currents, salinity, and, for intertidal species, desiccation can limit both growth and carbon sequestration (Millar *et al.*, 2020; Ross *et al.*, 2022)

Seaweed, often called macroalgae, is classified into three major groups based on color: brown, red, and green (Sawarkar *et al.*, 2024). Brown algae, known as Phaeophyta, are typically large and can vary significantly. They include species like giant kelp, which can reach lengths of up to 20 m, and smaller varieties that measure 30–60 cm. There are also thick, leathery types that grow 2–4 m long. Red algae, classified as Rhodophyta, tend to be smaller, with lengths ranging from just a few centimeters to about a meter. Interestingly, red algae are not always red; they can also appear purple or brownish-red. Green algae, or Chlorophyta, are similar in size to red algae, also generally being small (McHugh, 2002).

The three most important types of seaweed used for human consumption are various species of Porphyra, Laminaria, and Undaria. In recent years, Porphyra has ranked as the third most important catch in Japanese fishing statistics. These three algae were originally harvested from wild populations, but large-scale cultivation methods are now necessary to meet current demand. Porphyra is classified as a red algae, while Laminaria and Undaria are brown algae (McHugh, 2002; Miranda *et al.*, 2015).

8.10.1 Nutritional compounds

Algae are frequently cited as foods likely to gain in popularity in the coming years. This is largely due to their nutritional benefits, as they are rich in protein, essential amino acids, vitamins, Omega-3 fatty acids, minerals, dietary fibers, and polyphenols (Rajapakse and Kim, 2011). Edible algae contain approximately 81% carbohydrates based on dry weight. Red, green, and brown algae contain carbohydrate levels that range from 8.3% to 68.2%, from 4% to 79.9%, and from 12.8 to 81% of their dry weight, respectively (Sawarkar *et al.*, 2024). Additionally, the blue-green algae spirulina is renowned for its exceptionally high protein content, comprising nearly 70% of its dry weight, higher than legumes (Fleurence *et al.*, 2012).

The amino acid values of proteins from certain red algae, such as *Porphyra* spp. and *Undaria* spp., are 91 and 100, respectively, making them comparable to foods of animal origin (Murata and Nakazoe, 2001; Rajapakse and Kim, 2011). In *Palmaria palmata* L., the average composition of leucine, valine, and methionine is comparable to that of ovalbumin. Moreover, the relative concentrations of isoleucine and threonine are similar to those found in the major proteins of legumes. In contrast, *Ulva rigida* C. contains abundant amounts of leucine, phenylalanine, and valine, while its histidine content is comparable to that of eggs and legumes (Lordan *et al.*, 2011; Conde *et al.*, 2013; Gutiérrez Cuesta *et al.*, 2017). Dawczynski *et al.* (2007) analyzed the chemical composition of 34 edible algae and concluded that red algae, particularly, are significant sources of proteins containing all the essential amino acids.

Many species of macroalgae have low levels of histidine, while their methionine content can be relatively high. Additionally, macroalgae often contain significant amounts of glutamic acid, which contributes to their characteristic flavor. They also contain various bioactive amino acids and peptides, including taurine, carnosine, and glutathione. Taurine is essential in algae for osmoregulation. In humans, it plays a role in various physiological processes including immunomodulation, membrane stabilization, and the development of the ocular and nervous systems

(Holdt and Kraan, 2011; Gutiérrez Cuesta *et al.*, 2017; Overland *et al.*, 2018).

Research indicates that 100 g of seaweed exceeds the daily requirements for vitamins A, B2, and B12, providing two-thirds of the recommended vitamin C intake. Seaweed is also recognized as a natural source of water-soluble and fat-soluble vitamins, including thiamine, riboflavin, β-carotene, and tocopherols. Brown algae contain 14.5 mg of vitamin E per 100 g, significantly higher than found in peanuts. This elevated vitamin E content protects the algae's polyunsaturated fatty acids (PUFAs) and helps preserve their nutritional benefits. Both red and brown algae are also rich in carotenoids and vitamin C, with concentrations ranging from 20 to 170 ppm for carotenoids and from 500 to 3000 ppm for vitamin C. Additionally, they are considered good sources of vitamin B12, a nutrient typically absent in most land plants, although it can be found in considerable amounts in some vegetables (Rajapakse and Kim, 2011).

Seaweeds contain very low amounts of lipids, typically between 1% and 5% of their dry matter. The main classes of lipids found in all seaweeds are neutral lipids and glycolipids. The proportion of essential fatty acids in seaweeds is higher than in terrestrial plants, which may enhance their effectiveness as part of a balanced diet (Rajapakse and Kim, 2011). Seaweeds produce larger quantities of PUFAs, with concentrations varying depending on the specific type of seaweed. Additionally, seaweeds contain essential long-chain PUFAs from the omega-3 family, such as eicosapentaenoic acid, which constitutes 30% of the total fatty acid content (Khotimchenko *et al.*, 2002; Narayan *et al.*, 2006; Rajapakse and Kim, 2011). Phospholipids comprise about 4–10% of total lipids in seaweeds. Dietary phospholipids serve as emulsifiers that enhance the digestion and absorption of fatty acids, thereby improving the nutritional value of foods.

These nutrients may help reduce the risk of heart disease, thrombosis, and atherosclerosis, while also exhibiting antiviral properties. Additionally, seaweed is a significant source of fiber, particularly soluble fiber, which is crucial in preventing obesity, constipation,

cardiovascular diseases, and colon cancer, among others (Khotimchenko *et al.*, 2002; Ortiz *et al.*, 2006).

Seaweed is an excellent source of iodine, a trace mineral essential for thyroid health and helping regulate metabolism. Macroalgae are also rich in other important minerals, such as magnesium, phosphorus, potassium, calcium, zinc, sulfide, copper, selenium, molybdenum, fluoride, manganese, boron, nickel, cobalt, and iron. Additionally, macroalgae can accumulate significant amounts of heavy metals, including cadmium, arsenic, and lead, which may limit their use as food sources. Understanding the bioavailability of these metals is crucial for assessing toxicity risks. Many species of macroalgae naturally have heavy metal levels below food safety limits (Fleurence *et al.*, 2012; Wells *et al.*, 2017; Overland *et al.*, 2018). It is important to note that the mineral content in macroalgae can vary significantly among different species. Factors such as seasonal changes and environmental conditions can also affect the concentration of these minerals (Maehre *et al.*, 2014).

8.10.2 Bioactive compounds

Seaweeds have garnered significant attention as potential sources of bioactive compounds. Recent evidence indicates that these compounds exhibit a broad spectrum of bioactivity. Extracts obtained from seaweeds have been shown to have antioxidant, anti-inflammatory, and antimicrobial properties and effects on the central nervous system (Gutiérrez Cuesta *et al.*, 2017). Ethanolic extracts from seaweeds demonstrate antibacterial and anti-inflammatory properties and potential antioxidant activities. The effectiveness of these properties varies depending on the specific seaweed species. Research on the chemical defenses of marine organisms indicates considerable variability in the production of secondary metabolites. This production may be influenced by various factors, including physical conditions (such as light and temperature), biological factors (like community composition and developmental stages), seasonality, and geographic location (Lima *et al.*, 2002; Venkatesalu *et al.*, 2004; Frikha *et al.*, 2011).

8.10.3 Economic opportunities

Commercial seaweed farming includes harvesting naturally available wild seaweed or cultivating it on land or near the coast (Choudhary et al., 2021). Macroalgae production has been increasing at a rate of approximately 5.7% annually. The global harvest from natural sites has remained relatively stable, fluctuating between 1 million and 1.3 million tonnes per year since 2000 (Rebours et al., 2014; Mac Monagail et al., 2017).

Algae production in aquaculture is a vital economic and cultural industry, valued at $22.13 billion in 2024. When managed sustainably, it can enhance biodiversity, improve carbon sequestration, reduce acidification, and provide opportunities for social development. Most tropical seaweed aquaculture occurs near shore, using simple cultivation infrastructure, with 97% of production originating from South-east Asia (Ross et al., 2025). The seaweed industry plays a crucial role in providing income and support to rural, coastal, and remote communities globally. The primary economic benefit of wild harvesting is related to subsistence (Hart et al., 2014; Mac Monagail et al., 2017).

Due to its established nutritional and functional benefits, including seaweed in bakery products has become a key area of research. Studies indicate that adding seaweed can enhance the phytochemical content, antioxidant activity, and dietary fiber in these products. Additionally, incorporating seaweed can enhance the nutritional value, texture, and sensory qualities of various bakery items (Cox and Abu-Ghannam, 2013; Roohinejad et al., 2017; Hajare et al., 2024). About 25% of the food consumed in Japan is seaweed, which is prepared and served as sushi wrappers, seasonings, condiments, and vegetables. As a result, it has become a significant source of income for fishermen (Mwalugha et al., 2015).

Of approximately 200 commercially cultivated species, ten are actively grown for seaweed production. Saccharina japonica represents 33% of the total, while Kappaphycus alvararezi accounts for 17%; various other species make up the remaining percentage, such as Undaria pinnatifida, Porphyra spp., Sargassum fusiforme, Eucheuma spp., Gracilaria spp., Enteromorpha clathrate, Monsotroma nitidium, and Caluerpa spp. (Choudhary et al., 2021).

Offshore seaweed farming can impact marine life. The associated risks include the entanglement of marine mammals, competition for nutrients between algae and phytoplankton, effects on bird species, invasion of habitats by algae, interactions with shipping traffic, changes in marine biogeochemistry, visual and noise pollution, and industrialization of rural areas (Ross et al., 2022).

References

Abril, S., Pinzón, M., Hernández-Carrión, M. and Sánchez-Camargo, A. (2022) Edible insects in Latin America: A sustainable alternative for our food security. Frontiers in Nutrition 9, 1–16.

Akullo, J., Obaa, B.B., Acai, J.O., Nakimbugwe, D. and Agea, J.G. (2017) Knowledge, attitudes and practices on edible insects in Lango sub-region, northern Uganda. Journal of Insects as Food and Feed 3(2), 73–82.

Alamu, O.T., Amao, A.O., Nwokedi, C.I., Oke, O.A. and Lawa, I.O. (2013) Diversity and nutritional status of edible insects in Nigeria: A review. International Journal of Biodiversity and Conservation 5(4), 215–222.

Alkire, S., Oldiges, C. and Kanagaratnam, U. (2021) Examining multidimensional poverty reduction in India 2005/6–2015/16: Insights and oversights of the headcount ratio. World Development 142, 105454.

Ambrosio, A.F., Sotelo Díaz, I., Deaza Fernández, M.P. and Ramírez Pulido, B. (2021) Desde Cundinamarca Harina de Grillo Gastronomía y Sostenibilidad Para Colombia y el Mundo. Universidad de La Sabana, MinCiencias, ArthroFood S.A.S., and Gobernación de Cundinamarca, Chía, Colombia.

An, C., Sun, C., Li, N., Huang, B., Jiang, J. et al. (2022) Nanomaterials and nanotechnology for the delivery of agrochemicals: Strategies towards sustainable agriculture. Journal of Nanobiotechnology 20(1), 11.

Arévalo-Arévalo, H., Vernot, D. and Barragán-Fonseca, K. (2022) Prospects for the sustainable use of the tropical house cricket (*Gryllodes sigillatus*) for human consumption in Colombia. *Revista de La Facultad de Medicina Veterinaria y de Zootecnia* 69(3), 310–324.

Arroyo-Marlés, I. (2023) Insectos comestibles como modelo de negocio sostenible: Revisión sistemática. *UVserva* 16, 187–205.

Asociación Latinoamericana de Integración (ALADI) and Food and Agriculture Organization (FAO) (2015) *Desarrollo del Comercio Intrarregional de Alimentos y Fortalecimiento de la Seguridad Alimentaria en América Latina y el Caribe*. FAO, Santiago de Chile. Available at: http://www.fao.org/3/a-i4454s.pdf (accessed 24 June 2025).

Avendaño, C., Sánchez, M. and Valenzuela, C. (2020) Insectos: Son realmente una alternativa para la alimentación de animales y humano. *Revista Chilena de Nutrición* 47(6), 1029–1037.

Barennes, H., Phimmasane, M. and Rajaonarivo, C. (2015) Insect consumption to address undernutrition, a national survey on the prevalence of insect consumption among adults and vendors in Laos. *PLOS One* 10(8), e0136458.

Barragán–Fonseca, K.Y., Barragán-Fonseca, K.B., Verschoor, G., van Loon, J.J.A. and Dicke, M. (2020) Insects for peace. *Current Opinion in Insect Science* 40, 85–93.

Barsics, F., Caparros Megido, R., Brostaux, Y., Barsics, C., Blecker, C. *et al.* (2017) Could new information influence attitudes to foods supplemented with edible insects? *British Food Journal* 119(9), 2027–2039.

Batool, S. and Sheikh, M. (2024) Food security in developing countries: Role of agricultural exports, income inequality and economic growth. *Journal of Education and Social Studies* 5(2), 435–452.

Bermúdez-Serrano, I.M. (2020) Challenges and opportunities for the development of an edible insect food industry in Latin America. *Journal of Insects as Food and Feed* 6(5), 537–556.

Bermúdez-Serrano, I.M., Quirós-Blanco, A.M. and Acosta-Montoya, O. (2023) Production of edible insects: Challenges, opportunities, and perspectives for Costa Rica. *Agronomía Mesoamericana* 34(3), 1–19.

Bertuzzi, D.L., Becher, T.B., Capreti, N.M.R., Amorim, J., Jurberg, I.D. *et al.* (2018) General protocol to obtain D-glucosamine from biomass residues: Shrimp shells, cicada sloughs and cockroaches. *Global Challenges* 2(11), 1–6.

Bisconsin-Junior, A., Rodrigues, H., Behrens, J., Lima, V., Silva, M. *et al.* (2019) Examining the role of regional culture and geographical distances on the representation of unfamiliar foods in a continental-size country. *Food Quality and Preference* 79, 1–12.

Bucea-Manea-Tonis, R., Martins, O.M.D., Urdes, L., Coelho, A.S. and Simion, V.-E. (2023) Nudging consumer behavior with social marketing in Portugal: Can perception have an influence over trying insect-based food. *Insects* 14(6), 1–21.

Bug Burger (2022) Bugs Meet Meat: We Have Tasted the Swiss Flexiburger. Available at: https://www.bugburger.se/test/bugs-meet-meat-we-have-tasted-the-swiss-flexiburger/ (accessed 2 January 2025).

Cámara Hurtado, M.M., Conchello Moreno, P., Daschner, A., González Fandos, E., Palop Gómez, A. *et al.* (2018) Informe del Comité Científico de la Agencia Española de Consumo, Seguridad Alimentaria y Nutrición (AECOSAN) en relación a los riesgos microbiológicos y alergénicos asociados al consumo de insectos. *Revista Del Comité Científico de La AECOSAN* 27, 11–40.

Caparros Megido, R., Gierts, C., Blecker, C., Brostaux, Y., Haubruge, E. *et al.* (2016) Consumer acceptance of insect-based alternative meat products in Western countries. *Food Quality and Preference* 52, 237–243.

Carballo, A. (2005) Una revisión del modelo de crecimiento económico actual: Análisis de su problemática ambiental y desigualdades sociales. *Revista Luna Azul* 20, 1–15.

Cárcamo, R., Álvarez, A., Coral, C. and Santos, V. (2022) Mercado global-efectos locales: Un análisis coyuntural sobre el COVID-19, conflictos bélicos y cambio climático 2020-2022. *Sociedades Rurales, Producción y Medio Ambiente* 22(43), 16–16.

Cartay, R. (2018) Between shock and disgust: The consumption of insects in the Amazon basin. The case of *Rhynchophorus palmarum* (*Coleoptera curculionidae*). *Revista Colombiana de Antropología* 54(2), 143–169.

Casas, J.A. (2020) Develando el vínculo entre desigualdad y la pobreza. *Apuntes Del CENES* 39(69), 39–68.

Castillo, N. (2016) Desarrollo humano, desigualdad y pobreza. *Cultura De Paz* 22(68), 10–19.

Cerritos, R. (2009) Insects as food: An ecological, social and economical approach. *CAB Reviews: Perspectives in Agriculture, Veterinary Science, Nutrition and Natural Resources* 4(27), 1–10.

Chen, J., Zou, X., Zhu, W., Duan, Y., Merzendorfer, H. *et al.* (2022) Fatty acid binding protein is required for chitin biosynthesis in the wing of Drosophila melanogaster. *Insect Biochemistry and Molecular Biology* 149, 1–7.

Chen, X., Feng, Y. and Chen, Z. (2009) Common edible insects and their utilization in China. *Entomological Research* 39, 299–303.

Choudhary, P., Subhash, G.V., Khade, M. and Savant, S. (2021) Empowering blue economy: From under-rated ecosystem to sustainable industry. *Journal of Environmental Management* 291(312), 112697.

Christensen, D.L., Orech, F.O., Mungai, M.N., Larsen, T., Friis, H. *et al.* (2006) Entomophagy among the luo of Kenya: A potential mineral source. *International Journal of Food Sciences and Nutrition* 57(3–4), 198–203.

Cito, A., Botta, M., Francardi, V. and Dreassi, E. (2017) Insects as source of angiotensin converting enzyme inhibitory peptides. *Journal of Insects as Food and Feed* 3(4), 231–240.

Conde, E., Balboa, E.M., Parada, M. and Falqué, E. (2013) Algal proteins, peptides and amino acids. In: Dominguez, H. (ed.) *Functional Ingredients from Algae for Foods and Nutraceuticals*. Woodhead Publishing Series in Food Science, Technology and Nutrition, Oxford, UK, pp. 135–180.

Corporación Andina de Fomento (CAF) (2024) *Invertir Para la Seguridad Alimentaria y Nutricional en América Latina y el Caribe: Una vía Para Alcanzar el Hambre Cero*. CAF - Banco de Desarrollo de América Latina y El Caribe, Caracas.

Costa-Neto, E.M. (2005) Entomotherapy, or the medicinal use of insects. *Journal of Ethnobiology* 25, 93–114.

Cox, S. and Abu-Ghannam, N. (2013) Incorporation of *Himanthalia elongata* seaweed to enhance the phytochemical content of breadsticks using response surface methodology (RSM). *International Food Research Journal* 20, 1537–1545.

Cruz, M.A., Pelagio, R.C. and Vera, P.S. (2019) Propuesta de observatorios locales para el estudio de la pobreza y la seguridad alimentaria. In: Rubio, B. and Pasquier, A. (eds) *Inseguridad Alimentaria y Políticas de Alivio a La Pobreza: Una Visión Multidisciplinaria*. Universidad Nacional Autónoma do México. Instituto de Investigaciones Sociales, Mexico City, pp. 107–121.

Cruz-Fagua, D., Arévalo-Arévalo, H. and Vernot, D. (2021) *Artrópodos. Producción de Grillos de Forma Sustentable*. Universidad de La Sabana, MinCiencias, ArthroFood S.A.S., and Gobernación de Cundinamarca, Chía, Colombia.

Dawczynski, C., Schubert, R. and Jahreis, G. (2007) Amino acids, fatty acids, and dietary fibre in edible seaweed products. *Food Chemistry* 103(3), 891–899.

Dion-Poulin, A., Turcotte, M., Lee-Blouin, S. and Perreault, V. (2021) Acceptability of insect ingredients by innovative student chefs: An exploratory study. *International Journal of Gastronomy and Food Science* 24, 100362.

Dobermann, D., Swift, J.A. and Field, L.M. (2017) Opportunities and hurdles of edible insects for food and feed. *Nutrition Bulletin* 42(4), 293–308.

Durst, P.B. and Hanboonsong, Y. (2015) Small-scale production of edible insects for enhanced food security and rural livelihoods: Experience from Thailand and Lao People's Democratic Republic. *Journal of Insects as Food and Feed* 1(1), 25–31.

EFSA Scientific Committee (2015) Risk profile related to production and consumption of insects as food and feed. *EFSA Journal* 13(10), 1–60.

Entomo Farms (2024) Healthier for You & Our Planet. Available at: https://entomofarms.com/ (accessed 27 January 2025).

FAO (1996) *Rome Declaration on World Food Security and World Food Summit Plan of Action. Report of the 22nd Session of the Committee on World Food Security*. FAO, Rome. Available at: https://www.fao.org/4/w3613e/w3613e00.htm (accessed 23 June 2025).

FAO (2017) *Public Purchases of Food from Family Farming, and Food and Nutrition Security in Latin America and the Caribbean. Lessons Learned and Experiences*. FAO, Santiago de Chile.

FAO, International Fund for Agricultural Development (IFAD), and World Food Programme (WFP) (2013) *The State of Food Insecurity in the World 2013. The Multiple Dimensions of Food Security*. FAO, Rome. Available at: https://www.fao.org/4/i3434e/i3434e.pdf (accessed 17 June 2025).

FAO, Organización Panamericana de la Salud (OPS), WFP, and United Nations International Children's Emergency Fund (UNICEF) (2018) *Panorama of Food and Nutritional Security in Latin America and the Caribbean 2018*. FAO, Santiago de Chile.

FAO, IFAD, UNICEF, WFP, and WHO (2021) *The State of Food Security and Nutrition in the World 2021. Transforming Food Systems for Food Security, Improved Nutrition and Affordable Healthy Diets for*

All. FAO, Rome. Available at: https://openknowledge.fao.org/server/api/core/bitstreams/1c38676f-f5f7-47cf-81b3-f4c9794eba8a/content (accessed 24 June 2025).

FAO, IFAD, UNICEF, WFP, and WHO (2022) *The State of Food Security and Nutrition in the World 2022. Transforming Food Systems for Food Security, Improved Nutrition and Affordable Healthy Diets for All*. FAO, Rome. Available at: https://openknowledge.fao.org/server/api/core/bitstreams/67b1e9c7-1a7f-4dc6-a19e-f6472a4ea83a/content (accessed 24 June 2025).

FAO, IFAD, PAHO, UNICEF, and WFP (2023) *Latin America and the Caribbean – Regional Overview of Food Security and Nutrition 2023: Statistics and Trends*. FAO, Santiago de Chile.

FAO, IFAD, UNICEF, WFP, and WHO (2024) *The State of Food Security and Nutrition in the World 2024. Financing to End Hunger, Food Insecurity and Malnutrition in All Its Forms*. FAO, Rome.

Feng, Y., Chen, X.-M., Zhao, M., He, Z., Sun, L. *et al.* (2018) Edible insects in China: Utilization and prospects. *Insect Science* 25(2), 184–198.

Fernández-García, E., Carvajal-Lérida, I. and Pérez-Gálvez, A. (2009) *In vitro* bioaccessibility assessment as a prediction tool of nutritional efficiency. *Nutrition Research* 29(11), 751–760.

Finke, M.D. (2007) Estimate of chitin in raw whole insects. *Zoo Biology* 26(2), 105–115.

Fleta-Zaragozano, J. (2018) Entomofagia: ¿una alternativa a nuestra dieta tradicional? *Sanidad Militar* 74(1), 41–46.

Fleurence, J., Morançais, M., Dumay, J., Decottignies, P., Turpin, V. *et al.* (2012) What are the prospects for using seaweed in human nutrition and for marine animals raised through aquaculture? *Trends Food Science and Technology* 27, 57–61.

Fonseca-Carreño, N.E. and Bossa-Pabon, K.A. (2022) La agricultura y su incidencia en la seguridad y la soberanía alimentaria. Una Revisión. *Revista Científica Profundidad Construyendo Futuro* 17(17), 85–101.

Fontaneto, D., Tommaseo-Ponzetta, M., Galli, C., Risé, P., Glew, R.H. *et al.* (2011) Differences in fatty acid composition between aquatic and terrestrial insects used as food in human nutrition. *Ecology of Food and Nutrition* 50(4), 351–367.

Frikha, F., Kammoun, M., Hammami, N., Mchirgui, R.A., Belbahri, L. *et al.* (2011) Chemical composition and some biological activities of marine algae collected in Tunisia. *Ciencias Marinas* 37(2), 113–124.

Gahukar, R. (2011) Entomophagy and human food security. *International Journal of Tropical Insect Science* 31(3), 129–144.

Global Markets Insights Inc (2023) Edible Insects Market Size. Available at: https://www.gminsights.com/industry-analysis/edible-insects-market (accessed 20 December 2025).

Govorushko, S. (2019) Global status of insects as food and feed source: A review. *Trends in Food Science and Technology* 91, 436–445.

Grzelak, A. (2017) Income inequality and food security in the light of the experience of the OECD countries. In: *5th International Scientific Conference*, Vilnius Gediminas Technical University, May 11–12, Vilnius.

Guardiola, J. and González, F. (2010) La influencia de la desigualdad en la desnutrición de América Latina: Una perspectiva desde la economía. *Nutrición Hospitalaria* 25(3), 38–43.

Gutiérrez Cuesta, R., González García, K.L., Hernández Rivera, Y., Acosta Suárez, Y. and Marrero Delange, D. (2017) Marine algae, potential source of macronutrients. *Revista Investigaciones Marinas* 37(2), 16–28.

Haini, H., Musa, S., Loon, P. and Basir, K. (2023) Does unemployment affect the relationship between income inequality and food security? *International Journal of Sociology and Social Policy* 43(1/2), 48–66.

Hajare, R.B., Pagarkar, A.U., Desai, A.S., Koli, J.M., Shingare, P.E. *et al.* (2024) Seaweed-enriched cookies: A nutritional and functional perspective. *Asian Journal of Biotechnology and Biosource Technology* 10(4), 124–132.

Halloran, A., Hanboonsong, Y., Roos, N. and Bruun, S. (2017a) Life cycle assessment of cricket farming in North-Eastern Thailand. *Journal of Cleaner Production* 156, 83–94.

Halloran, A., Roos, N. and Hanboonsong, Y. (2017b) Cricket farming as a livelihood strategy in Thailand. *The Geographical Journal* 183(1), 112–124.

Hanboonsong, A. and Durst, P. (2020) *Guidance on Sustainable Cricket Farming: A Practical Manual for Farmers and Inspectors*. FAO, Bangkok.

Hanboonsong, Y., Jamjanya, T. and Durst, P.B. (2013) *Six-legged Livestock: Edible Insect Farming, Collection and Marketing in Thailand*. Food and Agriculture Organization of the United Nations, Regional Office for Asia and the Pacific, Bangkok. Available at: https://www.fao.org/4/i3246e/i3246e.pdf (accessed 17 June 2025).

Hart-Fredeluces, G.M., Ticktin, T., Kelman, D. and Wright, A.D. (2014) Contemporary gathering practice and antioxidant benefit of wild seaweed in Hawaii. *Economic Botany* 68, 30–43.

Hartmann, C. and Siegrist, M. (2017) Insects as food: Perception and acceptance: Findings from current research. *Ernahrungs Umschau International* 64(3), 44–50.

Hawkey, K.J., López-Viso, C., Brameld, J.M., Parr, T. and Salter, A.M. (2021) Insects: A potential source of protein and other nutrients for feed and food. *Annual Review of Animal Biosciences* 9, 333–354.

Holdt, S.L. and Kraan, S. (2011) Bioactive compounds in seaweed: Functional food applications and legislation. *Journal of Applied Phycology* 23, 543–597.

Hsieh, H.M. and Ju, Y.M. (2018) Medicinal components in termitomyces mushrooms. *Applied Mibrobiology and Biotechnology* 102(12), 4987–4994.

International Platform of Insects for Food and Feed (IPIFF) (2024). Available at: https://ipiff.org/ (accessed 25 January 2025).

IPCC (2021) *Climate Change 2021: The Physical Science Basis*. Intergovernmental Panel on Climate Change, Geneva.

IPCC (2022) *Climate Change 2022 – Impacts, Adaptation and Vulnerability: Contribution of Working Group II to the Sixth Assessment Report of the Intergovernmental Panel on Climate Change*. Cambridge University Press, Cambridge, UK.

Ishara, J., Cokola, M., Buzera, A., Mmari, M., Bugeme, D. *et al.* (2023) Edible insect biodiversity and anthropo-entomophagy practices in Kalehe and Idjwi territories, D.R. Congo. *Journal of Ethnobiology and Ethnomedicine* 19(3), 1–17.

Ites, S., Smetana, S., Toepfl, S. and Heinz, V. (2020) Modularity of insect production and processing as a path to efficient and sustainable food waste treatment. *Journal of Cleaner Production* 248, 1–17.

Jantzen da Silva Lucas, A., Menegon de Oliveira, L., da Rocha, M. and Prentice, C. (2020) Edible insects: An alternative of nutritional, functional and bioactive compounds. *Food Chemistry* 311, 1–11.

Juela-Tiban, E.V. and Chileno-Camacho, L.F. (2024) Prevención de la desnutrición infantil y educación sobre los hábitos alimenticios en las madres: Revisión sistemática. *Revista Científica Arbitrada En Investigaciones de La Salud GESTAR* 7(14), 546–570.

Junior, V.M.F. and Marzábal, Ó.R. (2012) La crisis financiera global en perspectiva: Génesis y factores determinantes. *Revista de Economía Mundial* 31, 199–226.

Karmakar, S. and Sarkar, D. (2014) Income inequality, poverty and food security in West Bengal, India. *Journal of Social Science Studies* 1(1), 31–43.

Kewuyemi, Y.O., Kesa, H., Chinma, C.E. and Adebo, O.A. (2020) Fermented edible insects for promoting food security in Africa. *Insects* 11(5), 1–16.

Khotimchenko, S.V., Vaskovsky, V.E. and Titlyanova, T.V. (2002) Fatty acids of marine algae from the Pacific Coast of North California. *Botanica Marina* 45, 17–22.

Kinyuru, J.N., Konyole, S.O., Roos, N., Onyango, C.A., Owino, V.O. *et al.* (2013) Nutrient composition of four species of winged termites consumed in Western Kenya. *Journal of Food Composition and Analysis* 30(2), 120–124.

Klunder, H.C., Wolkers-Rooijackers, J., Korpela, J.M. and Nout, M.J.R. (2012) Microbial aspects of processing and storage of edible insects. *Food Control* 26(2), 628–631.

Kourimská, L. and Adámková, A. (2016) Nutritional and sensory quality of edible insects. *NFS Journal* 4, 22–26.

Krongdang, S., Phokasem, P., Venkatachalam, K. and Charoenphun, N. (2023) Edible insects in Thailand: An overview of status, properties, processing, and utilization in the food industry. *Foods* 12(11), 1–24.

Lähteenmäki–Uutela, A., Grmelová, N., Hénault-Ethier, L., Deschamps, M.-H., Vandenberg, G.W. *et al.* (2017) Insects as food and feed: Laws of the European Union, United States, Canada, Mexico, Australia,and China. *Insects as Food and Feed* 3(2), 155–160.

Lange, K.W. and Nakamura, Y. (2021) Edible insects as future food: Chances and challenges. *Journal of Future Foods* 1(1), 38–46.

Latham, M. (2002) *Nutrición Humana en el Mundo en Desarrollo*, Organización de las Naciones Unidas para la Agricultura y la Alimentación, No. 29. FAO, Rome.

Latunde-Dada, G.O., Yang, W. and Vera-Aviles, M. (2016) *In vitro* iron availability from insects and sirloin beef. *Journal of Agricultural and Food Chemistry* 64(44), 8420–8424.

Lehtovaara, V.J., Valtonen, A., Sorjonen, J., Hiltunen, M., Rutaro, K. *et al.* (2017) The fatty acid contents of the edible grasshopper *Ruspolia differens* can be manipulated using artificial diets. *Journal of Insects as Food and Feed* 3(4), 253–262.

León, A., Martínez, R., Espíndola, E. and Schejtman, A. (2004) *Pobreza, Hambre y Seguridad Alimentaria en Centroamérica y Panamá.* Comisión Económica Para América Latina y El Caribe (CEPAL), Santiago de Chile.

Lima, J.V.M., Carvalho, A.F.F.U., Freitas, S.M. and Melo, V.M.M. (2002) Antibacterial activity of extracts of six macroalgae from the Northeastern Brazilian coast. *Brazilian Journal of Microbiology* 33, 311–313.

Liu, S., Sun, J., Yu, L., Zhang, C., Bi, J. *et al.* (2012) Antioxidant activity and phenolic compounds of *Holotrichia parallela* Motschulsky extracts. *Food Chemistry* 134(4), 1885–1891.

Lordan, S., Ross, R.P. and Stanton, C. (2011) Marine bioactives as functional food ingredients: Potential to reduce the incidence of chronic diseases. *Marine Drugs* 9(6), 1056–1100.

Mac Monagail, M., Cornish, L., Morrison, L., Araújo, R. and Critchley, A.T. (2017) Sustainable harvesting of wild seaweed resources. *European Journal of Phycology* 52(4), 371–390.

Macdiarmid, J.I. and Whybrow, S. (2019) Nutrition from a climate change perspective. *Proceedings of the Nutrition Society* 78(3), 380–387.

MacNabb, E. and Fletcher, B. (2021) Food insecurity and an economic crisis. In: Hoflund, A.B., Jones, J.C. and Pautz, M.C. (eds) *Administering and Managing the US Food System: Revisiting Food Policy and Politics.* Rowman & Littlefield, Lanham, Maryland, pp. 245–260.

Maehre, H.K., Malde, M.K, Eilertsen, K.-E., and Elvevoll, E.O. (2014) Characterization of protein, lipid and mineral contents in common Norwegian seaweeds and evaluation of their potential as food and feed. *Journal of the Science of Food and Agriculture* 94(15), 3281–3290.

Mahadevan, R. and Hoang, V. (2016) Is there a link between poverty and food security? *Social Indicators Research* 128, 179–199.

Mark, D. (2015) Complete nutrient content of four species of commercially available feeder insects fed enhanced diets during growth. *Zoo Biology* 34(6), 554–564.

McHugh, D.J. (2002) *Perspectivas para la Producción de Algas Marinas en los Países en Desarrollo,* Organización de las Naciones Unidas para la Agricultura y la Alimentación, FAO Circular de Pesca no.968. FAO, Rome.

Melgar-Lalanne, G., Hernández-Álvarez, A.J. and Salinas-Castro, A. (2019) Edible insects processing: Traditional and innovative technologies. *Comprehensive Reviews in Food Science and Food Safety* 18(4), 1166–1191.

Menasche, R. and Machado, C.J.B. (2019) Elementos para uma agenda de pesquisa em segurança alimentar e nutricional à luz da antropología. In: Rubio, B. and Pasquier, A. (eds) *Inseguridad Alimentaria y Políticas de Alivio a La Pobreza: Una Visión Multidisciplinaria.* Universidad Nacional Autónoma de México, Instituto de Investigaciones Sociales, Mexico City, pp. 227–249.

Mikulec, A., Platta, A., Radzymińska, M., Garbowska, B., Suwała, G. *et al.* (2024) Can sustainable food from edible insects become the food of the future? Exploring Poland's Generation Z. *Sustainability* 16(23), 1–19.

Millar, R.V., Houghton, J.D.R., Elsäβer, B., Mensink, P.J. and Kregting, L. (2020) Influence of waves and currents on the growth rate of the kelp *Laminaria digitata* (Phaeophyceae). *Journal of Phycology* 56(1), 198–207.

Miranda, A.F., Taha, M., Wrede, D., Morrison, P., Ball, A.S. *et al.* (2015) Lipid production in association of filamentous fungi with genetically modified cyanobacterial cells. *Biotechnology for Biofuels* 8(1), 179.

Miyashiro, M.J.K. (2022) Relación entre la especulación financiera con el hambre: Una revisión narrativa. *Spanish Journal of Community Nutrition* 28(4), 1–7.

Montowska, M., Kowalczewski, P.Ł., Rybicka, I. and Fornal, E. (2019) Nutritional value, protein and peptide composition of edible cricket powders. *Food Chemistry* 289, 130–138.

Morales-Ramos, J.A., Rojas, M.G., Dossey, A.T. and Berhow, M. (2020) Self-selection of food ingredients and agricultural by-products by the house cricket, *Acheta domesticus* (Orthoptera: Gryllidae): A holistic approach to develop optimized diets. *PLOS One* 15(1), e0227400.

Moreno, A. and Flores, J. (2011) *Agrobiodiversidad y Soberanía Alimentaria En Comunidades Shuar de Morona Santiago: Análisis de Impactos Del Programa GESOREN – Deutsche Gesellschaft Für Internationale Zusammenarbeit (GIZ).* Serie Estudios de Impacto, Fascículo 4. GIZ, Quito, Ecuador.

Mundo-Rosas, V., Unar-Munguía, M., Hernández, M., Pérez-Escamilla, R. and Shamah-Levy, T. (2019) La seguridad alimentaria en los hogares en pobreza de México: Una mirada desde el acceso, la disponibilidad y el consumo. *Salud Pública de México* 61(6), 866–875.

Murata, M. and Nakazoe, J. (2001) Production and use of marine algae in Japan. *Agricultural Research Quarterly* 35(4), 281–290.

Mutungi, C., Irungu, F.G., Nduko, J., Mutua, F., Affognon, H. *et al.* (2019) Postharvest processes of edible insects in Africa: A review of processing methods, and the implications for nutrition, safety and new products development. *Critical Reviews in Food Science and Nutrition* 59(2), 276–298.

Muzzarelli, R.A.A., Biagini, G., DeBenedittis, A., Mengucci, P., Majni, G. *et al.* (2001) Chitosan-oxychitin coatings for prosthetic materials. *Carbohydrate Polymers* 45(1), 35–41.

Mwalugha, H.M., Wakibia, J., Kenji, G.M. and Mwasaru, M. (2015) Chemical composition of common seaweeds from the Kenya coast. *Journal of Food Research* 4(6), 28–38.

Narayan, B., Miyashita, K. and Hosakawa, M. (2006) Physiological effects of eicosapentaenoic acid (EPA) and docosahexaenoic acid (DHA)—a review. *Food Reviews International* 22(3), 291–307.

Nelson, G.C., Rosegrant, M., Koo, J., Robertson, R., Sulser, T. *et al.* (2009) *Climate change: impact on agriculture and costs of adaptation.* Food Policy Report 21. International Food Policy Research Institute (IFPRI), Washington, DC.

Neves, J.A., Vasconcelos, F.D.A.G.D., Machado, M.L., Recine, E., Garcia, G.S., and Medeiros, M.A.T.D. (2022) The Brazilian cash transfer program (*Bolsa Família*): A tool for reducing inequalities and achieving social rights in Brazil. *Global Public Health* 17(1), 26–42.

Next Food (2025) ¿Comprar Insectos Comestibles en Línea? Available at: https://www.next-food.net/es/ (accessed 25 January 2025).

Nino, M.C., Reddivari, L., Osorio, C., Kaplan, I. and Liceaga, A. (2021) Insects as a source of phenolic compounds and potential health benefits. *Journal of Insects Food and Feed* 7(7), 1077–1087.

Niyonsaba, H.H., Höhler, J., Kooistra, J., Van der Fels-Klerx, H.J. and Meuwissen, M.P.M. (2021) Profitability of insect farms. *Journal of Insects as Food and Feed* 7(5), 923–934.

Nowakowski, A.C., Miller, A.C., Miller, M.E., Xiao, H. and Wu, X. (2022) Potential health benefits of edible insects. *Critical Reviews in Food Science and Nutrition* 62(13), 3499–3508.

Omuse, E.R., Tonnang, H.E.Z., Yusuf, A.A., Machekano, H., Egonyu, J.P. *et al.* (2024) The global atlas of edible insects: Analysis of diversity and commonality contributing to food systems and sustainability. *Scientific Reports* 14(5045), 1–17.

Oonincx, D.G.A.B., van Itterbeeck, J., Heetkamp, M.J.W., van den Brand, H., van Loon, J.J.A. *et al.* (2010) An exploration on greenhouse gas and ammonia production by insect species suitable for animal or human consumption. *PLOS One* 5(12), e14445.

Oonincx, D.G.A.B., van Broekhoven, S., van Huis, A. and van Loon, J.J.A. (2015) Feed conversion, survival and development, and composition of four insect species on diets composed of food by-products. *PLOS One* 10(12), e0222043.

Ortiz, J., Romero, N., Robert, P., Araya, J., Lopez-Hernández, J. *et al.* (2006) Dietary fiber, amino acid, fatty acid and tocopherol contents of the edible seaweeds *Ulva lactuca* and *Durvillaea antarctica*. *Food Chemistry* 99(1), 98–104.

Osimani, A., Milanović, V., Cardinali, F., Roncolini, A., Garofalo, C. *et al.* (2018) Bread enriched with cricket powder (*Acheta domesticus*): A technological, microbiological and nutritional evaluation. *Innovative Food Science & Emerging Technologies* 48, 150–163.

Ouma, O.M. (2017) The dynamics of poverty, inequality and economic well being in Uasin gishu County, Kenya. *Journal of Economics and Sustainable Development* 8(8), 115–116.

Overland, M., Mydland, L.T. and Skrede, A. (2018) Marine macroalgae as source of protein and bioactive compounds in feed for monogastric animals. *Journal of the Science of Food and Agriculture* 99, 13–24.

Panzella, L., Moccia, F., Nasti, R., Marzorati, S., Verotta, L. *et al.* (2020) Bioactive phenolic compounds from agri-food wastes: An update on green and sustainable extraction methodologies. *Frontiers in Nutrition* 7, 1–27.

Paoletti, M.G., Buscardo, E. and Dufour, D.L. (2000) Edible invertebrates among Amazonian Indians: A critical review of disappearing knowledge. *Environment, Development and Sustainability* 2(3–4), 195–225.

Paoletti, M.G., Norberto, L., Damini, R. and Musumeci, S. (2007) Human gastric juice contains chitinase that can degrade chitin. *Annals of Nutrition and Metabolism* 51(3), 244–251.

Park, S.J., Kim, K.Y., Kaik, M.Y. and Koh, Y.H. (2022) Sericulture and the edible-insect industry can help humanity survive: Insects are more than just bugs, food, or feed. *Food Science and Biotechnology* 31(6), 657–668.

Pasquier, A. (2019) Narrativas contrastantes en torno al concepto de "seguridad alimentaria". El caso del programa sin hambre. In: Rubio, B. and Pasquier, A. (eds) *Inseguridad Alimentaria y Políticas de Alivio a La Pobreza: Una Visión Multidisciplinaria*. Universidad Nacional Autónoma de México, Instituto de Investigaciones Sociales, Mexico City, pp. 95–130.

Payne, C.L.R., Sarborough, P., Rayner, M. and Nonaka, K. (2016) Are edible insects more or less "healthy" than commonly consumed meats? A comparison using two nutrient profiling models developed to combat over- and undernutrition. *European Journal of Clinical Nutrition* 70, 285–291.

Pérez De Armiño, K. (2013) La gobernanza global de la seguridad alimentaria: Debilidades, disparidades e iniciativas de reforma. In: Pons Rafols, X. (ed.) *Alimentación y Derecho Internacional: Normas, Instituciones y Procesos*. Marcial Pons, Madrid, pp. 83–115.

Pollard, C.M. and Booth, S. (2019) Addressing food and nutrition security in developed countries. *International Journal of Environmental Research and Public Health* 16(13), 2370. DOI: 10.3390/ijerph16132370.

Poore, J. and Nemecek, T. (2018) Reducing food's environmental impacts through producers and consumers. *Science* 360(6392), 987–992.

Poshadri, A., Palthiya, R., Shivacharan, G. and Butti, P. (2018) Insects as an alternate source for food to conventional food animals. *International Journal of Pure & Applied Bioscience* 6(2), 697–705.

Rahmanto, F., Purnomo, E.P. and Kasiwi, A.N. (2021) Food diversification: Strengthening strategic efforts to reduce social inequality through sustainable food security development in Indonesia. *Caraka Tani: Journal of Sustainable Agriculture* 36(1), 33–44.

Rainforest Alliance (2023) La Relación entre la Agricultura y el Cambio Climático. Available at: https://www.rainforest-alliance.org/es/en-el-campo/la-relacion-entre-la-agricultura-y-el-cambio-climatico/ (accessed 24 June 2025).

Rajapakse, N. and Kim, S.K. (2011) Nutritional and digestive health benefits of seaweed. *Advances in Food and Nutrition Research* 64, 17–28.

Ramos-Elorduy, J. (2008) Energy supplied by edible insects from Mexico and their nutritional and ecological importance. *Ecology of Food and Nutrition* 43(3), 280–297.

Ravagli, A.C. and Carolina, A. (2021) Prospección de los insectos comestibles como fuente de proteína animal para el consumo humano. Universidad Militar Nueva Granada, Colombia. Available at: http://hdl.handle.net/10654/38939 (accessed 24 June 2025).

Rebours, C., Marinho-Soriano, E., Zertuche-González, J.A., Hayashi, L., Vásquez, J.A. *et al.* (2014) Seaweeds: An opportunity for wealth and sustainable livelihood for coastal communities. *Journal of Applied Phycology* 26, 1939–1951.

Reverberi, M. (2020) Edible insects: Cricket farming and processing as an emerging market. *Journal of Insects as Food and Feed* 6(2), 211–220.

Ribeiro, J.C., Cunha, L.M., Sousa-Pinto, B., and Fonseca, J. (2018) Allergic risks of consuming edible insects: A systematic review. *Molecular Nutrition & Food Research* 62(1), 1–31.

Ritchie, H. and Roser, M. (2019) Half of the World's Habitable Land is Used for Agriculture. Our World in Data. Available at: https://ourworldindata.org/global-land-for-agriculture (accessed 25 January 2025).

Ritchie, H., Rosado, P. and Roser, M. (2023) Hunger and Undernourishment. Our World in Data. Available at: https://ourworldindata.org/hunger-and-undernourishment (accessed 25 January 2025).

Rodríguez-Peñaguirre, F.J. and González-Arellano, S. (2022) Revisión sistemática de los sistemas alimentarios en la transición a ciudades sostenibles. *Estudios Sociales. Revista de Alimentación Contemporánea y Desarrollo Regional* 32, 60.

Roohinejad, S., Koubaa, M., Barba, F.J., Saljoughian, S., Amid, M. *et al.* (2017) Application of seaweeds to develop new food products with enhanced shelf-life, quality and health-related beneficial properties. *Food Research International* 99(3), 1066–1083.

Roos, N. and van Huis, A. (2017) Consuming insects: Are there health benefits? *Journal of Insects as Food and Feed* 3(4), 225–229.

Ross, F.W.R., Tarbuck, P. and Macreadie, P.I. (2022) Seaweed afforestation at large-scales exclusively for carbon sequestration: Critical assessment of risks, viability and the state of knowledge. *Frontiers in Marine Science* 9, 1–15.

Ross, F.W.R., Malerba, M.E. and Macreadie, P.I. (2025) Global potential for seaweed aquaculture on existing offshore infrastructure. *Heliyon* 11(1), e41248.

Ruby, M.B., Rozin, P. and Chan, C. (2015) Determinants of willingness to eat insects in the USA and India. *Journal of Insects as Food and Feed* 1(3), 215–225.

Rumpold, B.A. and Schlüter, O.K. (2013a) Nutritional composition and safety aspects of edible insects. *Molecular Nutrition & Food Research* 57(5), 802–823.

Rumpold, B.A. and Schlüter, O.K. (2013b) Potential and challenges of insects as an innovative source for food and feed production. *Innovative Food Science & Emerging Technologies* 17(1), 1–11.

Rumpold, B.A. and Schluter, O.K. (2015) Insect-based protein sources and their potential for human consumption: Nutritional composition and processing. *Animal Frontiers* 5(2), 20–24.

Ruzengwe, F.M., Nyarugwe, S.P., Manditsera, F.A., Mubaiwa, J., Cottin, S. *et al.* (2022) Contribution of edible insects to improved food and nutrition security: A review. *International Journal of Food Science & Technology* 57(10), 6257–6269.

Sánchez-Muros, M., Barroso, F.G. and Manzano-Agugliaro, F. (2014) Insect meal as renewable source of food for animal feeding: A review. *Journal of Cleaner Production* 65, 16–27.

Sawarkar, K.T., Chaudhari, N.J. and Dhawade, A.G. (2024) Seaweed: A complete medical overview. *World Journal of Biology Pharmacy and Health Sciences* 20(2), 590–601.

Shelomi, M. (2016) The meat of affliction: Insects and the future of food as seen in expo 2015. *Trends in Food Science & Technology* 56, 175–179.

Siddiqui, F., Salam, R., Lassi, Z. and Das, J. (2020) The intertwined relationship between malnutrition and poverty. *Frontiers in Public Health* 8, 453. DOI: 10.3389/fpubh.2020.00453.

Simon, S.A., de Araujo, I.E., Gutierrez, R. and Nicolelis, M. (2006) The neural mechanisms of gustation: A distributed processing code. *Nature Reviews Neuroscience* 7(11), 890–901.

Smil, V. (2002) Eating meat: Evolution, patterns, and consequences. *Population and Development Review* 28(4), 599–639.

Soares de Castro, R.J., Ohara, A., Gonçalves dos, J. and Fontenele Domingues, M.A. (2018) Nutritional, functional and biological properties of insect proteins: Processes for obtaining, consumption and future challenges. *Trends in Food Science & Technology* 76, 82–89.

Ssepuuya, G., Kagulire, J., Katongole, J., Kabbo, D., Claes, J. *et al.* (2021) Suitable extraction conditions for determination of total anti-oxidant capacity and phenolic compounds in *Ruspolia differens* Serville. *Journal of Insects as Food and Feed* 7(2), 205–214.

Stewart, F. (2015) The sustainable development goals: A comment. *Journal of Global Ethics* 11(3), 288–293.

Stewart, R., Langer, L., Da Silva, N.R., Muchiri, F., Zaranyika, H. *et al.* (2015) The effects of training, innovation and new technology on African smallholder farmers' economic outcomes and food security: A systematic review. *Campbell Systematic Reviews* 11(1), 1–224.

St-Hilaire, S., Cranfill, K., McGuire, M.A., Mosley, E.E., Tomberlin, J.K. *et al.* (2007) Fish offal recycling by the black soldier fly produces a foodstuff high in Omega-3 fatty acids. *Journal of the World Aquaculture Society* 38(2), 309–313.

Tang, C., Yang, D., Liao, H., Sun, H., Liu, C. *et al.* (2019) Edible insects as a food source: A review. *Food Production, Processing and Nutrition* 1(8), 1–13.

Tonk, M. and Vilcinskas, A. (2017) The medical potential of antimicrobial peptides from insects. *Current Topics in Medicinal Chemistry* 17(5), 554–575.

Torres, B., Jadán Maza, O., Aguirre, P., Hinojosa, L. and Günter, S. (2014) Contribution of traditional agroforestry to climate change adaptation in the Ecuadorian Amazon: The chakra system. In: Filho, W.L. (ed.) *Handbook of Climate Change Adaptation*, 1st edn. Springer, Berlin, pp. 839–854.

Touch, V., Tan, D.K., Cook, B.R., Li Liu, D., Cross, R. *et al.* (2024) Smallholder farmers' challenges and opportunities: Implications for agricultural production, environment and food security. *Journal of Environmental Management* 370, 122536.

Tripathi, K. and Singh, A. (2018) Chitin, chitosan and their pharmacological activities: A review. *International Journal of Pharmaceutical Sciences and Research* 9(3), 2626–2635.

Trostle, R. (2008) *Global Agricultural Supply and Demand: Factors Contributing to the Recent Increase in Food Commodity Prices*. Economic Research Service WRS-0801. US Department of Agriculture, Washington, DC.

Tzompa-Sosa, D.A., Yi, L., van Valenberg, H.J.F., van Boekel, M.A.J.S. and Lakemond, C.M.M. (2014) Insect lipid profile: Aqueous versus organic solvent-based extraction methods. *Food Research International* 62, 1087–1094.

UNDP and FAO (2016) *Seguridad Alimentaria y Nutricional: Camino Hacia el Desarrollo Humano*. Programa de las Naciones Unidas para el Desarrollo (PNUD) / Organización de las Naciones Unidas para la Alimentación y la Agricultura (FAO). Cuaderno de Desarrollo Humano 12. UNDP and FAO, San Salvador.

Urdes, L., Simion, V.-E., Talaghir, L.-G. and Mindrescu, V. (2022) An integrative approach to healthy social-ecological system to support increased resilience of resource management in food-producing systems. *Sustainability* 14(22), 14830. DOI: 10.3390/su142214830.

van Broekhoven, S., Oonincx, D.G.A.B., van Huis, A. and van Loon, J.J.A. (2015) Growth performance and feed conversion efficiency of three edible mealworm species (Coleoptera: Tenebrionidae) on diets composed of organic by-products. *Journal of Insect Physiology* 73, 1–10.

van Huis, A. (2003) Insects as food in sub-saharan Africa. *International Journal of Tropical Insect Science* 23(3), 163–185.

van Huis, A. (2016) Edible insects are the future? *Proceedings of the Nutrition Society* 75(3), 294–305.

van Huis, A. (2019) Insects as food and feed, a new emerging agricultural sector: A review. *Journal of Insects as Food and Feed* 6(1), 27–44.

van Huis, A. (2020) Nutrition and health of edible insects. *Current Opinion in Clinical Nutrition & Metabolic Care* 23(3), 228–231.

van Huis, A. (2005) Insects eaten in Africa (Coleoptera, Hymenoptera, Diptera, Heteroptera, Homoptera). In: Paoletti, M.G. (ed.) *Ecological Implications of Minilivestock*. Science Publishers, New Hampshire, pp. 231–244.

van Huis, A., Van Itterbeeck, J., Klunder, H., Mertens, E., Halloran, A. *et al.* (2013) *Edible Insects: Future Prospects for Food and Feed Security*. FAO Forestry Paper 171. FAO, Rome.

van Huis, A., Halloran, A., Van Itterbeeck, J., Klunder, H. and Vantomme, P. (2021a) How many people on our planet eat insects: 2 billion? *Journal of Insects as Food and Feed* 8(1), 1–4.

van Huis, A., Rumpold, B., Maya, C., and Roos, N. (2021b) Nutritional qualities and enhancement of edible insects. *Annual Review of Nutrition* 41, 551–576.

Venkatesalu, V., Sundaramoorthy, P., Anantharaj, M., Gopalakrishnan, M. and Chandrasekaran, M. (2004) Studies on the fatty acid composition of marine algae of Rameswaram coast. *Seaweed Research and Utilization* 26, 83–86.

Verner, D., Roos, N., Halloran, A., Surabian, G., Tebaldi, E. *et al.* (2021) *Insect and Hydroponic Farming in Africa. The New Circular Food Economy*, Agriculture and Food Series. World Bank, Washington DC.

Vernot, D. (2021) *Nuevas Alternativas de Producción con Grillos G. sigillatus. Empoderamiento, Emprendimiento y Reconocimiento a Mujeres Rurales del Municipio de La Mesa, Cundinamarca–Colombia*. Universidad de La Sabana, MinCiencias, ArthroFood S.A.S., and Gobernación de Cundinamarca, Chía, Colombia.

Vigani, M. (2020) The bioeconomy of microalgae-based processes and products. In: Jacob-Lopes, E., Manzoni Maroneze, M., Queiroz, M.I. and Queiroz Zepka, L. (eds) *Handbook of Microalgae-Based Processes and Products*. Academic Press, London, pp. 799–821.

Vilar-Compte, M., Burrola-Méndez, S., Lozano-Marrufo, A., Ferré-Eguiluz, I., Flores, D. *et al.* (2021) Urban poverty and nutrition challenges associated with accessibility to a healthy diet: A global systematic literature review. *International Journal for Equity in Health* 20(1), 40.

Wahrendorf, M.S. and Wink, M. (2006) Pharmacologically active natural products in the defense secretion of *Palembus ocularis* (Tenebrionidae, Coleoptera). *Journal of Ethnopharmacology* 106(1), 51–56.

WCED (1987) *Our Common Future. Report of the World Commission on Environment and Development*. Oxford University Press, New York. Available at: https://sustainabledevelopment.un.org/content/documents/5987our-common-future.pdf (accessed 23 June 2025).

Wells, M.L., Potin, P., Craigie, J.S., Raven, J.A., Merchant, S.S. *et al.* (2017) Algae as nutritional and functional food sources: Revisiting our understanding. *Journal of Applied Phycology* 29, 949–982.

Wezel, A., Gemmill, B., Bezner, R., Barrios, E., Rodríguez, A. *et al.* (2020) Agroecological principles and elements and their implications for transitioning to sustainable food systems. A review. *Agronomy for Sustainable Development* 40, 1–13.

WFP (2024) *Climate Change Policy Update*. WFP, Rome. Available at: https://executiveboard.wfp.org/document_download/WFP-0000160977 (accessed 24 June 2025).

White, M.A. (2013) Sustainability: I know it when I see it. *Ecological Economics* 86, 213–217.

Willett, W., Rockström, J., Loken, B., Springmann, M., Lang, T. *et al.* (2019) Food in the Anthropocene: The EAT–Lancet commission on healthy diets from sustainable food systems. *The Lancet* 393(10170), 447–492.

Womeni, H.M., Linder, M., Tiencheu, B., Mbiapo, F.T., Villeneuve, P. *et al.* (2009) Oils of insects and larvae consumed in Africa: Potential sources of polyunsaturated fatty acids. *Oléagineux, Corps Gras, Lipides* 16(4), 230–235.

Woodhill, J., Kishore, A., Njuki, J., Jones, K. and Hasnain, S. (2022) Food systems and rural wellbeing: Challenges and opportunities. *Food Security* 14(5), 1099–1121.

World Bank (2024) *People in a Changing Climate: From Vulnerability to Action*. World Bank, Washington, DC. Available at: https://www.worldbank.org/en/topic/climatechange/publication/people-in-a-changing-climate-from-vulnerability-to-action (accessed 17 June 2025).

World Health Organization (WHO) (2007) Protein and Amino Acid Requirements in Human Nutrition. WHO Technical Report Series 935. Report of a Joint WHO/FAO/UNU Expert Consultation. United Nations University, Geneva.

World Wildlife Fund (WWF) (2024) *Living Planet Report 2024: A System in Peril*. WWF, Gland, Switzerland. Available at: https://wwflpr.awsassets.panda.org/downloads/2024-living-planet-report-a-system-in -peril.pdf (accessed 17 June 2025).

Yamada, Y., Matoba, N., Usui, H., Onishi, K. and Yoshikawa, M. (2002) Design of a highly potent anti-hypertensive peptide based on ovokinin(2–7). *Bioscience, Biotechnology, and Biochemistry* 66, 1213–1217.

Zhang, E., Ji, X., Ouyang, F., Lei, Y., Deng, S. *et al.* (2023) A minireview of the medicinal and edible insects from the Traditional Chinese Medicine (TCM). *Frontiers in Pharmacology* 16(14), 1–14.

Zhao, H., Li, H., Feng, Y., Zhang, Y., Yuan, F. *et al.* (2019) Mycelium polysaccharides from *Termitomyces albuminosus* attenuate CCl_4-induced chronic liver injury via inhibiting TGFβ1/smad3 and NF-κB signal pathways. *International Journal of Molecular Sciences* 20(19), 1–20.

Zielińska, E., Baraniak, B., Karaś, M., Rybczyńska, K. and Jakubczyk, A. (2015) Selected species of edible insects as a source of nutrient composition. *Food Research International* 77(3), 460–466.

Zimian, D., Yonghua, Z. and Xiwu, G. (2010) Medicinal insects in China. *Ecology of Food and Nutrition* 36(2–4), 209–220.

9 Bringing It All Together

Laura D. Urdes[1]*, Julius Tepper[2] and Chris Walster[3]

[1]*The World Aquatic Veterinary Medical Association, Romania; Faculty of Veterinary Medicine, University Spiru Haret Bucharest, Romania; Faculty of Management and Rural Development, University of Agricultural Sciences and Veterinary Medicine, Bucharest, Romania; [2]The World Aquatic Veterinary Medical Association, USA; Long Island Fish Hospital, Manorville, New York, USA; [3]The World Aquatic Veterinary Medical Association, UK*

Abstract

The concepts presented in the previous chapters are integrated and used in this chapter to indicate how One Health can be utilized to better protect public health and food and feed safety. Since the book addresses the One Health concept in relation to aquatic ecosystems, this chapter focuses on zoonoses of aquatic origin; in particular, a working example of a "neglected" zoonosis perceived by the authors as having considerable PH implications (i.e. mycobacteriosis) is discussed to indicate that using the One Health conceptual tools can support optimization of the decision-making process in health-related matters. The authors use the example of fish mycobacteriosis to highlight risks associated with the ornamental fish trade and how the One Health tools can be effectively used to mitigate these risks. In a separate section, the authors address the more neglected socio-political aspect required to ensure a practical implementation of One Health, and the sustainability of the concept, emphasizing the need for more effective operationalization and institutionalization of One Health.

9.1 How One Health Principles Integrate within Aquatic Ecosystems to Protect Public Health and Food and Feed Safety in a Healthy Environment

By now it should have become clear to the reader that One Health is a worldwide strategy, encouraging interdisciplinary collaboration and communication in relation to all aspects of health care for humans and animals. As explained in Chapter 2, 'Toward Sustainable One Ocean: Evaluating Life Below Water Through the Lens of One Health,' there is a close relationship between health and aquatic ecosystems, calling

for the One Health interdisciplinary approach to enable the development of effective disease control and water management policies.

According to the World Bank (Berthe *et al.*, 2018), there are five One Health disease threats involving aquatic ecosystems. These are:

* Emerging diseases—as explained in Chapter 5, 'Epidemiology, Biosecurity in Aquaculture, Statistics, and Health Economics,' and Chapter 7, 'Human–Wildlife Interface,' these are new diseases or diseases increasing in number and/or geographic area that are at risk of becoming pandemic (e.g. COVID-19, Ebola). Zoonotic spillover plays a fundamental role

*Corresponding author: urdeslaura@gmail.com

© CAB International 2025. *One Health Concepts and the Aquatic Ecosystem*
(eds L.D. Urdes *et al.*)
DOI: 10.1079/9781800623248.0009

in the emergence of new human infectious diseases such as COVID-19.

- Endemic diseases—these are constantly present, thus incurring economic and health expenses and consequences; some of these are zoonoses with reservoirs in wildlife and the environment (e.g. rabies, classical and African swine fever).
- Antimicrobial resistance (AMR)—as explained in detail in Chapter 6, 'Public Health,' infections resistant to antibiotics originate from veterinary and human practices, circulating via livestock and food production systems, and via wastewater.
- Vector-borne diseases—as shown in Chapter 7, animate vectors, such as ticks, mosquitos, or snails, play an important role in many (neglected) diseases with reservoirs in wildlife, causing around 700,000 human deaths per year globally (Berthe *et al.*, 2018). Commonly known vectors are closely linked to water bodies and waterborne pathogens.
- Chemical exposures—as explained in Chapter 4, 'Pollution and Mitigation of Environmental Risks,' Chapter 6, Public health, and Chapter 8, 'Food Security and Innovative Food Production,' chemical residues spread through wastewater from food-producing systems and agriculture, not only affecting the health of humans and animals, but also causing environmental degradation and loss of ecosystem services.

It has been pointed out in Chapter 7, Section 7.5.1 'Pathogen movement via trade,' that surveillance of aquatic animal diseases is greatly lacking compared with that of terrestrial animal diseases. In such cases, a One Health approach should be most effective in addressing this issue.

Trade of ornamental fish is a good example of risk factors for pathogen spread involving aquatic ecosystems. The related figures and facts presented in Chapter 7 should give an idea of the significant contribution this trade has made to the global economy. At the same time, it is placing pressure on the environment due to many species being overfished and their habitats being destroyed. This situation has triggered development of legislation aimed at protecting the environment and ensuring the welfare of the fish. The expanding interface between aquatic animals and humans has been suggested, however, as a possible threat to public health. To illustrate this threat, a brief introduction to zoonoses of aquatic animal origin is provided in the following section, where a One Health approach to *mycobacteriosis* in fish is used as a case study.

9.1.1   Zoonoses of aquatic animal origin

As indicated in previous chapters, zoonoses are diseases (or infections) transmitted from animals to people. In most cases, zoonotic diseases are connected with ecological dynamics, such as landscape and land-use change or climate and environmental changes. As shown in Chapter 7, these dynamics are facilitated by new or increased interactions between humans and wildlife. Most new and emerging diseases affecting humans seem to be originating from zoonoses connected with ecological changes.

Zoonotic diseases account for about one billion cases and millions of deaths every year (Berthe *et al.*, 2018). It is estimated that about 60% of emerging infectious diseases that are reported globally are zoonoses. According to the World Health Organization, over 30 new human pathogens have been detected in the last three decades, 75% of which have originated in animals (WHO, 2014).

Compared to their terrestrial counterparts, aquatic animals seem to cause a smaller number of zoonotic diseases. In most cases, amphibians and fish are cited sources of aquatic animal zoonotic pathogens and vectors of infectious agents with potential to cause diseases in humans. However, the current problem lies in the recognition and diagnosis of these diseases in humans, as there is poor knowledge of the zoonotic potential of these pathogens and diseases in humans under natural conditions (Haenen *et al.*, 2013). As in the case of zoonoses transmitted by terrestrial animals, zoonotic pathogens of fish can transmit to humans through contact (i.e. via broken, abraded, or chapped skin) with these organisms, which may be present on the skin, fins, or gills of fish, as well as through consumption of infected or contaminated fish (foodborne zoonoses). Some pathogens can also be transmitted through accidental ingestion of water contaminated with feces, skin mucus, or other physiological products of the fish.

Another source of disease is toxins produced by harmful algal blooms. These toxins are directly toxic, via ingestion or through the respiratory system, to humans as well as fish, aquatic birds, and marine mammals. Indirectly, there can be trophic transfer of these toxins from algivorous species, such as anchovies, crustaceans, or mollusks, when ingested by humans and other aquatic species (Urdes *et al.*, 2023).

Viral zoonoses are more commonly transmitted by shellfish (mollusks) which, as filter feeding organisms, can bio-accumulate environmental pathogens within their water habitats. Hepatitis A virus and enteroviruses are the most commonly found zoonotic viruses, which can be transmitted by the oysters, mussels, and clams from aquatic systems closely related with human habitats. When the mollusks are consumed, humans develop a wide range of pathologies, such as hepatitis (in hepatitis A infection), gastroenteritis, meningitis, nephritis, pneumonia, and myocarditis (in enteroviral infections).

Parasitic zoonoses are mostly caused by digenetic trematodes (families: Opisthorchiidae and Heterophyidae), nematodes (families: Anisakidae, Dioctophymatidae, and Gnathostomatidae), and cestodes (families: Diphyllobothriidae).

Bacterial zoonoses are commonly transmitted by fish, reptiles, and amphibians kept as pets. In the case of fish, these zoonoses include mycobacteriosis/fish tuberculosis (*Mycobacterium* spp.), botulism (*Clostridium botulinum*), streptococcosis (*Streptococcus iniae*), vibrioses (*Vibrio vulnificus*, *V. parahemolyticus*, and *V. cholera*), aeromonosis (*Aeromonas hydrophila*), edwardsiellosis (*Edwardsiella tarda* and *E. ictaluri*), and swine erysipelas (*Erysipelothrix rhusiopathiae*) (Boylan, 2011; Haenen *et al.*, 2013; Foyle, 2023). Salmonella infections may be transmitted by affected aquatic birds, reptiles, and marine mammals (Urdes *et al.*, 2023).

9.2 One Health Assessment of a Known Fish Disease

A working example of fish tuberculosis (mycobacteriosis—'fish handlers' disease') is provided below to:

- present the public health risks posed by a zoonotic disease originating from an aquatic ecosystem; and
- show how the One Health concept works in practice.

Fish tuberculosis is caused by atypical mycobacteria, which is found in nature in many varied environments. Fish mycobacteria are Gram-positive, aerobic, and non-motile. *Mycobacterium. marinum*, *M. fortuitum*, and *M. ulcerans* are among the most frequently reported species found in fish.

Transmission to humans can occur either directly from infected fish via the lacerated or abraded skin on the hands of the handler, or indirectly, by contact with the water contaminated with the infectious agent (Parent *et al.*, 1995; Akram and Aboobacker, 2023). Even though the disease is only occasionally diagnosed in humans, in susceptible fish the prevalence is 10–100% (Hashish *et al.*, 2018).

The benefit of using the One Health approach is that it ensures awareness and accurate diagnosis and treatment of such diseases in humans. It also illustrates issues revolving around AMR and the presence of active or residual pharmaceutical substances resulting in the environment due to the choice of specific antibiotics used in treating fish as well as the extended treatment time period.

9.2.1 The epidemiology of the disease

In fish, the infection may present in a clinically manifest, acute, or chronic form, or it may evolve subclinically (without clinical signs). The subclinical form is more commonly found. Many species of fish are susceptible to infection with mycobacteria, but only the most susceptible species develop the clinical form. In aquarium fish, the disease can evolve without symptoms or chronically, which means that these infected fish can spread the mycobacterium undetected into the water. Fish with chronic forms also remain carriers for a long time. After carrier fish excrete the bacteria into the water, it remains in the aquatic system for up to several months, due to its resistance and ability to stick to the biofilm. Stress and overpopulation are usually factors favoring the spread of the bacteria in an aquatic system (Foyle, 2023). There are no effective clinical tests to identify infected fish and no effective treatments or vaccines to break the chain of transmission of the infection in susceptible fish. Fish with clinical disease manifest a complex symptomatology, such as exophthalmia, abnormal swimming, lethargy, scale loss, abdominal distention, skin pigmentation changes, loss of body condition, and skin erosions and ulcers (Urdes and Loh, 2021). In fish, the fecal-oral route is the main route of transmission, followed by the gill and cutaneous routes (necrotic detritus from ulcerated areas) (Niemeyer-Corbellini *et al.*, 2017). Because fish mycobacteria can persist for months in water, and because common disinfectants such as dilute sodium hypochlorite and quaternary ammonia compounds are ineffective in this case, contaminated water remains the most common source of infection for both susceptible fish and humans who come into contact with it. On the other hand, the biofilms of contaminated tanks act as a protective layer for mycobacteria, making them resistant to these disinfectants (Boylan, 2011). Phenols, high concentrations of alcohol, and strong sodium chlorite solutions are the most effective disinfectants in this case. Infected fish and contaminated water resulting from tanks containing fish infected with mycobacteria also pose risks to public health and the environment.

9.2.2 Fish tuberculosis—a One Health assessment framework

The concept of the following framework is to ensure consideration of all relevant facts.

- Action 1: Consider the information about fish tuberculosis given in the above section and in the review paper by Hashish *et al.* (2018).
- Action 2: Identify the risks—you can use the biosecurity representation introduced in Chapter 6 as a template (see Fig. 6.10).
- Action 3: Identify the stakeholders to consider, such as immunocompromised humans, fish farmers, aquarists, veterinary and medical staff, economists, ecologists.

  ○ Identify who the stakeholders are: Who holds pertinent information and who might be impacted by any measures taken?

- Action 4: Carry out a risk assessment. This should indicate whether there is a high probability of exposure to fish tuberculosis in the water and which equipment was used in contact with these fish. The risk of infection is currently considered low but the public health consequences are high in terms of diagnosis and treatment costs.

  ○ List the areas of concern: Who and what is impacted by the disease?

- Action 5: Having determined there is sufficient risk to merit further investigation, the next step is to conduct an assessment based on the following points:

  ○ The cost–benefit of the ornamental trade: in considering these economic arguments, you may want to also include the cost of fish tuberculosis in food fish.
  ○ As the majority of wild-caught fish are marine fish, you may also want to consider the impact on marine ecosystems' degradation and conduct an environmental cost–benefit assessment.
  ○ Human health and welfare: conduct a health cost–benefit assessment to include such issues as AMR, prolonged duration of treatment, and possible misdiagnoses.
  ○ Fish health and welfare: consider the effects of a lack of effective treatment leading to depopulation, as well as inefficient sterilization procedures leading to the pathogen's persistence in the life support systems and the environment.

- Action 6: Is there a way to implement beneficial changes?

  ○ What, if any, would be the benefits of new or additional controls? The costs of any measures must be less than the benefits gained.

○ Allow for interdisciplinary training for professionals across all relevant disciplines to help them understand and address the risks incurred by the disease.

○ Disseminate new knowledge about ongoing disease research.

○ Raise the awareness of communities at risk.

○ Provide necessary funding and resource allocation to include disease prevention programs, infrastructure for health data modeling, etc.

The actions above are for illustrative purposes. It should be clear that to outline new policies and create awareness on public health issues requires multiple disciplines. It can also be extremely complex to investigate and obtain satisfactory results without the cooperation of relevant stakeholders.

9.3 Operationalization and Institutionalization of One Health

9.3.1 Operationalization

Without adequate biosecurity measures, both the occurrence and the impact of communicable diseases are likely to increase. Factors such as changes in agricultural practices and land use, demographic and urbanization expansion, and climate change are leading to an increased risk of spillover and spread of pathogens. Addressing these factors as public health challenges requires a whole-of-government approach, with inputs from all sectors related to human, animal, and environmental health (Ortenzi *et al.*, 2022; European Commission, 2023). At this point, the reader should look again at Figure 3.7, in Chapter 3, 'Biodiversity Conservation,' where good governance is illustrated in relation to sustainable development, preservation of aquatic ecosystems, and adaptation to climate change.

Operationalizing One Health implies collaborative work that responds to emerging diseases and outbreaks of major importance. It involves establishing multisectoral programs at the governmental level and strong international partnerships to provide concrete and sustainable grounds for better health outcomes across the sectors of human and veterinary medicine and the environmental area (FAO-OIE-WHO, 2010; Berthe *et al.*, 2018). Currently, public institutions operate in a fractured manner, working independently and drawing only on the knowledge and perspectives specific to their respective area. This often results in gaps or overlaps in effort. For practitioners serving in the One Health area, regular communication should start taking place between practitioners from different disciplines and sectors. This allows for a better understanding of the significance of a finding or an event with regard to their own field by practitioners in the other fields, and makes professionals from these sectors more likely to collaborate to improve health outcomes.

A number of initiatives to operationalize the implementation of the One Health concept, by setting up workable strategies with concrete and measurable objectives, have started to crystallize. These refer to: (i) *trans-sectorial collaborations* between public health, veterinary medicine, environmental sciences, animal husbandry, agriculture, and other stakeholders, aimed at addressing health threats in an integrated manner; (ii) *new policies* that align human, animal, and environmental health objectives; (iii) *surveillance and data modeling* designed for health issues across all three poles, human, animal, and environment, to share information in order to detect and respond to emerging diseases, as explained in the study case about mycobacteriosis; (iv) *interdisciplinary training* for professionals across disciplines to understand and address health threats in a holistic manner; (v) *science-based preventive measures* and responses that address health challenges at the human–animal–environment interface; (vi) *funding and resource allocation* to include supporting research on zoonotic diseases, funding for disease prevention programs, and building infrastructure for health data modeling; (vii) *community active participation*, such as public awareness campaigns and community-driven initiatives addressing health challenges by using all levels of understanding (Ceptureanu and Ceptureanu, 2019).

In essence, it's about making the One Health concept a practical and operational part of day-to-day health management and policy to better protect and improve health across human, veterinary, and environmental sectors.

9.3.2 Institutionalization

New research approaches (transdisciplinary and operational research), specific training, and appropriate curricula are required to enable development of sustainable health systems to tackle public health issues.

As shown in the case study of risk management of mycobacteriosis in the ornamental fish trade, engagement of broad One Health professionals from suitable disciplines and sectors is of paramount importance to ensure relevance, avoid duplication, and maximize the health security outcome.

To set up the One Health practical approach, the operationalization initiatives presented in the previous section require that the practicing of One Health is established as a generally accepted norm within the society. It means, therefore, that operationalization of One Health requires an a priori institutionalization of One Health.

Higher education establishments play a crucial role in this process, as they act as training and research hubs, 'multipliers' and disseminators of scientific results relevant to the health security area. Institutionalization of One Health requires implementation of the following actions by higher education institutions (Shyaka *et al.*, 2024):

- benchmarking, developing, and improving the One Health curricula at the master's level;
- supporting researchers to deliver One Health science-based solutions for sustainable development;
- training educators in the field of One Health;
- supporting One Health professional development for One Health practitioners; and
- working with primary and secondary schools to integrate One Health into their curricula.

References

Akram, S.M. and Aboobacker, S. (2023) *Mycobacterium marinum* infection. In: *StatPearls*. StatPearls Publishing, Treasure Island, Florida. Available at: https://www.ncbi.nlm.nih.gov/books/NBK441883/ (accessed 4 February 2025).

Berthe, F.C.J., Bouley, T., Karesh, W.B., Legall, I.C., Machalaba, C.C. *et al.* (2018) *One Health Operational Framework for Strengthening Human, Animal, and Environmental Public Health Systems at their Interface*. World Bank Group, Washington, DC. Available at: http://documents.worldbank.org/curated /en/961101524657708673 (accessed 18 June 2025).

Boylan, S. (2011) Zoonoses associated with fish. *Veterinary Clinics of North America. Exotic Animal Practice* 14, 427–438.

Ceptureanu, S.I. and Ceptureanu, E.G. (2019) Community-based healthcare programs sustainability impact on the sustainability of host organizations: A structural equation modeling analysis. *International Journal of Environmental Research and Public Health* 16(20), 4035. DOI: 10.3390/ijerph16204035.

European Commission (2023) *Mutual Learning Exercise on the Whole of Government Approach in Research and Innovation. First Thematic Report: Introduction and Overview of the Whole of Government Approaches in Research and Innovation*. Publications Office of the European Union, Luxembourg City. Available at: https://data.europa.eu/doi/10.2777/92395 (accessed 4 February 2025).

FAO, OIE, WHO (2010) *Sharing Responsibilities and Coordinating Global Activities to Address Health Risks at the Animal-Human-Ecosystems Interfaces—A Tripartite Concept Note*. Available at: https:// cdn.who.int/media/docs/default-source/ntds/neglected-tropical-diseases-non-disease-specific/ tripartite_concept_note_hanoi_042011_en.pdf?sfvrsn=8042da0c_1&download=true (accessed 18 June 2025).

Foyle, L. (2023) Epidemiology of aquatic animal diseases. In: Urdes, L., Walster, C. and Tepper, J. (eds) *Fundamentals of Aquatic Veterinary Medicine*. Wiley, Oxford, UK, pp. 135–150.

Haenen, O.L., Evans, J.J. and Berthe, F. (2013) Bacterial infections from aquatic species: Potential for and prevention of contact zoonoses. *Revue Scientifique et Technique* 32, 497–507.

Hashish, E., Merwad, A., Elgaml, S., Amer, A., Kamal, H. *et al*. (2018) *Mycobacterium marinum* infection in fish and man: Epidemiology, pathophysiology and management; a review. *Veterinary Quarterly* 38(1), 35–46. DOI: 10.1080/01652176.2018.1447171.

Niemeyer-Corbellini, J.P., Lupi, O., Klotz, L., Montelo, L., Elston, D.M. *et al*. (2017) Environmental causes of dermatitis. In: Tyring, S.K., Lupi, O. and Hengge, U.R. (eds) *Tropical Dermatology*. Elsevier, Amsterdam, pp. 443–470.

Ortenzi, F., Marten, R., Valentine, N.B., Kwamie, A. and Rasanathan, K. (2022) Whole of government and whole of society approaches: Call for further research to improve population health and health equity. *BMJ Global Health* 7(7), e009972. DOI: 10.1136/bmjgh-2022-009972.

Parent, L.J., Salam, M.M., Appelbaum, P.C. and Dossett, J.H. (1995) Disseminated *Mycobacterium marinum* infection and bacteremia in a child with severe combined immunodeficiency. *Clinical Infectious Diseases* 21(5), 1325–1327.

Shyaka, A., Igihozo, G., Tegli, M., Ntiyaduhanye, E., Ndizeye, E. *et al*. (2024) Rwanda one health achievements and challenges: The final push toward policy implementation. *One Health Cases* ohcs20240010. DOI: 10.1079/onehealthcases.2024.0010.

Urdes, L. and Loh, R. (2021) A case report on fish tuberculosis ("fish handlers' disease") in rainbowfish (Fam. Melanotaeniidae) and rosy barb (Pethia conchonius). *Scientific Papers. Series D. Animal Science* LXIV(2). Available at: https://animalsciencejournal.usamv.ro/index.php/scientific-papers/ 22-articles-2021-issue-2/1005-a-case-report-on-fish-tuberculosis-fish-handlers-disease-in-rainbowfish-fam-melanotaeniidae-and-rosy-barb-pethia-conchonius (accessed 18 June 2025).

Urdes, L., Walster, C. and Tepper, J. (2023) *Pathology and Epidemiology of Aquatic Animal Diseases for Practitioners*. Wiley, Oxford, UK.

World Health Organization (WHO) (2014) Zoonotic Disease: Emerging Public Health Threats in the Region. WHO, Geneva. Available at: https://www.emro.who.int/about-who/rc61/zoonotic-diseases.html (accessed 18 June 2025).

Further Reading

Adeyemo, K.O. and Foyle, L. (2022) Public health, zoonoses, and seafood safety. In: Urdes, L., Walster, C. and Tepper, J. (eds) *Fundamentals of Aquatic Veterinary Medicine*. Wiley, Oxford, UK.

WHO Commission on Macroeconomics and Health & World Health Organization (2001) *Macroeconomics and Health: Investing in Health for Economic Development: Executive Summary / Report of the Commission on Macroeconomics and Health*. WHO, Geneva. Available at: https://apps.who.int/iris /handle/10665/42463

Glossary of Terms

_____

Aminoglycosides: a class of antibiotics that inhibit protein synthesis in bacteria. They are commonly used to treat serious infections caused by Gram-negative bacteria.

Aminopenicillins: a subclass of penicillin antibiotics that are effective against a wider range of bacteria than traditional penicillin. They are often used to treat infections caused by Gram-positive and some Gram-negative bacteria.

Antimicrobial resistance (AMR): the ability of microorganisms to become increasingly resistant to an antimicrobial to which they were previously susceptible, making infections harder to treat and increasing the risk of disease spread, severe illness, and death.

Antimicrobial resistance gene (ARG): genes that confer resistance to antimicrobial drugs. These genes can be located on chromosomes, plasmids, or transposons.

Antimicrobial-resistant bacteria (ARB): bacteria that have developed resistance to one or more antimicrobial drugs.

Approved establishments: producing plants listed by the EU (also named EU-approved establishments). These establishments are allowed to export consignments to the EU.

Aquatic Ecosystem: the total sum of the varied aquatic ecosystems on earth, as understood from a One Health perspective.

Aquatic ecosystems: habitats for aquatic organisms that also provide water for drinking, irrigation, and recreational activities for people. Aquatic ecosystems are vital components of the environment; these ecosystems contribute to biodiversity and formation of biomass in the ecosystem (also known as "ecological productivity"). There are several types of aquatic ecosystems: saltwater (marine) and freshwater (inland) ecosystems, coral reefs (in tropical and subtropical regions), kelp forests (in colder coastal waters), wetlands (including marshes and swamps), and mangroves (in coastal areas).

Beta-lactams: a broad class of antibiotics characterized by a beta-lactam ring in their molecular structure. This ring is essential for their antibacterial activity, which involves inhibiting bacterial cell wall synthesis. Beta-lactams include penicillins, cephalosporins, carbapenems, and monobactams.

Biosecurity: a set of measures aimed at reducing the risk of the introduction and spread of pathogens onto and between populations, and to contain the spread of a disease within a population. It comprises policy and regulatory frameworks that analyze and manage risks in the sectors of food safety, animal and plant life, and health, including associated environmental risk. Biosecurity covers the introduction of plant pests, animal pests and diseases, and zoonoses; the introduction and release of genetically modified organisms (GMOs) and their products; and the introduction and management of invasive alien species and genotypes. Biosecurity is a holistic concept of direct relevance to the sustainability of agriculture, food safety, and the protection of the environment, including biodiversity.[1]

blaKPC-2: a gene that encodes for a carbapenemase enzyme, which can inactivate carbapenem antibiotics.

blaTEM: a gene that encodes for a beta-lactamase enzyme, which can inactivate penicillin and other beta-lactam antibiotics.

Centrifugal microfluidics: a technology that uses centrifugal force to manipulate small volumes of fluids on a microfluidic platform. It has applications in various fields, including diagnostics and drug discovery.

Chromosome: a DNA molecule that carries the genetic material of an organism.

Composite products: food commodities containing mixed ingredients of processed products of animal origin and plant-based products.

Conjugation: the transfer of genetic material from one bacterium to another through direct contact.

Disease Contingency Plan: is a documented work plan designed to ensure that all needed actions are taken, requirements met, and resources provided in order to eradicate or bring under control outbreaks of infectious diseases of significance. As an example, the FAO Fisheries Technical Paper 486[2] refers to aquatic animal productivity and/or market access.

EcoHealth and Planetary Health: *ecohealth* emphasizes science at the intersection of ecology and health through an ecosystem approach to address a wide range of topics in relation to health, including waterborne and water-related disease, while *planetary health* calls for solutions to address the drivers of global environmental change, such as those deriving from climate change, biodiversity loss, and soil erosion.[3]

Ecosystem approach: integrated management of soil, water, and air comprised in an ecosystem; this approach promotes conservation and sustainable use of the ecosystem's resources in an equitable way.

Emerging disease: a new disease appearing in a population.

Environment: the natural world inhabited by people, animals, and plants. It is composed of air, land, and water, and includes all known ecosystems: forest, grassland, desert, tundra, and aquatic.

Epidemiological unit (EpiUnit): the basic unit relevant for disease control measures. EpiUnits are geographically defined areas in which a disease can be transmitted and maintained among individuals of a population.

Epidemiology: a branch of public health studying the distribution and occurrence of disease. Epidemiology is explained in Chapter 5.

Epigenetic modifications: heritable changes in gene expression that do not involve alterations to the underlying DNA sequence.

***erm*B**: a gene that encodes for a ribosomal methylase enzyme, which can confer resistance to macrolide antibiotics.

EU: European Union

Extended-spectrum beta-lactamase (ESBL): enzymes that can inactivate a wide range of beta-lactam antibiotics, including penicillins, cephalosporins, and monobactams.

Food safety: refers to handling, preparing, and storing food. The main objective of food safety measures is to reduce the risk of individuals becoming sick from foodborne illnesses.

Gene: a unit of heredity that is responsible for a particular trait.

Genetic mutation: a permanent alteration in the DNA sequence of an organism.

Glycopeptides: a class of antibiotics that inhibit cell wall synthesis in bacteria. They are commonly used to treat infections caused by Gram-positive bacteria, including methicillin-resistant *Staphylococcus aureus* (MRSA).

Hazard identification: recognizes those organisms which pose a disease risk to others, and aids in deciding which diseases should be controlled.

Health (clinical) care: defined as the actions taken by qualified and authorized professionals to preserve, promote, and restore a person's physical, mental, and emotional well-being. With evolving technological advancements, a relatively new approach has emerged in the field of health care, namely clinical care management (CCM). CCM is a collective and anticipatory approach to personalized care (patient-centered care) of patients who present with complex medical needs, particularly those with chronic diseases. CCM involves personalized services designed to improve health outcomes and reduce hospitalizations and medical costs while ensuring the efficient use of healthcare resources.

Health disparities: the Centers for Disease Control and Prevention (CDC)[4] defines health disparities as the preventable differences in the burden of disease, which are related to unequal distribution of socio-political, socio-economic, and socio-environmental resources. The existence of health disparities (mainly caused by poverty and lack of good quality health care) is perceived as the main limitation in improving the health of people.

Health outcome: improved access to quality essential health services. To allow for measurement of the progress toward achieving health outcomes, WHO uses outcome indicators,[5] which are also milestones to the targets of the United Nations' Sustainable Development Goals.

Health security (global health security): protection of public health against threats posed by communicable diseases and environmental hazards. The means used for health security are: surveillance and early warning, risk assessment, crisis/disease outbreak management, preparedness, and response planning.

Huanglongbing (HLB): a devastating bacterial disease of citrus trees, also known as citrus greening disease. It is caused by bacteria of the genus *Candidatus liberibacter* and spread by psyllids (small insects). HLB affects all citrus varieties, causing yellowing of leaves, stunted growth, and bitter, inedible fruit, eventually leading to tree death. There is no cure for HLB, and it poses a significant threat to citrus production worldwide.

Macrolides: a class of antibiotics that inhibit protein synthesis in bacteria. They are commonly used to treat infections caused by Gram-positive bacteria and some Gram-negative bacteria.

Metagenomics: the study of genetic material recovered directly from environmental samples.

Microbiomes are complex and dynamic systems, modulating host immunity, digestion, and resilience against pathogens. The diversity and stability of microbiomes are essential for health and are influenced by numerous factors, including diet, environmental pollutants, antibiotics, and climate changes.[6]

Mobile genetic elements: DNA sequences that can move around the genome. They can include plasmids, transposons, and integrons.

One Health: unifying multiple sectors, disciplines, and communities to sustain the health of people, animals, and ecosystems through the application of measures aimed at ensuring balance and sustainable development, as reflected in the Sustainable Development Goals (SDG) of the United Nations.[7] In operational terms, One Health is defined as a collaborative effort aimed at strengthening the systems to optimally prevent (i.e. prepare for and detect) and respond to health risks occurring at the human–animal–environment interface, by using tools such as surveillance and reporting, with the ultimate goal to achieve global health security.

Operational research: mathematically based analysis method (quantitative data) used to improve management and decision making.

Phytosanitary measures: actions aimed at ensuring plant health.

Plasmid: a small, circular DNA molecule that is separate from the chromosome. Plasmids can carry genes that confer resistance to antibiotics or other traits.

Public health surveillance: the World Health Organization[8] defines public health surveillance as "the continuous and systematic collection, orderly consolidation and evaluation of pertinent data with prompt dissemination of results to those who need to know, particularly those who are in a position to take action." Public health surveillance response and support systems are intrinsic constituents of disease control programs. Different diseases may have specific surveillance needs, but surveillance programs should use similar structures, processes, and resources to achieve public health objectives. It is important to remember that the effectiveness of disease control programs is directly proportional with the effectiveness of public health surveillance.

Quinolones: a class of antibiotics that inhibit DNA synthesis in bacteria. They are commonly used to treat a variety of infections.

Registered establishments: those establishments under the control of a state's authority which are not listed to export composite products to the EU (i.e. these establishments are registered within their national competent authorities).

Reservoir: a source of pathogens or antimicrobial resistance genes.

Risk analysis: investigates both the likelihood and the consequences of undesirable events, known as hazards.[9]

Sanitary and Phytosanitary (SPS) Agreement of the World Trade Organization (WTO): the agreement on the application of sanitary and phytosanitary measures by the WTO, available since January 1, 1995, concerning the application of food safety and animal and plant health regulations. As of August 2024, the WTO has 166 members.[10] The EU is a full member of the organization.

Sanitary measures: actions aimed at ensuring human and animal health.

Selection pressure: any factor that favors the survival and reproduction of organisms with certain traits.

***sul*1**: a gene that encodes for a dihydropteroate synthase enzyme, which can confer resistance to sulfonamide antibiotics.

Sulfonamides: a class of antibiotics that inhibit folic acid synthesis in bacteria. They are commonly used to treat urinary tract infections and other infections.

***tet*(A)xz**: genes that encode for efflux pumps, which can confer resistance to tetracycline antibiotics.

Tetracycline: an antibiotic that inhibits protein synthesis in bacteria. It is commonly used to treat a variety of infections.

Third-generation cephalosporin: subclass of cephalosporin antibiotics that have a broader spectrum of activity than earlier generations. They are often used to treat serious infections caused by Gram-negative bacteria.

Transduction: the transfer of genetic material from one bacterium to another by a bacteriophage.

Transformation: the uptake of genetic material from the environment by a bacterium.

Transposon: a DNA sequence that can move from one location to another in the genome. Transposons can carry genes that confer resistance to antibiotics or other traits.

Vector: an organism or vehicle that transmits a pathogen from one host to another.

Zoonoses: diseases transmissible to humans from animals.

Notes

[1] https://documents1.worldbank.org/curated/en/961101524657708673/pdf/122980-REVISED-PUBLIC-World-Bank-One-Health-Framework-2018.pdf

[2] https://www.fao.org/4/a0090e/A0090E10.htm

[3] https://documents1.worldbank.org/curated/en/961101524657708673/pdf/122980-REVISED-PUBLIC-World-Bank-One-Health-Framework-2018.pdf

[4] https://www.cdc.gov/nchs/healthy_people/hp2020/health-disparities.htm

[5] https://data.who.int/indicators

[6] Stanhope, J., Breed, M., and Weinstein, P. (2022) Biodiversity, microbiomes, and human health. In: Rook, G. and Lowry, C. (eds) *Evolution, Biodiversity and a Reassessment of the Hygiene Hypothesis.* Springer International Publishing, Cham, Switzerland, pp. 67–104.

[7] https://www.who.int/news/item/01-12-2021-tripartite-and-unep-support-ohhlep-s-definition-of-one-health

[8] https://www.emro.who.int/health-topics/public-health-surveillance/index.html

[9] Peeler, E.J., Murray, A.G., Thebault, A., Brun, E., Giovaninni, A., and Thrush, M.A. (2007) The application of risk analysis in aquatic animal health management. *Preventive Veterinary Medicine* 81(1–3).

[10] https://www.wto.org/english/thewto_e/whatis_e/tif_e/org6_e.htm

Index